T0257837

Modern Concepts of Flow Cytometry

Modern Concepts
of Flow Cytometry

Edited by **Barbara Roth**

New York

Published by Callisto Reference,
106 Park Avenue, Suite 200,
New York, NY 10016, USA
www.callistoreference.com

Modern Concepts of Flow Cytometry
Edited by Barbara Roth

International Standard Book Number: 978-1-63239-463-7 (Hardback)

Printed in the United States of America.

Contents

Preface

This book is a compilation of complete reviews and unique technical researches in modern concepts of flow cytometry. The data demonstrates the continuously developing application of flow cytometry in a varied number of technical fields as well as its wide-scale utilization evident from the global composition of the group of contributing authors. The book deals with the employment of the flow cytometry methodology in fundamental sciences and covers many diverse fields of microbiology, biotechnology etc. This book offers useful information to students and experts dealing with flow cytometry.

After months of intensive research and writing, this book is the end result of all who devoted their time and efforts in the initiation and progress of this book. It will surely be a source of reference in enhancing the required knowledge of the new developments in the area. During the course of developing this book, certain measures such as accuracy, authenticity and research focused analytical studies were given preference in order to produce a comprehensive book in the area of study.

This book would not have been possible without the efforts of the authors and the publisher. I extend my sincere thanks to them. Secondly, I express my gratitude to my family and well-wishers. And most importantly, I thank my students for constantly expressing their willingness and curiosity in enhancing their knowledge in the field, which encourages me to take up further research projects for the advancement of the area.

Editor

Stem Cell Characterization

Arash Zaminy

Department of Anatomy & Cell Biology,
Shahid Beheshti University of Medical Sciences, Tehran,
Iran

1. Introduction

One of the main problems in stem cell studies is how researchers can identify and characterize a stem cell.

Identification and characterization of stem cells is a difficult and often evolving procedure. Stem cells not only must exhibit the appropriate markers, but also a healthy and robust stem cell population must also lack specific markers. In addition to the difficulty of this area of stem cell biology markers, profiles change based on the species, site of origin and maturity (totipotent vs. multipotent) of a given population. Furthermore, stem cell populations may consist of several specific phenotypes which are often indicators of the population's general health. Flow cytometry employs instrumentation that scans single cells flowing past excitation sources in a liquid medium. It is a widely used method for characterizing and separating individual cells.

This chapter tries to explain what stem cells are, as well as to summarize current knowledge on stem cell characterization and usage of stem cells markers.

2. Stem cells

All life forms initiate with a stem cell, which is defined as a cell that has the dual capacity to self-renew and to produce progenitors and different types of specialized cells in the organism. Scientists mostly work with two kinds of stem cells from animals and humans: embryonic stem cells and non-embryonic "somatic" or "adult" stem cells.

At the beginning of human life, one fertilized egg cell – the zygote – divides into two and two becomes four (Carlson, 1996). Within 5 to 7 days, some 40 cells are produced which build up the inner cell mass encircled by an outer cell layer subsequently forming the placenta. At this phase, each of these cells in the inner cell mass has the potential to give rise to all tissue types and organs – that is, these cells are pluripotent. Finally, the cells forming the inner cell mass will give rise to the some 1013 cells that constitute a human body, organized in 200 differentiated cell types (Sadler, 2002). Many somatic, tissue-specific or adult stem cells are produced during the foetal period. These stem cells have a more limited ability than the pluripotent embryonic stem cells (ESCs) and they are multipotent – that is, they have the potential to give rise only to a limited number of cell lineages. These adult stem cells keep on in the related organs to varying degrees over the whole of a person's lifetime.

Stem cells are well-known from other cell types because of two important characteristics. First, they are unspecialized cells and the have ability to renew themselves through cell division, sometimes after long periods of inactivity. Second, under certain physiologic or experimental conditions, they can be induced to become tissue- or organ-specific cells with special abilities (Thrasher, 1966; Merok & Sherley, 2001). In some organs, such as the gut and bone marrow, stem cells divide regularly to repair and restore exhausted or damaged tissues. In other organs, however, such as the pancreas and the heart, stem cells only divide under special conditions.

3. Stem cells markers

In recent years researchers have revealed a broad range of stem cells that have unique capabilities to self-renew, grow indefinitely and differentiate or develop into multiple kinds of cells and tissues. Researchers now know that many different types of stem cells exist, but they are all found in very small populations in the human body, in some cases one stem cell in 100,000 cells in circulating blood. In addition, when scientists study these cells under a microscope, they are similar to other cells in the tissue where they are found. So, like the search for a needle in a haystack, how do scientists recognize these uncommon types of cells found in many different cells and tissues? The answer is rather simple, thanks to stem cell "markers."

What are stem cell markers? The surface of every cell in the body has specialized proteins called receptors that have the capability of selectively binding or adhering to other "signalling" molecules. There are many different types of receptors that differ in their composition and affinity for the signalling molecules. Generally, cells use these receptors and the molecules that bind to them as a way of communicating with other cells, and to perform their correct functions in the body. These cell surface receptors are commonly used as cell markers. Each cell type, for example a liver cell, has a certain combination of receptors on their surface that makes them distinguishable from other types of cells. Scientists have taken advantage of the biological exclusivity of stem cell receptors and the chemical properties of certain compounds to label or mark cells. Researchers owe much of the past success in finding and characterizing stem cells to the use of markers.

Stem cell markers are given shorthand names based on the molecules that attach to the stem cell surface receptors. For example, a cell that has the receptor stem cell antigen -1 on its surface is known as Sca-1. In many cases, a mixture of multiple markers is used to identify a particular stem cell type. So now, researchers often identify stem cells in shorthand by a combination of marker names reflecting the presence (+) or absence (-) of them. For example, a special type of haematopoietic stem cell from blood and bone marrow is described as (CD34$^{-/low}$, c-Kit$^+$, Sca-1$^+$) (Jackson et al., 2001).

Researchers employ antibody molecules that selectively bind with the receptors on the surface of the cell as a way to identify stem cells. In former years a method was developed to attach to the antibody molecule another molecule (or tag) that has the ability to fluoresce or emit light energy when triggered by an energy source such as an ultraviolet light or laser beam. Now, multiple fluorescent labels are available with emitted light that differ in colour and intensity.

Researchers exploit the combination of the chemical properties of fluorescence and unique receptor patterns on cell surfaces to identify specific numbers of stem cells. One approach for using markers is a technique known as fluorescence-activated cell sorting (FACS) (Bonner et al., 1972; Herzenberg, 2000; Julius, 1972). Researchers frequently use a FACS instrument to sort out the rare stem cells from the millions of other cells. By this method, a suspension of tagged cells (i.e. fluorescently-labelled antibodies are bound to the cell surface markers) is sent under pressure through a very fine nozzle. Upon exiting the nozzle cells pass through a light source, usually a laser, and then through an electric field. Operators apply a series of criteria. If the cell stream meets the criteria, they become negatively or positively charged. When cells are passing among an electric field, the charge difference permits the desired cells to be separated from other cells. The researchers now have a population of cells that have all of the same marker characteristics and with these cells they can conduct their research.

A second method uses stem cell markers and fluorescent antibodies to visually assess cells as they exist in tissues. Often researchers want to assess how stem cells appear in tissues and in doing so they use a microscope to evaluate them rather than the FACS instrument. In this case, a thin slice of tissue is prepared and the stem cell markers are tagged by the antibodies that have the fluorescent tag attached. The fluorescent tags are then activated either by special light energy or a chemical reaction. The stem cells will emit a fluorescent light that can easily be seen under the microscope.

4. Embryonic stem cells

An embryonic stem cell (ESC) is described by its origin. It is obtained from the blastocyst stage of the embryo. Embryonic stem cells are unique cell populations with the capability of both self-renewal and differentiation, and thus ESCs can give rise to any adult cell type. Pluripotent embryonic stem (ES) cells, like embryonal carcinoma cells, were first used as a tool to examine thoroughly early differentiation. However, the properties of ESCs identify them as being highly appropriate for making specific cell lineages in vitro. The ability of embryonic stem cells to almost limitless self-renewal and differentiation capacity has opened up the panorama of widespread applications in biomedical research and regenerative medicine.

ESCs are harvested from the inner cell mass of the pre-implantation blastocyst and have been derived from rodents (Martin, 1981; Evans & Kaufman, 1981; Doetschman et al., 1988; Graves & Moreadith, 1993), primates (Thomson et al., 1995) and humans (Thomson et al., 1998; Reubinoff et al., 2000).

4.1 Embryonic stem cell markers

Some approaches have been applied to characterize ESCs, but the most widely used approach is analysis of cell surface antigens by flow cytometry and evaluation of gene expression profile by RT-PCR or microarrays. Many cell surface antigens used to identify hESCs were first detected with antibodies prepared against pre-implantation mouse embryos and/or against mouse or human embryonal carcinoma cells (Pera et al., 2000).

Although the functions of those antigens in the continuance of undifferentiated human embryonal carcinoma cells are not necessarily clear, they may represent helpful markers for the recognition of pluripotent stem cells. These antigens include the globo-series glycolipid antigens, stage-specific embryonic antigen-3 (SSEA-3) and -4 (SSEA-4), keratan sulphate antigens TRA-1-60, TRA-1-81, GCTM2 and GCTM343, a set of various protein antigens comprising the two liver alkaline phosphatase antigens TRA-2-54 and TRA-2-49, Thy1, CD9, HLA class 1 antigens, Oct3/4, Nanog and the absence of hESC negative markers, such as SSEA-1 (Pera et al., 2000; Carpenter et al., 2003; Chambers et al., 2003; Draper et al., 2004; Heins et al., 2004; Nichols et al., 1998).

Some cell surface biomarkers are also listed in Table 1. The International Stem Cell Initiative (ISCI) established by the International Stem Cell Forum (http://www.stemcellforum.org.uk) carried out a comparative study of a large and different set of hESC lines derived from and maintained in different laboratories worldwide (Adewumi et al., 2007). Fifty-nine independent hESC lines derived from 17 laboratories in 11 countries were investigated for the expression of 17 cell surface antigens and 93 genes, which have been chosen as potential markers of undifferentiated stem cells or their differentiated derivatives (Adewumi et al., 2007). All of the independent hESC lines displayed a common expression profile for a specific set of marker antigens, despite the fact that they had different genetic backgrounds and were produced by different techniques in each laboratory. All examined cell lines expressed a comparable spectrum of cell surface marker antigens characteristic of hESCs, suggesting that there is a common set of markers that can be used to monitor, in general, the presence of pluripotent stem cells. SSEA-3 and SSEA-4 were expressed in all hESCs tests, indicating that these molecules are valuable operational markers of this cell type; however, a study revealed that they are not necessary for the pluripotency of hESCs (Brimble et al., 2007).

	Mouse ES cells	Human ES cells
SSEA-1	+	-
SSEA-3	-	+
SSEA-4	-	+
TRA-1-60	-	+
TRA-1-81	-	+
GCTM-2	-	+
Alkaline phosphatase	+	+
Oct-4	+	+
GDF-3	+	?

Table 1. Marker expression and growth properties of mouse and primate pluripotent cells

5. Haematopoietic stem cells

Blood cells are responsible for continuous preservation and immune protection of every cell type of the body. This persistent and brutal work requires blood cells, along with skin cells, to have the greatest power of self-renewal of any adult tissue. The stem cells that form blood and immune cells are known as haematopoietic stem cells (HSCs). HSCs are among the best characterized adult stem cells and the only stem cells being regularly used in clinics.

A haematopoietic stem cell is a cell isolated from the blood or bone marrow that can renew itself, can differentiate a variety of specialized cells and can mobilize out of the bone marrow into circulating blood. Since HSCs look and behave in culture like ordinary white blood cells, it has been a challenge to identify them by morphology (size and shape). Even now, scientists must rely on cell surface proteins, which generally serve as markers of white blood cells.

5.1 Haematopoietic stem cell markers

HSCs have an identity problem. First, the ones with long-term replicating ability are rare. Second, there are multiple types of stem cells. Third, the stem cells look like many other blood or bone marrow cells. So how do researchers find the desired cell populations? The most common approach is through markers that emerge on the surface of cells.

A variety of markers has been found to help distinguish and separate HSCs. Early marker efforts focused on cell size, density and recognition by lectins (carbohydrate-binding proteins derived largely from plants) (Bauman et al., 1988), but more recent attempts have focused mostly on cell surface protein markers, as defined by monoclonal antibodies. For mouse HSCs, these markers contain panels of 8 to 14 different monoclonal antibodies that recognize cell surface proteins present on differentiated haematopoietic lineages, such as the red blood cell and macrophage lineages (thus, these markers are collectively referred to as Lin) (Spangrude et al., 1988; Uchid & Weissman, 1992) as well as the proteins Sca-1 (Spangrude et al., 1988; Uchid & Weissman, 1992), CD27 (Weissman et al., 2000), CD34 (Osawa et al., 1996), CD38 (Randall et al., 1996), CD43 (Moore et al., 1994), CD90.1 (Thy-1.1) (Spangrude et al., 1988; Uchid & Weissman, 1992), CD117 (c-Kit) (Ikuta & Weissman, 1992), AA4.1 (Jordan et al., 1996) and MHC class I (Bauman et al., 1988), and CD150 (Kiel et al., 2005).

Human HSCs have been described with respect to staining for CD34 (Civin et al., 1984), CD38 (Kiel et al., 2005), CD43 (Moore et al., 1994), CD45RO (Lansdorp et al., 1990), CD45RA (Lansdorp et al., 1990), CD59 (Hill et al., 1996), CD90 (Bauman et al., 1988), CD109 (Sutherland et al., 1996), CD117 (Gunji et al., 1993), CD133 (Miraglia et al., 1997; Yin et al., 1997), CD166 (Uchida et al., 1997), HLA DR (human) (Srour et al., 1992; Tsukamoto et al., 1995) and lacking expression of lineage (Lin) markers (Baum et al., 1992). It is important to note that lineage markers are cell surface antigens that can be used for immunophenotyping cells of a particular developmental lineage. Cells that do not express these marker antigens, or express them at very low levels, are said to be lineage marker negative [lin(-)].

While none of these markers recognize functional stem cell activity, combinations (typically with 3 to 5 different markers, see examples below) led to the purification of near-homogenous populations of HSCs. The ability to obtain pure preparations of HSCs, albeit in limited numbers, has greatly facilitated the functional and biochemical characterization of these important cells. However, now there has been limited impact of these discoveries on clinical practice, as highly purified HSCs have only rarely been used to treat patients. The irrefutable advantages of using purified cells (e.g., the absence of contaminating tumour cells in autologous transplantations) have been offset by practical difficulties and increased purification costs.

HSC assays, when combined with the ability to purify HSCs, have provided increasingly detailed insight into the cells and the early steps involved in the differentiation process.

Several marker combinations have been developed that describe murine HSCs, including [CD117^{high}, CD90.1^{low}, Lin$^{neg/low}$, Sca-1^{pos}] (Morrison & Weissman, 1994), [CD90.1^{low}, Lin^{neg}, Sca-1^{pos}Rhodamine123^{low}] (Kim et al., 1998), [CD34$^{neg/low}$, CD117^{pos}, Sca-1^{pos}, Lin^{neg}] (Osawa et al., 1996), [CD150 ^{pos}, CD48^{neg}, CD244^{neg}] (Kiel et al., 2005) and side-population; cells using Hoechst-dye (Goodell et al., 1996). Each of these combinations allows purification of HSCs to near-homogeneity. Similar strategies have been widened to purify human HSCs, employing markers such as CD34, CD38, Lin, CD90, CD133 and fluorescent substrates for the enzyme, aldehyde dehydrogenase. The use of highly purified human HSCs has been mostly experimental and clinical use normally employs enrichment for one marker, usually CD34. CD34 enrichment yields a population of cells enriched for HSC and blood progenitor cells, but still contains many other cell types. However, limited trials in which highly FACS-purified CD34^{pos} CD90^{pos} HSCs were used as a source of reconstituting cells have demonstrated that rapid reconstitution of the blood system can reliably be obtained using only HSCs (Negrin et al., 2000; Vose et al., 2001).

None of the HSC markers currently used are directly linked to crucial HSC function, and consequently, even within species, markers can differ depending on genetic alleles (Spangrude & Brooks, 1992), mouse strains (Spangrude & Brooks, 1993), developmental stages (Morrison et al., 1995) and cell activation stages (Randall & Weissman, 1997; Sato et al., 1999). In spite of this, there is an obvious connection with HSC markers between divergent species such as humans and mice. However, unless the current efforts at defining the complete HSC gene expression patterns will yield usable markers that are linked to essential functions for maintaining the stemness of the cells (Ramalho et al., 2002; Ivanova et al., 2002), functional analysis will remain necessary to identify HSCs clearly (Domen et al., 1999).

Mouse	Human
CD34low/-	CD 34+
SCA-1+	CD59+
Thy1+/low	Thy1+
CD38+	CD38low/-
C-kit+	C-kit -/low

Table 2. Proposed cell surface markers of undifferentiated haematopoietic stem cells

5.2 Side population

When in adult mouse haematopoietic tissue, unpurified bone marrow cells are labelled with the membrane-permeate DNA binding dye Hoechst 33342, a very small fraction of cells extrudes this dye via a membrane pump (Goodell et al., 1996, 1997, 2001). Analysis of these cells on a flow cytometer equipped with an ultraviolet (UV) laser source allows finding of these cells; when Hoechst-labelled cells are analysed simultaneously through blue and red emission filters, the SP forms a dim tail extending from the normal G1 cell populations. These cells can reconstitute the bone marrow of lethally irradiated mice at an ED$_{50}$ (Effective Dose 50) of fewer than 100 cells, indicating that they are highly enriched for totipotent stem cells. The SP cell subpopulation is also enriched for cells expressing the murine stem cell markers Sca-1 and c-kit, further suggesting that they contain very early haematopoietic progenitors (Goodell et al., 1996).

The SP fraction expresses an ABC transporter, Bcrp-1(ABCG2), on the cell surface and this transporter contributes to efflux of the Hoechst dye from the cells, leading to low levels of staining (Zhou et al., 2001). Interestingly, the bone marrow and peripheral immune system in ABCG2 transporter knockouts animals, is normal, suggesting that the capability to efflux Hoechst 33342 is characteristic of stem cells, but not essential for function (Uchida et al., 2002). Similar SP subpopulations have been observed in primates and humans (Kim et al., 2002; Allen et al., 2002). The SP phenotype, therefore, has become a significant marker for stem cell activity in the identification of these cells and in their physical isolation by fluorescence-activated cell sorting.

6. Mesenchymal stem cells

Stem cells from adult tissues are an interesting source for cell therapy, gene therapy and tissue engineering. These cells normally have limited lineage potential in comparison to embryonic stem cells and this can be advantageous from the viewpoint of controlling cell growth and differentiation in certain therapeutic applications (Barrilleaux et al., 2006; Barry & Murphy, 2004; Haynesworth et al., 1998).

In 1961, bone marrow was shown to have haematopoietic progenitor cells (Till & McCulloch, 1961). In the early 1970s, many investigators confirmed that bone marrow also had cells with fibroblastic morphology that could differentiate into bone, cartilage, fat and muscle (Prockop, 1997). These cells have been variously designated as marrow stromal cells or mesenchymal stem cells, and abbreviated as "MSCs." It has been demonstrated that individual cells from the bone marrow stromal population possessed multilineage potential (Pittenger et al., 1999). Since the recognition of MSCs in bone marrow, cells with the same multilineage potential have been isolated from other tissues, including trabecular bone (Noth et al., 2002; Sottile et al., 2002) , adipose tissue (Lee et al., 2004; Zuk et al., 2001) and umbilical cord (Secco, 2008). The presence of MSCs in adipose tissue has generated special interest because harvesting fat tissue is generally less invasive to the donor than harvesting bone marrow and larger quantities may be available. Adipose-derived MSCs are also called adipose-derived stem cells (ADSCs) and adipose-derived adult stromal or stem cells (ADAS cells). But in addition to ADSCs, umbilical cord is also another interesting source for MSCs and has recently gained some attention.

6.1 Mesenchymal stem cell markers

Cell surface proteins may characterize particular cell types or lineages. In some cases, the role of a specific surface protein and its role in the biology of the cell type is known. However, often the function of the protein has not been determined, but the protein has been shown to be related to a certain type of cell and can serve as a marker. Exclusive diagnostic surface markers for human MSCs have not been identified, however, several surface markers have been found to be commonly associated with hMSCs, including STRO-1, CD105 (endoglin), CD166 (activated leukocyte cell adhesion molecule, ALCAM) (Barry & Murphy, 2004; Gronthos et al., 2001) and more recently CD271 (low affinity nerve growth factor receptor, LNGFR) (Buhring et al., 2007; Quirici et al., 2002). Surface marker antigens can be used to distinguish the cells in a specific preparation and monitor their differentiation. Surface markers that are exclusively positive for a different cell type, for example, the haematopoietic surface markers CD45 and CD34, can be used to search for contamination of MSC preparations with other cell types. Surface markers have also been used for positive and negative immunoselection of MSC cell populations (Buhring et al., 2007; Simmons & Torok-Storb, 1991).

The expressed genes that appear on the hMSC surface include receptors for growth factors, matrix molecules and other cells, and point out how the hMSC will interact with its environment. The flow cytometry analysis also indicates the homogeneity of the hMSC population or whether it is a mixture of different cell types. A wide-ranging, yet incomplete, list of the surface molecules on hMSCs is provided in Table 3.

Surface antigens	
Positive	CD13, CD29, CD44, CD49b(Integrin alpha 2,5), CD54(ICAM1), cd71(Transferrin Rec), CD73(SH-3), CD105(Endoglin.SH-2), CD106(VCAM), CD166(ALCAM)
Negative	CD3, CD4, CD6, CD9, CD10, CD11a, b, CD14, CD15, CD34, CD45, D18 (Integrin beta 2), CD31 (PECAM), CD49d (Intergrin alpha 4), CD50 (ICAM3), CD62E (E-Selectin), CD117(c-kit), CD133

Table 3. Mesenchymal stem cells markers

7. Neural stem cells

Neurogenesis is defined as the procedure of generating new neurons from neural stem cells (NSCs),which consists of the proliferation and fate determination of NSCs, migration and survival of young neurons, and maturation and integration of recently matured neurons (Ming & Song, 2005).

NSCs are defined as undifferentiated cells that developmentally originate from the neuroectodermal layer during early embryogenesis. After neural tube closing, these undifferentiated precursor cells and their immediate progeny compose the neuroepithelial layer that surrounds the lateral, third and fourth ventricles in the midbrain and forebrain, and the central canal in the spinal cord. They are the main source of cells that later form all major structures of the brain and spinal cord (Maric & Barker, 2004).

NSCs have recently attracted a great deal of attention because of their inherent ability to generate all major classes of cells of the nervous system. NSCs have therefore been supposed as a useful resource for potentially repairing and restoring the physiological functions to damaged, diseased or aging neural tissues (Gang, 2000; Anderson, 2001; Temple, 2001; Vaccarino et al., 2001; Vescovi et al., 2001; Weissman et al., 2001).

However, with the accelerated interest in and growth of the NSC field, there has been growing uncertainty around the understanding of what cell phenotype actually makes up a neural stem cell. NSCs in their undifferentiated shape are characterized by a unique bipolar morphology that can help identify them from the heterogeneity associated with early culture. Derivation from human foetal material gives rise to an apparently mixed population of NSCs, exhibiting both classic bipolar NSC morphology and other cell morphologies.

7.1 Neural stem cell markers

The major research limitation is that the cellular preparations used as a source of NSCs are themselves naturally heterogeneous and consisting of both NSCs and self-renewing, but more lineage restricted, progenitors; accordingly making the retrospective studies of NSC biology skewed to an unknown degree. Adding to this is the increasing evidence that implies clear functional differences between neural stem and progenitor cells (Galli et al., 2003; Cai & Rao, 2002). Consequently, there is a critical need to use strategies to identify and isolate pure populations of NSCs and other type cells with the aim of resolving their shared or unique biological properties with respect to cell-fate determination and lineage progression.

NSCs are immunoreactive for a range of neural precursor/radial glia markers such as Nestin, Vimentin, RC2, 3CB2, Sox-2 and brain lipid-binding protein (BLBP). However, subtle differences exist between mouse and human NSCs. For example, hNS cells display moderate levels of glial fibrillary acidic protein (GFAP) expression unlike mouse NSCs (Conti et al., 2005), reflecting the differences between the species in vivo (Malatesta et al., 2000; Rakic, 2003).

So, as mentioned above, the cells which are gathered from neural tissue are heterogeneous and identifying cells is required. Therefore, some markers that are used in studies are listed below:

Neural stem cells	GFAP, Nestin, Prominin, SOX-2
Proliferating cells	Ki-67,BrdU, PCNA
Immature neurons	beta Tubulin,DCX,PSA-NCAM
Radial glia	GLAST, RC2
Mature neurons	NeuN, MAP-2, NF, BLBP
Oligodendrocyte precursors	NG2
Oligdendrocytes	O4, MBP, RIP

Table 4. Neural stem cells markers

8. Spermatogonial stem cells

Germ cells are specific cells that transfer the genetic information of an individual to the next generation. Making functional germ cells is vital for continuation of the germ line of the species. Spermatogenesis, the process of male germ cell production, takes place in the seminiferous tubules of the postnatal testis and is an extremely productive system in the body. In the mammalian testis, more than 20 million sperms per gram of tissue are created daily (Amann, 1986). The high productivity relies on spermatogonial stem cells (SSCs). Similar to other kinds of stem cells in adult tissues, SSCs are self-renewing and produce daughter cells that assign to differentiate throughout the life of the male (Meistrich & van Beek, 1993). In addition, in mammals, SSCs are unique among stem cells in the adult body, because they are the only cells that undergo self-renewal and transmit genes to subsequent generations. Furthermore, SSCs provide an excellent model to study stem cell biology due to the availability of a functional assay that clearly identifies the stem cell (Weissman et al., 2001).

Spermatogonial stem cells derive from primordial germ cells (PGCs), which in turn originate from epiblast cells (embryonal ectoderm) (Lawson KA et al., 1992). Soon after the development of the PGCs, they migrate from the base of the allantois, along the hindgut, finally reaching the genital ridges. The PGCs increase in number during migration, when these cells have reached the genital ridges; their number increases to about 10,000 per gonad (Tam and Snow, 1981). PGCs are single cells that under certain culture conditions can make colonies of cells which morphologically are similar to undifferentiated embryonic stem cells (ESCs) (Resnick JL et al., 1992).When they have arrived in the genital ridges, the PGCs are surrounded by the differentiating Sertoli cells, so seminiferous cords are formed.

The germ cells present within the seminiferous cords are different morphologically from PGCs and are called gonocytes (Clermont and Perey, 1957; Sapsford CS et al., 1962; Huckins & Clermont, 1968) or various subsequent types of pro-spermatogonia (Hilscher B et al;1974). Shortly after birth, the gonocytes restart proliferation to give rise to adult types of spermatogonia (Sapsford CS et al., 1962; Huckins & Clermont, 1968; Vergouwen RPFA et al., 1991; Novi & Saba, 1968; Bellye AR et al., 1977). This happening indicates the start of spermatogenesis.

8.1 Spermatogonial stem cell markers

Since the establishment of the transplantation technique, several new markers and characteristics of spermatogonial stem cells have been identified that can be used to isolate a population from the testis that is enriched for spermatogonial stem cells - Tables 5 and 6.

Markers for positive selection of spermatogonial stem cells	CD9 (Kanatsu-Shinohara M et al., 2004), integrin alpha 6 (Shinohara T et al., 1999), integrin beta 1 (Shinohara T et al., 1999), THY-1, CD24 (Kubota H et al., 2003)
Markers for negative selection of spermatogonial stem cells	c-kit, MHC1, Ly6A(Sca-1), CD34 (Kubota H et al., 2003)

Table 5. Overview of markers that have been successfully used to isolate spermatogonial stem cell populations from the testis by either positive or negative selection

8.2 Testicular side population

So far, four groups have separated a side population of testicular cells; meanwhile, different results were drawn as to whether these were spermatogonial stem cells. The first group reported the existence of a testicular side population. Amazingly, they did not find this population to be capable of colonizing a recipient testis after transplantation and concluded that it did not contain spermatogonial stem cells (Kubota et al., 2003).

Then, two other groups found the testicular side population to be enriched for spermatogonial stem cells (Falciatori et al., 2004; Lassalle et al., 2004). A fourth group then explained that testicular side population cells contain Leydig cell progenitors (Lo et al., 2004; de Rooij.,2004) and later failed to find spermatogonial stem cells in this population (Lo et al., 2005).

The controversial results can probably be explained by the strictness of the FACS gating and the different procedures used to separate the side population. It may be possible to isolate a very pure population of spermatogonial stem cells from the testis using the side population technique, alone or in combination with membrane markers (van Bragt et al., 2005), however, for this to be possible more research needs to be performed to determine the optimal procedures and combinations of markers.

A(s) and	GFRalpha-1(Von Schonfeldt et al., 2004, Hofmann et al, 2005) FC, MACS, IHC, ISH, WM
A(s), A (pr) and A(al)	PLZF (Buaas et al., 2004) (Costaya et al., 2004) FC, ISH, IHC, WM, Mu; OCT4 (Pesce et al., 1998) FC, ISH, IHC, WM TG; NGN3 (Yoshida et al., 2004) ISH, TG, WM, ISH; NOTCH1 (Von Schonfeldt et al., 2004) RT–PCR, IHC, SOX3 (Raverot et al., 2005) KO, IHC; c-RET (Meng et al., 2000) IHC, MACS
A spermatogonia	RBM (Jarvis et al., 2005) RT–PCR, IHC
Spermatogonia	EP-CAM (Anderson et al., 1999) FC, IHC, MACS
Premeiotic germ cells	STRA8 (Oulad Abdelghani et al., 1996) RT–PCR, ISH, IHC, WM EE2 (Koshimizu et al., 1995) WB , IHC
Cells on basal membrane and interstitium	CD9 (Kanatsu-Shinohara et al., 2004) FC, IHC, MACS
Spermatogonia, spermatocytes and round spermatids	GCNA1 (Enders & May., 1994) FC, WB, IHC
Premeiotic spermatogonia and postmeiotic spermatid	TAF4B (Falender et al., 2005) FC, KO, IHC

A(s), A-single; A (pr), pair of spermatogonia; A (al), A-aligned spermatogonia; FC, flow cytometry (including FACS); Mu, mutant mouse; TG, transgenic mouse; KO, Knockout mouse; IHC, immunohistochemistry; WM, whole mount immunostaining; WB, Western blot; ISH, in situ hybridization; RT–PCR, reverse transcriptase– PCR; MACS, magnetic-activated cell sorting

Table 6. Overview of markers used to identify spermatogonial stem cells

9. Epidermal stem cells

The skin is the body's strong outer cover that maintains the inside of the body being moist and protects the body from outside assaults by physical, environmental and biological factors. Skin and its associated hair follicles and glandular structures, sebaceous and sweat glands, are made by a stratified epithelium where the position of the cell within the tissue relates to its state of differentiation. The terminally differentiated stratum corneum, hairs and oil-filled sebocytes have a limited lifespan and are constantly shed from the body throughout the adult life. This continual shedding requires that the epithelium is replenished and restored by a stem cell population during normal maintenance of the skin and also in response to injury (Fuchs & Horsley, 2008; Watt et al., 2006). By definition, adult stem cells (ASCs) have the ability to both self-renew and make differentiated progeny (Lajitha, 1979). In healthy skin, epidermal stem cells divide uncommonly, but upon skin injury, stem cells quickly divide to repair the lesion.

There has been important progress in the recognizing of epidermal stem cells (ESCs) since the 1970s, when the idea of interfollicular epidermis was firstly suggested; later, much work was focused on the specific region of the hair follicle outer root sheath, mainly the bulge region. Hair follicle stem cells are multipotent, capable of giving rise to all cell types of the hair, the epidermis and the sebaceous gland (Morris et al., 2004).

9.1 Epidermal stem cells markers

Recognizing the ESCs is major progress in the field of skin biology which lets scientists examine their biochemical properties, lineage and their relation to other cells. There is evidence of ESCs in the bulge region of the hair follicles (Myung et al., 2009a, 2009b; Zhang et al, 2009), as well as in the interfollicular epidermis (Abbas & Mahalingam, 2009; Ambler & Maatta, 2009). When the epidermis undergoes severe damage, it may fully regenerate from the ESCs of the bulge (Watt, 2006). The ESCs present in the bulge and interfollicular epidermis are potentially interconvertible, but under normal conditions they only differentiate a more confined progeny.

ESCs can be identified in vivo by label retention or in vitro by clonogenicity, but neither of these methods allows easy isolation of stem cells for analysis. Therefore, there is a strong need for specific ESCs markers to be identified.

Identifying stem cells by their cell cycling properties has limited potential. Therefore, several research groups have undertaken wide attempts to characterize a set of stem cell specific markers. Much of this research has focused on the bulge region, as this is the most clearly defined stem cell niche in the skin (Fuchs & Horsley, 2008; Watt et al., 2006).

Many efforts have been made in recent years to recognize ESCs. The potential candidate hair follicle stem cells markers include integrin beta 1, keratin 15, keratin 19, CD71, transcription factor p63 and CD34 (Ma et al., 2004). Keratinocyte shows the characteristics of keratin intermediate filaments. In the epidermis, keratins 5 and 14 are expressed in the basal layer, while keratins 1 and 10 are found in the suprabasal layer. The hair follicle stem cells expressed the above keratins and keratins 6, 16 and 17 (Al-Refu et al., 2009; Hoang et al., 2009), and desmosomal proteins, including desmoglein, may serve as negative markers of ESCs (Wan et al., 2003).

In 2001, p63 was identified as a marker for ESCs; p63 is a transcription factor belonging to a family that contains an additional two structurally-related proteins, p53and p73 (Pellegrini et al., 2001). Although p53 fulfils an important role in tumour suppression, p63 and p73 participate in morphogenetic processes (Klein et al., 2010). Their expression is evidenced in ESCs.

CD34 is also a specific marker for bulge keratinocytes. The mouse bulge marker CD34, often used for isolating murine bulge cells, is expressed below the bulge region in human hair follicles (Ohyama et al., 2006). As mentioned before, CD34 is also a specific marker for haematopoietic stem and progenitor cells, however, much more work is needed to clarify specific markers for ESCs.

10. Conclusion

Flow cytometry is able to rapidly check thousands of cells stained with antibodies conjugated to fluorescent dyes. Each cell is individually assessed for a mixture of features such as size and biochemical and/or antigenic composition. High accuracy and sensitivity, combined with the large numbers of cells that can be examined, allows resolution of even very minor subpopulations from complex mixtures with high levels of statistical validity.

As mentioned earlier, the main problem with stem cell research is that a specific marker for each stem cell is not available for researchers and markers usually are common between some cell populations. Therefore, it is clear that we should wait to hear more from future studies to resolve this issue and introduce new and specific markers for each individual stem cell.

11. References

Abbas, O. & Mahalingam, M. (2009). Epidermal stem cells: practical perspectives and potential uses. *Br J. Dermatol.*, 161: 228-236.

Adewumi, O., Aflatoonian, B., Ahrlund-Richter, L. & Amit, M. (2007). Characterization of human embryonic stem cell lines by the International Stem Cell Initiative. International Stem Cell Initiative. *Nat. Biotechnol.*, 25, 803–816.

Allen, JD., van Loevezijn, A., Lakhai, JM., van der Valk, M., van Tellingen, O., Reid, G., Schellens, JH., Koomen, GJ. & Schinkel, AH. (2002). Potent and specific inhibition of the breast cancer resistance protein multidrug transporter in vitro and in mouse intestine by a novel analogue of fumitremorgin C. *Mol Cancer Ther.*, 1:417– 425.

Al-Refu, K., Edward, S., Ingham, E. & Goodfield, M. (2009). Expression of hair follicle stem cells detected by cytokeratin 15 stain: implications for pathogenesis of the scarring process in cutaneous lupus erythematosus. *Br J. Dermatol.*, 160: 1188-1196.

Amann, RP. (1986). Detection of alterations in testicular and epididymal function in laboratory animals. *Environ. Health Perspect.*, 70, 149–158.

Ambler, CA. & Maatta, A. (2009). Epidermal stem cells: location, potential and contribution to cancer. *J. Pathl* 217: 206-216.

Anderson, DJ. (2001). Stem cells and pattern formation in the nervous system: the possible versus the actual. *Neuron*, 30, 19-35.

Anderson, R., Schaible, K., Heasman, J. & Wylie, C. (1999). Expression of the homophilic adhesion molecule, Ep-CAM, in the mammalian germ line. *J. Reprod. Fertil.*, 116:379–84.

Barrilleaux, B., Phinney, DG., Prockop, D. J. & O'Connor, KC. (2006). Review: Ex vivo engineering of living tissues with adult stem cells. *Tissue Eng.*, 12, 3007–3019.

Barry, FP. & Murphy, JM. (2004). Mesenchymal stem cells: clinical applications and biological characterization. *Int. J. Biochem. Cell Biol.*, 36, 568–584.

Baum, CM., Weissman, IL., Tsukamoto, AS., Buckle, AM. & Peault, B. (1992). Isolation of a candidate human hematopoietic stem-cell population. *Proc Natl Acad Sci USA*, Vol 89,2804-2808.

Bauman, JG., de Vries, P., Pronk, B. & Visser, JW. (1988). Purification of murine hemopoietic stem cells and committed progenitors by fluorescence activated cell sorting using wheat germ agglutinin and monoclonal antibodies. *Acta Histochem Suppl.*, Vol 36,241-253.

Bellve, AR., Cavicchia, JC., Millette, CF., O'Brien, DA., Bhatnagar, YM. & Dym,M. (1977). Spermatogenic cells of the prepuberal mouse. Isolation and morphological characterization. *J. Cell Biol.*, 74:68– 85.

Bonner, WA., Hulett, HR., Sweet, RG. & Herzenberg, LA. (1972). Fluorescence activated cell sorting. *Rev. Sci. Instrum.*, 1972:43, 404–409.

Brimble, SN., Sherrer, ES., Uhl, EW. & Wang, E. (2007). The cell surface glycosphingolipids SSEA-3 and SSEA-4 are not essential for human ESC pluripotency. *Stem Cells*, 25, 54–62.

Buaas, FW., Kirsh, AL., Sharma, M., McLean, DJ., Morris, JL. & Griswold, MD. (2004). Plzf is required in adult male germ cells for stem cell self-renewal. *Nat Genet.*, 36:647–52.

Buhring, H-J., Battula, V L., Treml, S., Schewe, B., Kanz, L. & Vogel, W. (2007). Novel markers for the prospective isolation of human MSC. *Ann. N. Y. Acad. Sci.*, 1106, 262–271.

Cai, J. & Rao, MS. (2002) Stem cell and precursor cell therapy. *Neuromolecular Med.* 2, 233–249.

Carlson, BM. (1996). *Patten's Foundations of Embryology*. 6th edn. New York, McGraw-Hill.

Carpenter, MK., Rosler, E. & Rao, MS. (2003).Characterization and differentiation of human embryonic stem cells.*Cloning Stem Cells*,5(1):79–88.

Chambers, I., Colby, D., Robertson, M., Nichols, J., Lee, S. & Tweedie, S. (2003). Functional expression cloning of Nanog, a pluripotency sustaining factor in embryonic stem cells. *Cell*, 113(5):643–655.

Civin, CI., Strauss, LC., Brovall, C., Fackler, MJ., Schwartz, JF. & Shaper, JH. (1984). Antigenic analysis of hematopoiesis. III. A hematopoietic progenitor cell surface antigen defined by a monoclonal antibody raised against KG- 1a cells. *J. Immunol.*, Vol 133,157–165.

Clermont, Y. & Perey, B. (1957). Quantitative study of the cell population of the seminiferous tubules of immature rats. *Am J. Anat*, 100:241–68.

Conti, L., Pollard, SM., Gorba,T., Reitano, E., Toselli, M., Biella, G., Sun, Y., Sanzone, S., Ying Q-L, Cattaneo E. & Smith, A. (2005). Niche-independent symmetrical self-renewal of a mammalian tissue stem cell. *PLoS Biol*, 3:e283.

Costoya, JA., Hobbs, RM., Barna, M., Cattoretti, G., Manova, K. & Sukhwani M. (2004). Essential role of Plzf in maintenance of spermatogonial stem cells. *Nat Genet*, 36:653–9.

de Rooij, DG. & Van Bragt, MP. (2004). Leydig cells: testicular side population harbors transplantable leydig stem cells. *Endocrinology*, 2004, 145:4009–10.

Doetschman, T., Williams, P., Maeda, N. (1988). Establishment of hamster blastocyst-derived embryonic stem cells. *Dev. Biol*, 127: 224–227.

Domen, J. & Weissman, IL. (1999). Self-renewal, differentiation or death: regulation and manipulation of hematopoietic stem cell fate. *Mol Med Today*. Vol 5,201–208.

Draper, JS., Pigott, C., Thomson, JA & Andrews, PW. (2002). Surface antigens of human embryonic stem cells: changes upon differentiation in culture. *J. Anat*, 200(Part 3):249–258.

Eiges, R., Schuldiner, M., Drukker, M., Yanuka, O., Itskovitz-Eldor, J. & Benvenisty, N. (2001). Establishment of human embryonic stem cell-transduced clones carrying a marker of undifferentiated cells. *Curr. Biol.* 11, 514–518.

Ellis, P., Fagan, B. M., Magness, S. T., Hutton, S., Taranova, O., Hayashi, S., McMahon, A., Rao, M. & Pevny, L. (2004). SOX2, a persistent marker for multipotential neural stem cells derived from embryonic stem cells, the embryo or the adult. *Dev. Neurosci* 26,148-65.

Enders, GC, May JJ, 2nd. Developmentally regulated expression of a mouse germ cell nuclear antigen examined from embryonic day 11 to adult in male and female mice. *Dev. Biol.,*1994, 163:331-40.

Evans, MJ., Kaufman, MH. (1981). Establishment in culture of pluripotential stem cells from mouse embryos. *Nature*, 291:154-156.

Falciatori, I., Borsellino, G., Haliassos, N., Boitani, C., Corallini, S. & Battistini, L. (2004). Identification and enrichment of spermatogonial stem cells displaying side-population phenotype in immature mouse testis. *Faseb J*,18:376-8.

Falender, AE., Freiman, RN., Geles, KG., Lo, KC., Hwang, K. & Lamb, DJ. (2005). Maintenance of spermatogenesis requires TAF4b, a gonad-specific subunit of TFIID. *Genes Dev*,19:794-803.

Fuchs, E. & Horsley, V. (2008). More than one way to skin. *Genes Dev*, 22(8):976-985.

Gage, F.H. (2000) Mammalian neural stem cells. *Science* 287, 1433-1438.

Galli R., Gritti A., Bonfanti L., & Vescovi A.L. (2003). Neural stem cells: an overview. *Circ. Res.*92, 598-608.

Goodell, MA., Brose, K., Paradis, G., Conner, AS. & Mulligan, RC. (1996). Isolation and functional properties of murine hematopoietic stem cells that are replicating in vivo. *J. Exp Med.* Vol 183,1797-1806.

Goodell, MA, Rosenzweig M, Kim H, Marks DF, De Maria M, Paradis G, Grupp S, Seiff CA, Mulligan RC, Johnson RP. (1997). Dye efflux studies suggest that hematopoietic stem cells expressing low or undetectable levels of CD34 antigen exist in multiple species. *Nat Med*, 3:1337-1345.

Goodell, MA. (2001). Stem cell identification and sorting using the Hoechst 33342 side population (SP). In: Robinson JP, Darzynkiewicz Z, Dean PN, Hibbs AR, Orfao A, Rabinovitch PS, Wheeless LL, editors. *Current protocols in cytometry*. New York: John Wiley & Sons; P 9.18.1–9.18 -21.

Graves, KH., Moreadith, RW. (1993). Derivation and characterization of putative pluripotential embryonic stem cells from preimplantation rabbit embryos. *Mol Reprod Devel*, 36:424-433.

Gronthos, S., Franklin, DM., Leddy, HA., Robey, PG., Storms, RW. & Gimble, JM. (2001). Surface protein characterization of human adipose tissue-derived stromal cells. *J. Cell. Physiol.* 189, 54-63.

Gunji, Y., Nakamura, M. & Osawa, H. (1993). Human primitive hematopoietic progenitor cells are more enriched in KITlow cells than in KIThigh cells. *Blood,* Vol 82,3283-3289.

Haynesworth, S. E., Reuben, D. & Caplan, A. I. (1998). Cell-based tissue engineering therapies: the influence of whole body physiology. *Adv. Drug Deliv. Rev.* 33, 3–14.

Heins, N., Englund, MC., Sjöblom, C., Dahl, U., Tonning, A., Bergh, C & Lindah, L. (2004). Derivation, characterization and differentiation of human embryonic stem cells. *Stem Cells*, 22(3):367-376.

Herzenberg, LA. & De Rosa, SC. (2000). Monoclonal antibodies and the FACS: complementary tools for immunobiology and medicine. *Immunol. Today.* 21, 383-390.

Hill, B., Rozler, E. & Travis, M. (1996). High-level expression of a novel epitope of CD59 identifies a subset of CD34& bone marrow cells highly enriched for pluripotent stem cells. *Exp Hematol.* Vol 24,936–943.

Hilscher, B., Hilscher, W., Bulthoff-Ohnolz, B., Kramer, U., Birke, A. & Pelzer, H. (1974). Kinetics of gametogenesis. I. Comparative histological and autoradiographic studies of oocytes and transitional prospermatogonia during oogenesis and prespermatogenesis. *Cell Tissue Res* ,154: 443–70.

Hoang, MP., Keady, M. & Mahalingam, M. (2009).Stem cell markers (cytokeratin 15, CD34 and nestin) in primary scarring and nonscarring alopecia. *Br J. Dermatol* 160: 609–615.

Hofmann, MC., Braydich-Stolle, L. & Dym, M. (2005). Isolation of male germ-line stem cells, influence of GDNF. *Dev. Biol,* 279:114–24.

Huckins, C. & Clermont, Y. (1968). Evolution of gonocytes in the rat testis during late embryonic and early post-natal life. *Arch Anat Histol Embryol,* 51:341–54.

Ikuta, K. & Weissman, IL. (1992) Evidence that hematopoietic stem cells express mouse c-kit but do not depend on steel factor for their generation. *Proc Natl Acad Sci USA.* Vol 89,1502–1506.

Ivanova, NB., Dimos, JT., Schaniel, C., Hackney, JA., Moore, KA. & Lemischka, IR. (2002). A stem cell molecular signature. *Science,* Vol 298,601–604.

Jackson, K., Majka, SM., Wang, H., Pocius, J., Hartley, CJ., Majesky, MW., Entman, ML., Michael, LH., Hirschi, KK. & Goodell, MA. (2001). Regeneration of ischemic cardiac muscle and vascular endothelium by adult stem cells. *J. Clin. Invest.* 107(11), 1395–1402.

Jarvis, S., Elliott, DJ., Morgan, D., Winston, R. & Readhead, C. (2005). Molecular markers for the assessment of postnatal male germ cell development in the mouse. *Hum Reprod* 2005,20:108–16.

Jorda, CT., McKearn, JP. & Lemischka, IR. (1990). Cellular and developmental properties of fetal hematopoietic stem cells. *Cell,* Vol 61,953–963.

Julius, MH., Masuda, T. & Herzenberg, LA. (1972). Demonstration that antigen-binding cells are precursors of antibody-producing cells after purification with a fluorescence-activated cell sorter. *Proc. Natl. Acad. Sci. U. S. A.* 69, 1934–1938.

Kanatsu-Shinohara, M., Toyokuni, S. & Shinohara, T. (2004). CD9 is a surface marker on mouse and rat male germline stem cells. *Biol. Reprod,* 70:70–5.

Kempermann, G., Jessberger, S., Steiner, B. & Kronenberg, G. (2004) Milestones of neuronal development in the adult hippocampus. *Trends Neurosci* 27, 447–52.

Kiel, MJ., Yilmaz, OH., Iwashita, T.,Terhorst , C. & Morrison, SJ. (2005) SLAM family receptors distinguish hematopioetic stem and progenitor cells and reveal enothelial niches for stem cells. *Cell,* 121:1109–1121.

Kim, M., Cooper, DD., Hayes, SF. & Spangrude, GJ. (1998). Rhodamine123 staining in hematopoietic stem cells of young mice indicates mitochondrial activation rather than dye efflux. *Blood,* Vol 91,4106–4117.

Kim, M., Turnquist, H., Jackson, J., Sgagias, M., Yan,Y., Gong, M., Dean, M., Sharp, JG. & Cowan, K. (2002). The multidrug resistance transporter ABCG2 (breast cancer resistance protein 1) effluxes Hoechst 33342 and is over expressed in hematopoietic stem cells. *Clin Cancer Res,* 8:22–28.

Klein, AM., Brash, DE., Jones, PH. & Simons, BD. (2010). Stochastic fate of p53-mutant epidermal progenitor cells is tilted toward proliferation by UVB during preneoplasia. *Proc Natl Acad Sci USA* 107: 270-275.

Koshimizu, U., Nishioka, H., Watanabe, D., Dohmae, K. & Nishimune, Y. (1995). Characterization of a novel spermatogenic cell antigen specific for early stages of germ cells in mouse testis. *Mol Reprod Dev*, 40:221-7.

Kubota, H., Avarbock, MR. & Brinster, RL. (2003). Spermatogonial stem cells share some, but not all, phenotypic and functional characteristics with other stem cells. *Proc Natl Acad Sci U S A* ,100:6487-92.

Lajtha, LG. (1979). Stem cell concepts. *Differentiation*, 14(1-2):23-34.

Lansdorp, PM., Sutherland, HJ. & Eaves, CJ. (1990). Selective expression of CD45 isoforms on functional subpopulations of CD34& hemopoietic cells from human bone marrow. *J. Exp Med.*. Vol 172,363-366.

Lassalle, B., Bastos, H., Louis, JP., Riou, L., Testart, J. & Dutrillaux, B. (2004). 'Side Population' cells in adult mouse testis express Bcrp1 gene and are enriched in spermatogonia and germinal stem cells. *Development*, 131:479-87.

Lawson, KA. & Pederson, RA. (1992). Clonal analysis of cell fate during gastrulation and early neurulation in the mouse. In: Ciba Foundation Symposium 165. Post implantation development in the mouse. New York: John Wiley & Sons, 3-26.

Lee, RH., Kim, B., Choi, I., Kim, H., Choi, HS., Suh, K., Bae, Y. C. & Jung, J. S. (2004). Characterization and expression analysis of mesenchymal stem cells from human bone marrow and adipose tissue. *Cell Physiol. Biochem.* 14, 311-324.

Lo, KC., Brugh,VM., Parker, M. & Lamb, DJ. (2005). Isolation and enrichment of murine spermatogonial stem cells using rhodamine 123 mitochondrial dye. *Biol. Reprod.*, 72:767-71.

Lo, KC., Lei, Z., Rao, Ch, V., Beck, J. & Lamb, DJ. (2004). De novo testosterone production in luteinizing hormone receptor knockout mice after transplantation of Leydig stem cells. *Endocrinology*, 145:4011-5.

Ma, DR., Yang , EN. & Lee, ST. (2004). A Review: The Location, Molecular Characterisation and Multipotency of Hair Follicle Epidermal Stem Cells. *Ann Acad Med Singapore*, 33:784-8

Malatesta, P., Hartfuss, E. & Gotz, M. (2000) Isolation of radial glial cells by fluorescent-activated cell sorting reveals a neuronal lineage. *Development*, 127: 5253-5263.

Maric, D. & Barker, J. L. (2004). Neural Stem Cells Redefined A FACS Perspective. *Molecular Neurobiology*: 04:30(1): 49-76.

Martin, G. Isolation of a pluripotent cell line from early mouse embryos cultured in medium conditioned by teratocarcinoma cells. *Proc Natl Acad Sci U S A* 1981, 78: 7634-7638.

Meistrich, M. L. & van Beek, M. E. A. B. (1993). Spermatogonial stem cells. In *Cell and Molecular Biology of the Testis* (C. Desjardins, and LL. Ewing, eds.), pp. 266-295. Oxford University Press, New York.

Meng, X., Lindahl, M., Hyvonen, ME., Parvinen, M., de Rooij, DG. & Hess, MW. (2000). Regulation of cell fate decision of undifferentiated spermatogonia by GDNF. *Science* ,287:1489-93.

Merok, J. R & Sherley, JL. (2001). Breaching the kinetic barrier to in vitro somatic stem cell propagation. *J. Biomed. Biotech.* 1:25-27.

Ming, G. L. & Song, H. (2005) Adult neurogenesis in the mammalian central nervous system. *Annu Rev Neurosci.*, 28, 223–50.

Minguell, JJ., Erices, A. & Conget, P. (2001). Mesenchymal stem cells. *Exp. Biol. Med.*, 226:506–520.

Miraglia, S., Godfrey, W. & Yin, AH. (1997). A novel five-transmembrane hematopoietic stem cell antigen: isolation, characterization, and molecular cloning. *Blood*, Vol 90,5013–5021.

Moore, T., Huang, S., Terstappen, LW., Bennett, M. & Kumar, V. (1994). Expression of CD43 on murine and human pluripotent hematopoietic stem cells. *J. Immunol.* Vol 153,4978–4987.

Morris, RJ., Liu, Y., Marles, L., Yang, Z., Trempus, C. & Li, S. (2004). Capturing and profiling adult hair follicle stem cells. *Nat Biotech*, 22:411-7.

Morrison, SJ. & Weissman, IL. (1994). The long-term repopulating subset of hematopoietic stem cells is deterministic and isolatable by phenotype. *Immunity*, 1:661–673.

Morrison, SJ., Hemmati, HD., Wandycz, AM. & Weissman, IL. (1995). The purification and characterization of fetal liver hematopoietic stem cells. *Proc Natl Acad Sci USA*. Vol 92,10302–10306.

Myung, P., Andl T & Ito, M. (2009b). Defining the hair follicle stem cell (Part II). *J. Cutan Pathol.*, 36: 1134-1137.

Myung, P., Andl T. & Ito, M. (2009a). Defining the hair follicle stem cell (Part I). *J. Cutan Pathol.*, 36: 1031-1034.

Negrin, RS., Atkinson, K. & Leemhuis T. (2000). Transplantation of highly purified CD34&Thy-1& hematopoietic stem cells in patients with metastatic breast cancer. *Biol. Blood Marrow Transplant*, Vol 6,262–271.

Nichols, J., Zevnik, B., Anastassiadis, K., Niwa, H., Klewe-Nebenius, D & Chambers, I. (1998). Formation of pluripotent stem cells in the mammalian embryo depends on the POU transcription factor Oct 4. *Cell*, 95(3):379–391.

Noth, U., Osyczka, AM., Tuli, R., Hickok, NJ., Danielson, KG. &Tuan, RS. (2002). Multilineage mesenchymal differentiation potential of human trabecular bone-derived cells. *J. Orthop. Res.*, 20, 1060–1069.

Novi, AM., & Saba, P. (1968). An electron microscopic study of the development of rat testis in the first 10 postnatal days. *Z Zellforsch Mikrosk Anat* , 86:313–26.

Ohyama, M., Terunuma, A., Tock, CL., Radonovich, MF., Pise- Masison, CA. & Hopping, SB. (2006). Characterization and isolation of stem cell-enriched human hair follicle bulge cells. *J. Clin Invest*, 116(1):249–260.

Osawa M., Hanada, K., Hamada, H., Nakauchi, H. (1996). Long-term lymphohematopoietic reconstitution by a single CD34low/negative hematopoietic stem cell. *Science*, Vol 273,242–245.

Oulad Abdelghani, M., Bouillet, P., Decimo, D., Gansmuller, A., Heyberger, S. & Dolle, P. (1996). Characterization of a premeiotic germ cell-specific cytoplasmic protein encoded by Stra8, a novel retinoic acid-responsive gene. *J. Cell Biol*, 135: 469–77.

Pellegrini, G., Dellambra, E., Golisano, O., Martinelli, E., Fantozzi, I., Bondanza, S., Ponzin, D., McKeon, F. & De Luca, M. (2001). P63 identifies keratinocyte stem cells. *Proc Natl Acad Sci USA* 98: 3156-3161.

Pera, M. F., Reubinoff, B. & Trounson, A. (2000). Human embryonic stem cells. *J. Cell Sci.*, 113, 5–10.

Pesce, M., Wang, X., Wolgemuth, DJ. & Scholer, H. (1998). Differential expression of the Oct-4 transcription factor during mouse germ cell differentiation. *Mech Dev*, 71:89–98.

Pittenger, MF., Mackay, AM., Beck, SC., Jaiswal, RK., Douglas, R., Mosca, JD., Moorman, MA., Simonetti, DW., Craig, S. & Marshak, DR. (1999). Multilineage potential of adult human mesenchymal stem cells. *Science*, 284, 143–147.

Prockop, DJ. (1997). Marrowstromal cells as stemcells for nonhematopoietic tissues. *Science*, 276, 71–74.

Quirici, N., Soligo, D., Bossolasco, P., Servida, F., Lumini, C. & Deliliers, GL. (2002). Isolation of bone marrow mesenchymal stem cells by anti-nerve growth factor receptor antibodies. *Exp. Hematol.*, 30, 783–791.

Rakic, P. (2003) Elusive radial glia cells: historical and evolutionary perspective. *Glia* 43: 19–32.

Ramalho-Santos, M., Yoon, S., Matsuzaki, Y., Mulligan, RC. & Melton, DA. (2002). Stemness:transcriptional profiling of embryonic and adult stem cells. *Science*, Vol 298,597–600.

Randall, TD., Lund, FE., Howard, MC. & Weissman, IL. (1996). Expression of murine CD38 defines a population of long-term reconstituting hematopoietic stem cells. *Blood*, Vol 87,4057–4067.

Randall, TD. & Weissman, IL. (1997). Phenotypic and functional changes induced at the clonal level in hematopoietic stem cells after 5-fluorouracil treatment. *Blood*, Vol 89,3596–3606.

Raverot, G., Weiss, J., Park, SY., Hurley, L. & Jameson, JL. (2005). Sox3 expression in undifferentiated spermatogonia is required for the progression of spermatogenesis. *Dev. Biol.*, 283:215–25.

Resnick, JL., Bixler, LS., Cheng, L. & Donovan, PJ. (1992). Long-term proliferation of mouse primordial germ cells in culture. *Nature*, 359:550–1.

Reubinoff, BE., Pera, MF., Fong, CY., Trouson, A., Bongso, A. (2000). Embryonic stem cell lines from human blastocysts: somatic differentiation in vitro. *Nature Biotechnol.*, 18: 399–404

Sadler, TW. (2002). *Langman's Medical Embryology*, 8th edn. Philadelphia, Lippincott Williams and Wilkins.

Sapsford, CS. (1962). Changes in the cells of the sex cords and the seminiferous tubules during development of the testis of the rat and the mouse. *Austr J. Zool*,10:178–92.

Sato, T., Laver, JH., & Ogawa, M. (1999). Reversible Expression of CD34 by Murine Hematopoietic Stem Cells. *Blood*, Vol 94,2548–2554.

Secco, M., Zucconi, E., Vieira, NM., Fogaça, LL., Cerqueira, A.,Carvalho, MD., Jazedje, T., Okamoto, OK., Muotri, AR. & Zatz, M. (2008). Multipotent stem cells from umbilical cord: cord is richer than blood! *Stem Cells*, 26: 146–150.

Shibata, T., Yamada, K., Watanabe, M., Ikenaka, K., Wada, K., Tanaka, K. & Inoue, Y. (1997) Glutamate transporter GLAST is expressed in the radial glia-astrocyte lineage of developing mouse spinal cord. *J. Neurosci.*, 17, 9212–9.

Shinohara, T., Avarbock, MR. & Brinster, RL. (1999). Beta(1)- and alpha(6)-integrin are surface markers on mouse spermatogonial stem cells. *Proc Natl Acad Sci U S A,* 1999, 96:5504–9.

Shinohara, T., Orwig, KE., Avarbock, MR. & Brinster, RL. (2000). Spermatogonial stem cell enrichment by multiparameter selection of mouse testis cells. *Proc Natl Acad Sci U S A,* 97:8346–51.

Simmons, P. J. & Torok-Storb, B. (1991). Identification of stromal cell precursors in human bone marrow by a novel monoclonal antibody, STRO-1. *Blood,* 78, 55–62.

Sottile, V., Halleux, C., Bassilana, F., Keller, H., & Seuwen, K. (2002). Stem cell characteristics of human trabecular bone-derived cells. *Bone* 30, 699–704.

Spangrude, GJ. & Brooks, DM. (1993). Mouse strain variability in the expression of the hematopoietic stem cell antigen Ly-6A/E by bone marrow cells. *Blood,* Vol 82,3327–3332.

Spangrude, GJ., Heimfeld, S. & Weissman, IL. (1988). Purification and characterization of mouse hematopoietic stem cells. *Science,* Vol 241,58–62.

Spangrude, G. & Brooks, DM. (1992). Phenotypic analysis of mouse hematopoietic stem cells shows a Thy-1- negative subset. *Blood,* Vol 80,1957–1964.

Srour, EF., Brandt, JE., Leemhuis, T., Ballas, CB. & Hoffman, R. (1992). Relationship between cytokine-dependent cell cycle progression and MHC class II antigen expression by human CD34& HLA-DR- bone marrow cells. *J. Immunol.* Vol 148,815–820.

Sutherland, DR., Yeo, EL. & Stewart, AK. (1996). Identification of CD34& subsets after glycoprotease selection: engraftment of CD34&Thy-1&Lin– stem cells in fetal sheep. *Exp Hematol.* Vol 24,795–806.

Tam, PP. & Snow, MH. (1981).Proliferation and migration of primordial germ cells during compensatory growth in mouse embryos. *J. Embryol Exp Morphol* 64:133–47.

Temple, S. (2001) Stem cell plasticity—building the brain of our dreams. *Nat. Rev. Neurosci.* 2, 513–520.

Terstappen, LW., Huang, S., Safford, M., Lansdorp, PM. & Loken, MR. (1991).Sequential generations of hematopoietic colonies derived from single nonlineage-committed CD34&CD38– progenitor cells. *Blood,* Vol 77,1218–1227.

Thomson, JA., Itskovitz-Eldor, J., Shapiro SS, et al. (1998). Embryonic stem cell lines derived from human blastocysts. *Science,* 282: 1145–1147.

Thomson, JA., Kalishman, J., Golos, TG., et al. (1995). Isolation of a primate embryonic stem cell line. *Proc Natl Acad Sci U S A,*92: 7844–7848

Thrasher, J. D. (1966) Analysis of renewing epithelial cell populations. In: *Methods in Cell Physiology,* vol. 2 (Prescott, D. M., ed.), Academic, New York, pp. 323–357.

Till, J. E. & McCulloch, C. E. (1961). A direct measurement of the radiation sensitivity of normal mouse bone marrow cells. *Radiat. Res.* 14, 213–222.

Toresson, H., Mata de Urquiza, A., Fagerstrom, C., Perlmann, T. & Campbell, K. (1999). Retinoids are produced by glia in the lateral ganglionic eminence and regulate striatal neuron differentiation. *Development,* 126, 1317–26.

Tsukamoto, A., Weissman, I. & Chen, B. (1995). Phenotypic and functional analysis of hematopoietic stem cells in mouse and man. In: Mertelsman La, ed. *Hematopoietic Stem Cells, Biology and Therapeutic Applications: Phenotypic and functional analysis of hematopoietic stem cells in mouse and man.*

Uchida, N., Leung, FY. & Eaves, CJ. (2002). Liver and marrow of adult mdr-1a/ 1b (-/-) mice show normal generation, function, and multi-tissue trafficking of primitive hematopoietic cells. *Exp Hematol.*, 30:862– 869.

Uchida, N. & Weissman, IL. (1992). Searching for hematopoietic stem cells: evidence that Thy-1.1lo Lin- Sca-1& cells are the only stem cells in C57BL/Ka-Thy-1.1 bone marrow. *J. Exp Med.*, Vol 175,175–184.

Uchida, N., Yang, Z. & Combs, J. (1997). The characterization, molecular cloning, and expression of a novel hematopoietic cell antigen from CD34& human bone marrow cells. *Blood,* Vol 89,2706–2716.

Vaccarino, FM., Ganat, Y., Zhang, Y., & Zheng W. (2001) Stem cells in neurodevelopment and plasticity. *Neuropsychopharmacology,* 25, 805–815.

van Bragt, MP., Ciliberti, N., Stanford, WL., de Rooij, DG. & van Pelt, AM. (2005). LY6A/E (SCA-1) expression in the mouse testis. *Biol. Reprod.,* 2005, 73:634–8. Epub,Jun 1.

Vergouwen, RPFA., Jacobs, SG., Huiskamp, R., Davids, JAG. & de Rooij, DG. (1991). Proliferative activity of gonocytes, Sertoli cells and interstitial cells during testicular development in mice. *J. Reprod. Fertil.,* 93:233–43.

Vescovi, A.L., Galli R., & Gritti A. (2001) The neural stem cells and their transdifferentiation capacity. *Biomed. Pharmacother.,* 55, 201–205.

von Schonfeldt, V., Wistuba, J. & Schlatt, S. (2004). Notch-1, c-kit and GFRalpha-1 are developmentally regulated markers for premeiotic germ cells. *Cytogenet Genome Res,* 105:235–9.

Vose, JM., Bierman, PJ. & Lynch, JC. (2001). Transplantation of highly purified CD34&Thy-1& hematopoietic stem cells in patients with recurrent indolent non-Hodgkin's lymphoma. *Biol. Blood Marrow Transplant,* Vol 7,680–687.

Wan, H., Stone, M., Simpson, C., Reynols, L., Marshall, J., Hart, I., Dilke, K. & Eady, R. (2003). Desmosomal proteins, including desmoglein 3, serve as novel negative markers for epidermal stem cell-containing population of keratinocytes. *J. Cell Sci.,* 116: 4239-4248.

Watt, FM., Lo Celso, C. & Silva-Vargas, V. (2006). Epidermal stem cells: an update. *Curr Opin Genet Dev.,* 16(5):518–524.

Wcissman, IL., Anderson, DJ, & Gage, F. (2001) Stem and progenitor cells: origins, phenotypes, lineage commitments, and transdifferentiations. *Annu. Rev. Cell. Dev. Biol.,* 17, 387–403.

Wiesmann, A., Phillips, RL. & Mojica, M. (2000). Expression of CD27 on murine hematopoietic stem and progenitor cells. *Immunity,* Vol 12,193–199.

Yin, AH., Miraglia, S. & Zanjani, ED. (1997). AC133, a novel marker for human hematopoietic stem and progenitor cells. *Blood,* Vol 90,5002–5012.

Yoshida, S., Takakura, A., Ohbo, K., Abe, K., Wakabayashi, J. & Yamamoto, M. (2004). Neurogenin3 delineates the earliest stages of spermatogenesis in the mouse testis. *Dev. Biol.,* 269:447–58.

Zhang, YV., Cheong, J., Ciapurin, N., McDermitt, DJ. & Tumbar, T. (2009). Distinct self-renewal and differentiation phases in the niche of infrequently dividing hair follicle stem cells. *Cell Stem Cell,* 5: 267-278.

Zhao, X., Ueba, T., Christie, BR., Barkho, B., McConnell, MJ., Nakashima, K., Lein, ES., Eadie, BD., Willhoite, AR., Muotri, AR., Summers, RG., Chun, J., Lee, KF. & Gage,

FH. (2003) Mice lacking methyl-CpG binding protein 1 have deficits in adult Neurogenesis and hippocampal function. *Proc Natl Acad Sci U S A,* 100, 6777–82.

Zhou, S., Schuetz JD., Bunting KD., Colapietro AM., Sampath J., Morris JJ., Lagutina I., Grosveld GC., Osawa M., Nakauchi H., Sorrentino BP. (2001). The ABC transporter Bcrp1/ABCG2 is expressed in a wide variety of stem cells and is a molecular determinant of the side-population phenotype. *Nat Med.,* 7:1028 –1034.

Zuk, PA., Zhu, M., Mizuno, H., Huang, J., Futrell, JW., Katz, AJ., Benhaim, P., Lorenz, HP., & Hedrick, MH. (2001). Multilineage cells from human adipose tissue: implications for cell-based therapies. *Tissue Eng.,* 7, 211–228.

Experimental Conditions and Mathematical Analysis of Kinetic Measurements Using Flow Cytometry – The FacsKin Method

Ambrus Kaposi[1], Gergely Toldi[1], Gergő Mészáros[1],
Balázs Szalay[2], Gábor Veress[3] and Barna Vásárhelyi[2,4]
[1]*First Department of Pediatrics, Semmelweis University*
[2]*Department of Laboratory Medicine, Semmelweis University*
[3]*Analytix Ltd.*
[4]*Research Group of Pediatrics and Nephrology, Hungarian Academy of Sciences*
Hungary

1. Introduction

Flow cytometry is in use for the assessment of different cell subsets' prevalence for decades. However, the development of specific dyes sensitive for the quickly changing intracellular analytes provided an opportunity for the real-time monitoring of intracellular processes with flow cytometry. In these kinetic measurements additionally to cell prevalence values the time as a novel variable is introduced. This enables researchers to gather data on cellular functionality from a new perspective, as the recording of a kinetic parameter with flow cytometry provides information on intracellular processes in several cell subtypes in a simultaneous manner, where physiological cell-cell interactions (such as cytokines) are still present. This approach may be a useful tool in many different areas of immune research. However, the mathematical formulae to characterize the several millions of data recorded during one measurement are still missing.

During past year our team made efforts to develop algorithms to extract the biologically relevant information from these measurements.

2. Possible ways of analysis of kinetic data obtained with flow cytometry

We define here the term 'kinetic measurement' as when the distribution of a parameter measured varies over time (Figure 1). For commonly used parameters (like FSC, SSC and most fluorescent parameters) this does not apply in general and just simply observing the distribution of these parameters regardless of the time parameter is suitable for statistical comparisons of different measurements. For the description and comparison of kinetic measurements however approaches different from the standard ones should be taken. Table 1 shows a review of the analytical methods already reported in papers on kinetic flow cytometric measurements.

2.1 Dot plot

The simplest way of presenting kinetic information is using a scatter plot (or dot plot) with time parameter and measured analyte on axis x and y, respectively, when each dot represents an individual cell. This approach and variants such as density plot (Figure 1, left), contour plot or 3D densitiy plot (June et al., 1986) are widely used (Table 1). The careful selection of cutoff values enables users to compare measurements qualitatively. The use of this technique is limited by its subjective nature. It does not provide data about the magnitude or shape of the kinetic process either.

2.2 Smoothing method

A more complex approach is to average the kinetic parameter at given small time intervals and to replace the dots in the scatter plot with these averaged values (red or green curves on Figure 1, left). This approach replaces the distribution of the kinetic parameter at each different time interval with one average value calculated from that distribution (red or green lines on Figure 1, right). Another way of explaining this method is that it is essentially a smoothing of the spiked curve that goes through all the dots in the scatter plot. There are several averaging methods that could be used: mean (used in Omann et al., 1990; Lund-Johansen et al., 1992; Rijkers et al., 1993; do Céu Monteiro et al., 1999; Jakubczak et al., 2006; Schepers et al., 2009), geometric mean (Bailey et al., 2006), median (Szalay et al., 2012). The mean is sensitive to outliers and is suitable for the characterization of a normal distribution. The median separates the lower and upper 50% range of the data.

There are also different approaches to define time intervals where the averaging takes place. These include partitioning the whole time-frame into intervals of the same length (used in the method described in Section 3); partitioning the whole time-frame into intervals all containing the same number of cells; having a fixed-length time-frame and shifting it through the whole measurement and calculating the average in these (overlapping) intervals (moving average, used in Rijkers et al., 1993, Bailey et al., 2006); using local regression (lowess method, published in (Cleveland, 1979), used in (Szalay et al., 2012)); using cubic splines etc.

The result of a smoothing method is a curve describing the measurement. If one has more measurements from the same type these curves can be averaged. A curve is an easy way to understand the graphical representation of a measurement and it shows information on how a hypothetical "average" cell behaves during the measurement. When using median as averaging method 50% of the cells showed a kinetic reaction that was higher than that of the "average" cell and 50% of the cells showed a lower reaction[1].

Furthermore, different parameters with possible biological meaning can be calculated (or simply read) from the curve such as maximum (highest value), time to reach maximum, maximal slope, value at given time points, slope at given time points, AUC (area under curve) etc. Note that these calculated parameters differ from the parameters that would be recorded during the measurement for each cell: from a single measurement one can only derive a single value for each such calculated parameter. It is possible to compare the

---

[1] The precise mathematical formulation of "higher" and "lower" in this sense is not within the scope of this chapter.

reference	investigated cell types	dyes used	software used	graphical presentation of kinetic data	standardization against	averaging method	parameters calculated	statistical comparison
June et al., 1986	lymphocytes	Indo-1		3D intensity against time density plot; mean instensity in 100 equal-length time intervals; % of responding cells in 100 equal-length time intervals	first part	Mean	time to 50% maximal response; % of responding cells	-
Omann et al., 1990	leukocytes	Fluo-3	BD Chronys	mean intensity of 2 s intervals with 8 s gaps against time	previous calibration of [Ca^{2+}]	Mean	intensity 10 s after stimulation	-
Lund-Johansen et al., 1992	monocytes, leukocytes	Fura-Red	Cyclops (Cytomation Inc)	intensity against time scatter plot	first 40 s	Mean	intensity at a given time point	t-test
1990, Norgauer et al., 1993	neutrophils	Fluo-3, Indo-1		histogram of intensity values before and 10 s after stimulation; smoothed [Ca^{2+}] against time; averaged [Ca^{2+}] at 10 s time points	previous calibration of [Ca^{2+}]	mean (?)	intensity 10 s after stimulation; maximal intensity	-
Rijkers et al., 1993	lymphocytes	Fluo-3, SNARF-1 (Ca^{2+}), Mag-indo-1 (Mg^{2+})		intensity against time scatter plot; moving mean of instensity against time	?	mean	maximal intensity	-
do Céu Monteiro et al., 1999	platelets	Fluo-3	EPICS XL-MCL, System II software	intensity against time scatter plot; mean intensity every 25 s	first 20 s	mean	maximal intensity	?

Table 1. (continues on next page) Analysis methods used in a selection of articles publishing results of kinetic flow cytometric measurements. *Intensity* refers to the fluorescent intensity value of the measured kinetic parameter. These values are usually given relative to the begininng values so as to show the fold increase of intensity. We call the the method of transforming raw values into fold-increase values (or sometimes actual concentration values) *standardization*

reference	investigated cell types	dyes used	software used	graphical presentation of kinetic data	standardization against	averaging method	parameters calculated	statistical comparison
Jakubczak et al., 2006	Neutrophils	Fluo-3, Fura-Red	Statistica 6.0	mean intensity every 60 s	first 40 s	mean	slope at given time points; intensity at given time points	Mann-Whitney U test
Bailey et al., 2006	PBMC, Jurkat	Fluo-3, Fura-Red, Indo-1	FCSPress 1.3	moving geometric mean intensity against time	first part	geometric mean	maximum	-
Stork et al., 2007	B lymphocytes	Indo-1	FlowJo	moving median (?) intensity against time	none	?	-	-
Demkow et al., 2009	Neutrophils	Fluo-3, Fura-Red	Statistica 6.0	-	first 40 s	mean (?)	intensity after 60s	Mann-Whitney U test
Schepers et al., 2009	granulocytes, monocytes, lymphocytes	Fluo-3		intensity against time scatter plot; mean intensity every 30 s	first 30 s	mean	intensity at every 30 s	paired Wilcoxon-test
Szalay et al., 2012	T cells	Fluo3AM (cytoplasmic Ca^{2+}), Rhod2-AM (mitochondrial $Ca2+$ levels), Dihydroethidium (superoxide), DAF-FM diacetate (NO)	R, Statistica 7	moving median intensity against time in representative measurements	first 5 s	median	AUC; maximum; time to reach maximum	paired Wilcoxon-test, Mann-Whitney U test

Table 1. (continued) Analysis methods used in a selection of articles publishing results of kinetic flow cytometric measurements. *Intensity* refers to the fluorescent intensity value of the measured kinetic parameter. These values are usually given relative to the begininng values so as to show the fold increase of intensity. We call the the method of transforming raw values into fold-increase values (or sometimes actual concentration values) *standardization*

corresponding parameters of different groups of measurements with regular statistical methods, eg. Mann-Whitney U test for 2 groups (used in Jakubczak et al.; 2006, Demkow et al.; 2009, Toldi et al.; 2010b, Szalay et al., 2012), t-test (used in Lund-Johansen et al., 1992), paired Wilcoxon-test (used in Schepers et al., 2009; Toldi et al., 2010a, Szalay et al., 2012).

The fitting of these curves should be standardized as they usually should start at value 1.0. Hence the values represent relative parameter values (rpv). This can be done by dividing the

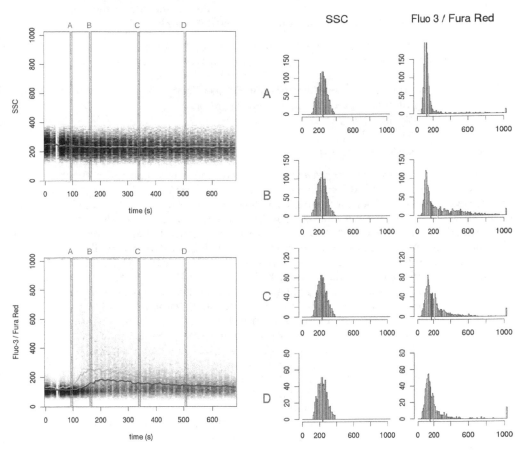

Fig. 1. Left side: 2D density plots of SSC over time and Fluo 3 / Fura Red over time in a representative Ca^{2+}-flux measurement (human CD4+ lymphocytes stimulated with PHA). Black areas represent the highest density while white areas represent zero density. The measurement interval was splitted up to 100 time intervals of equal length and medians (red curve) and means (green curve) in each interval were calculated. Right side: histograms of SSC and Fluo 3 / Fura Red parameters at 4 representative time intervals named A, B, C and D. The red line shows the place of the median and the green line that of the mean. Note that the distribution of SSC is constant over time while that of Fluo 3 / Fura Red changes: there is a sudden shift towards higher values at approx. 110 s and a slow continuous shift towards lower values afterwards

values at each time point by the value at the beginning of the experiment. This changes the values of those calculated parameters that depend on axis y (i.e. the maximum, slope parameters) but not those dependent only on axis x (i.e. time to reach maximum). Sometimes a calibration is made and the measured kinetic parameter value is converted into real biochemical/biophysical units (as in Omann et al., 1990, Norgauer et al., 1993). Another kind of standardization can be done by shifting the whole measurement by subtracting or adding a time value from/to all time points. This is useful when the exact beginning of the kinetic reaction is not known and the goal is to calculate distances from the maximal value or some well-defined time point inside the measured time-frame.

However, the value of curve parameters derived by smoothing method depends very much on the exact method used and the adjustments that are made. Adjustments have to be done for each measurement manually. Moreover, these methods are very sensitive to experimental conditions and provide no qualitative feedback on the shape of the smoothed curve. Smoothing methods are well-established for presenting but not for analyzing the data (Motulsky et al., 1987).

2.3 Fitting a model to the smoothed values

By selecting a mathematical model the assumptions that one would make about a kinetic process are made explicit (Motulsky et al., 2004). By fitting the model to the smoothed values one can test whether the measured values really follow the selected model; if so, the parameters that describe the model can be calculated. The model can be empirically determined or mechanistic. The former describes the general shape of the data and its parameters do not necessarily correspond to a biological process, while the latter is specifically formulated to describe a biological process with parameters such as dissociation constants, catalytic velocities etc. In our case the model can be formulated by a function that takes a time value and some other parameters describing the exact shape of the function and returns a numeric value that estimates the smoothed kinetic parameter value (an example of an empirically determined function is shown in Figure 2). The parameters describing the exact shape correspond to those calculated from the smoothed curve in 2.2. A series of empirical models is described in (Kaposi et al., 2008, Mészáros et al., 2011). We will describe a version improved further in Section 3. Some applications of this method are described in Section 5 (Toldi et al., 2010a-b-c, 2011a-b). Mechanistic models for calcium flux kinetics are also available (Tang et al., 1996; Politi et al., 2006). Fitting a function to the smoothed kinetic parameter values could be done by non-linear least-squares regression (Bates et al., 1988) or by one of the several robust regression methods (Motulsky et al., 2004). A robust way of testing whether the model fits the dataset is cross-validation (Picard et al., 1984).

2.4 Fitting to quantiles

An extension of the smoothing method using median averaging (method 2.2) which is orthogonal to method 2.3 comes from the idea of quantile regression (Koenker et al., 2001). One limitation of methods 2.2 and 2.3 is that they replace the distributions of the kinetic parameter at each time interval with one single value forgetting about the deviation around this average value. Replacing such a distribution with eg. 100 percentile values preserves the shape of the distribution. Afterwards, one could fit models to values corresponding to the same percentile just as in method 2.3. This is the essence of our method "FacsKin" described precisely in Section 3.

$$\text{logist+}(t;\, y0,\, y2,\, x1,\, m1) = \frac{y0 + (y2 - y0)}{1 + e^{\frac{4*m1*(-t-m1)}{y2-y0}}} \quad (y0,\, y2,\, x1,\, m \geq 0 \text{ and } y0 < y2)$$

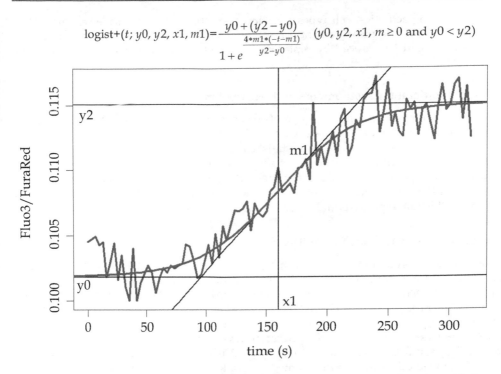

Fig. 2. A kinetic model describing kinetic reactions following an increasing sigmoid shape curve. The corresponding mathematical function is called *logist+* function and has 4 parameters that fully describe the shape of a particular logistic function (*y0, y2, x1* and *m1*, black lines with captions). The time parameter is denoted *t*. The 320 s long kinetic flow cytometric measurement time-frame was divided into 100 time intervals of equal length and the median of the kinetic parameter was calculated in each interval (red curve). The *logist+* function was fitted (blue curve) to these median values with Nelder-Mead optimization method minimizing the sum of absolute deviances (Nelder et al., 1965). The result of the current fit was: *y0=0.1018, y2=0.1151, x1=160.29, m1=0.0001026*

2.5 Fitting to the whole measurement at the same time

Replacing a distribution with percentile (or more generally, quantile) values is unnecessary if the type of the probability distribution is known. In this case fitting the parameters of the probability distribution is a more robust method than fitting functions separately to quantiles. It also helps to avoid overfitting (Hawkins, 2004). Then, it is necessary to describe the change of the parameters of the probability distribution over time by fitting a function to this change as well, one function per parameter for the probability distribution. Having a probability distribution with 2 parameters and two functions each having 8 parameters to describe their changes would result in a model containing *2*8=16* parameters (compare it with the *8*100* parameters for the functions for all 100 percentiles). The same result could be achieved by creating a mechanistic model based on the biological characteristics of the kinetic process but with parameters having exact biological meaning like dissociation constants, enzyme activities, cell types etc. However, this is a great challenge due to the large variety of cells and measurement conditions. The computational capacity required for this approach could be reduced by replacing the whole measurement with percentiles at each time interval.

2.6 Fitting to different measurements at the same time

To statistically compare different (groups of) measurements one can take the (empirical or estimated) distributions of parameters derived from fitting models separately to these measurements and compare them with statistical methods such as Kruskall-Wallis test. In this case, p values should be adjusted for multiple comparisons because each parameter has to be compared separately. A more robust method would be to define a common model for several measurements selecting parameters that are common in different measurements according to the null hypothesis and other parameters that are tested for differences.

3. Description of the FacsKin method

We developed a method for describing and comparing kinetic flow cytometry measurements using method 2.4 described in the previous section. Our aim was to provide a readily usable standard way of analyzing and comparing kinetic measurements. Previous versions of the method were published in (Kaposi et al., 2008) and (Mészáros et al., 2011). A computer program that implements the method is available at the website http://www.facskin.com. The implementation was done in Java (Oracle Inc., Redwood Shores, USA.) and R (R Development Core Team, 2006). This section describes the technical details of the method. A more user-oriented description and tutorial is the User's Guide on the website: http://www.facskin.com/node/3.

3.1 Input data

The input of the method are the time (seconds) and kinetic parameter value (raw measured intensity value or calculated ratio in case of eg. Fluo 3 / Fura Red fluorescent dyes) for each cell in the gated cell population that the user is interested in.

We divide the whole measurement into 100 equal-length time intervals. Sometimes the resolution of the time parameter isn't high enough and the cells are not evenly distributed in the time intervals. To prevent this, we recalculate time values so as to make time points evenly distributed:

$$(t'_i, t'_{i+1}, t'_{i+2}, t'_{i+3}, \ldots t'_{j-1}) := (t_i, t_i + d, t_i + 2*d, t_i + 3*d, \ldots, t_i + (j-i-1)*d),$$

where t_k is the old time value of the kth cell, t'_k is the new time value of the kth cell, $t_{i-1} \neq t_i = t_{i+1} = \ldots = t_{j-1} \neq t_j$, $d = (t_j - t_i) / (j - i)$. In the special case when t_{j-1} is the time value of the last cell, $t_j = t_i + (t_i - t_{i-1})$.

In each time interval we calculate the following 201 quantiles distributed equally: $1/402$, $1/402 + 1/201$, $1/402 + 2/201$, $1/402 + 3/201$, $\ldots 1/402 + 200/201$. We use 201 quantiles so that the quartiles (0.25, 0.5, 0.75 corresponding to 51th, 101th, 151th quantile) can be obtained directly. From now on, we will use these 201 quantiles in each time interval instead of the original measurement data (we replace the original measurement data with 201*100 + 100 = 20,200 values (quantiles + time values)).

3.2 Kinetic models

We defined 5 kinetic models (Figure 3) each for different kind of kinetic measurement:

1. **constant:** the value of the kinetic parameter is constant during the measurement timeframe
2. **logist+:** the kinetic parameter starts at a given value, increases during the measurement timeframe and reaches a given value
3. **logist-:** same as *logist+*, but instead of increasing the kinetic parameter value decreases during the measurement timeframe
4. **dlogist+:** the kinetic parameter starts at a given value, increases, reaches a maximum value and then decreases and reaches a given value during the measurement timeframe
5. **dlogist-:** same as *dlogist+*, but instead of increasing and then decreasing the kinetic parameter value first decreases, reaches a minimum and then increases

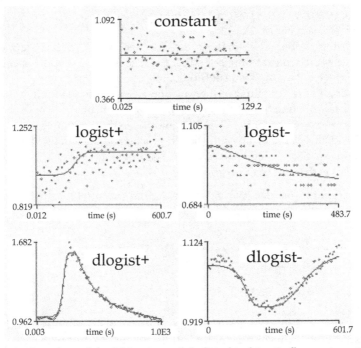

Fig. 3. Representative plots of the 5 kinetic models fitted to kinetic flow cytometric measurements (calcium flux). The dots represent the median values (quantile 0.5) at each of the 100 time intervals, the curves the functions fitted on these dots

Each model corresponds to a function. Here follows the formula and parameters for each function. Most of the parameters are constrained which means that the domain of the function is restricted to such values. *Logist+* and *logist-* differ only in their constraints and the meaning of their parameters, this is true for *dlogist+* and *dlogist-* as well. The reason for separating these functions is that we would like to avoid fitting increasing functions to some quantiles while decreasing functions to other quantiles of the same measurement.

1. **constant:** The function is a horizontal line having the same value (y) all the time. Formula:

$$constant\ (t;\ y) = y$$

Constraints: $y \geq 0$.
Parameter: y: constant value

2. **logist+:** an S-shape function that starts at a given value $(y0)$ increases and reaches a higher given value $(y2)$. Formula:

$$logist + (t;\ y0,\ y2,\ x1,\ m1) = y0 + \frac{y2 - y0}{1 + e^{\frac{(-t+x1)*4*m}{y2-y0}}}$$

Constraints: $y0 \geq 0$, $y2 \geq 0$, $x1 \geq 0$, $m1 \geq 0$, $y0 < y2$.
Parameters (see also Figure 2):
* $y0$: starting value. The limit of the function at $-\infty$ (minus infinity). It is not necessarily the value at time point 0. If the function begins with a steep, the starting value is lower than the value at time point 0.
* $y2$: ending value. The limit of the function at $+\infty$ (positive infinity). Not necessarily the value of the function at the end of the measurement.
* $x1$: time to reach 50% value. The time point when the function reaches the 50% value. The 50% value is the mean of the starting value and the ending value (unit: s).
* $m1$: slope at 50% value. The slope of the function at the 50% value (unit: int/s where int is the unit of the vertical axis).

3. **logist-:** an S-shape function that starts at a given value $(y0)$ decreases and reaches a lower given value $(y2)$. Formula:

$$logist - (t;\ y0,\ y2,\ x1,\ m1) = y0 + \frac{y2 - y0}{1 + e^{\frac{(-t+x1)*4*m}{y2-y0}}}$$

Constraints: $y0 \geq 0$, $y2 \geq 0$, $x1 \geq 0$, $m1 \geq 0$, $y0 < y2$.
Parameters:
* $y0$: starting value. The limit of the function at $-\infty$ (minus infinity). It is not necessarily the value at time point 0. If the function begins with a steep, the starting value is higher than the value at time point 0.
* $y2$: ending value. The limit of the function at $+\infty$ (positive infinity). Not necessarily the value of the function at the end of the measurement.
* $x1$: time to reach 50% value. The time point when the function reaches the 50% value. The 50% value is the mean of the starting value and the ending value (unit: s).
* $m1$: slope at 50% value. The slope of the function at the 50% value (unit: int/s where int is the unit of the vertical axis.).

4. **dlogist+:** a function that starts at a given value $(y0)$, has an increasing phase, reaches a maximum $(y1)$, has a decreasing phase and reaches a given ending value $(y2)$. Formula:

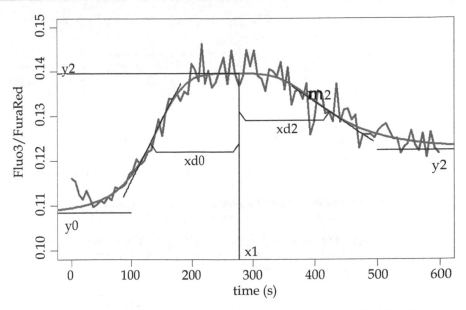

Fig. 4. Parameters of the *dlogist+* function (the formula is given in 3.2.5). The 600 s long
kinetic flow cytometric measurement time-frame was divided into 100 time intervals of
equal length and the median of the kinetic parameter was calculated in each interval (red
curve). The *dlogist+* function was fitted (blue curve) to these median values with Nelder-
Mead optimization method minimizing the sum of absolute deviances (Nelder et al., 1965).
The result of the current fit was: $y0=0.1083$, $y1=0.1395$, $y2=0.1222$, $x1=278.4$, $xd0=146.7$,
$xd2=145.0$, $m0=2.657*10^{-4}$, $m2=-9.519*10^{-5}$

$$\text{dlogist} + (t; y0, y1, y2, x1, xd0, xd2, m0\ m2) = \begin{cases} y0 + \dfrac{y1-y0}{1+\left(\dfrac{x1-t}{xd0}\right)^{\frac{4*xd0*m0}{y1-y0}}}, & t<x1 \\[20pt] y2 + \dfrac{y1-y2}{1+\left(\dfrac{t-x1}{xd2}\right)^{\frac{4*xd2*m2}{y2-y1}}}, & t\geq x1 \end{cases}$$

Constraints: $y0 \geq 0$, $y1 \geq 0$, $y2 \geq 0$, $m0 \geq 0$, $x1 \geq 0$, $xd0 \geq 0$, $xd2 \geq 0$, $m2 \leq 0$, $xd0 \leq x1$, $y1 >$
$y0$, $y1 > y2$.
Parameters (see also Figure 4):
- $y0$: starting value. The limit of the function at $-\infty$ (minus infinity). It is not
  necessarily the value at time point 0. If the function begins with a steep, the starting
  value is lower than the value at time point 0.
- $y1$: maximum value. The maximum of the function. It is possible that the maximum
  point is not in the measurement timeframe (usually meaning that the measurement
  does not follow this kinetic model).
- $y2$: ending value. The limit of the function at $+\infty$ (positive infinity). Not necessarily
  the value of the function at the end of the measurement.

- x1: time to reach maximum value. The time point when the function reaches the maximum value (unit: s).
- xd0: time from the first 50% value to maximum. The distance between the time point where the function reaches the first 50% value and where the function reaches the maximum. The first 50% value is the mean of the starting value and the maximum (unit: s).
- m0: slope at first 50% value. The slope of the function at the first 50% value (unit: int/s where int is the unit of the vertical axis).
- xd2: time from maximum to the second 50% value: the distance between the time point where the function reaches the maximum and where the function reaches the second 50% value. The second 50% value is the mean of the maximum and the ending value (unit: s)
- m2: slope at second 50% value: the slope of the function at the second 50% value (unit: int/s where int is the unit of the vertical axis).

5. **dlogist-:** a function that starts at a given value ($y0$), has a decreasing phase, reaches a minimum ($y1$), has an increasing phase and reaches a given ending value ($y2$). Formula:

$$\text{dlogist} - x(t; y0, y1, y2, x1, xd0, xd2, m0\ m2) = \begin{cases} y0 + \dfrac{y1 - y0}{1 + \left(\dfrac{x1 - t}{xd0}\right)^{\frac{4*xd0*m0}{y1-y0}}}, & t < x1 \\[4ex] y2 + \dfrac{y1 - y2}{1 + \left(\dfrac{t - x1}{xd2}\right)^{\frac{4*xd2*m2}{y2-y1}}}, & t \geq x1 \end{cases}$$

Constraints: $y0 \geq 0$, $y1 \geq 0$, $y2 \geq 0$, $m2 \geq 0$, $x1 \geq 0$, $xd0 \geq 0$, $xd2 \geq 0$, $m0 \leq 0$, $xd0 \leq x1$, $y1 < y0$, $y1 < y2$.

Parameters:

- $y0$: starting value. The limit of the function at $-\infty$ (minus infinity). It is not necessarily the value at time point 0. If the function begins with a steep, the starting value is higher than the value at time point 0.
- $y1$: minimum value. The minimum of the function. It is possible that the minimum point is not in the measurement timeframe (usually meaning that the measurement does not follow this kinetic model).
- $y2$: ending value. The limit of the function at $+\infty$ (positive infinity). Not necessarily the value of the function at the end of the measurement
- $x1$: time to reach minimum value. The time point when the function reaches the minimum value (unit: s).
- $xd0$: time from the first 50% value to minimum. The distance between the time point where the function reaches the first 50% value and where the function reaches the minimum. The first 50% value is the mean of the starting value and the minimum (unit: s).
- $m0$: slope at first 50% value: the slope of the function at the first 50% value (unit: int/s where int is the unit of the vertical axis).
- $xd2$: time from maximum to the second 50% value. The distance between the time point where the function reaches the minimum and where the function reaches the

second 50% value. The second 50% value is the mean of the minimum and the ending value (unit: s).
- $m2$: slope at second 50% value. The slope of the function at the second 50% value (unit: int/s where int is the unit of the vertical axis).

The parameters dependent on axis y ($y0$, $y1$, $y2$, $m0$, $m1$, $m2$) can be standardized by dividing each value by the $y0$ parameter of the same function so that the functions start at value 1.0 and the $y0$, $y1$ and $y2$ parameters become relative parameter values (rpv) while $m0$, $m1$ and $m2$ parameters become values of rpv/s units. The standardization of the parameters dependent only on axis x in case of *dlogist+* and *dlogist-* functions is done by having $xd0$ and $xd2$ parameters which are measured as a distance from the maximum, not as the time to reach the 50% values.

3.3 Function fitting

Before fitting a function, we estimate the parameters by applying the lowess smoothing method (Cleveland, 1979) several times with different smoother spans. We fit models with the Nelder-Mead optimisation method (Nelder et al., 1965) to quantile values starting from the estimated parameters minimizing the sum of absolute deviances.

For all 5 functions we do the following:

1. First we use 10-fold cross validation on the three quartiles separately to get an estimation of how well the function fits the measurement data. We summate the sum of absolute distances between the function fitted to the training set and the test set (Picard et al., 1984). The test set is one tenth of the whole dataset (which is 100 time and 100 intensity values for each quartile).
2. Then we fit the function to the 201 quantiles mentioned earlier.

The result of fitting the functions is 201*(1 + 4 + 4 + 8 + 8) = 5025 parameters and the 10-fold cross validation values for the 5 functions.

3.4 Comparison

To describe a measurement, we select the best function (by looking at how well the median function fits the median (0.5 quantile) values and looking for low 10-fold cross validation values) and we create distributions of the parameters by using the corresponding parameters of the 201 functions, hence each parameter will be represented by a distribution of 201 values. For example, in case of the *dlogist+* function we can talk about the distribution of the maximum value (Figure 5). The distributions can be summarized by giving median [range] or median [quartiles] values or visually by drawing a histogram or box plot.

To compare different (groups of) measurements we select a common function that describes every measurement well. Then, we summate the distributions of each parameter by group so that each group will have one single distribution for every parameter containing (measurement count in group)*201 values. We can compare distributions of parameters between different groups by probability binning method (Roederer et al., 2001) or by calculating the overlap of the middle 50% of the distributions. The T value given by probability binning and the overlap percentage give a measure of difference between the distributions and the former method also gives a p-value (corresponding to the null hypothesis that the two distributions are equal). There is the possibility of extracting only 1 value for each parameter from one measurement, ie. the parameters of the median function (which was fitted to the 0.5 quantile values) and in this case usual nonparametric statistical methods like Mann-Whitney U test (or Kruskall-Wallis test, for more than 2 groups) can be used to compare parameters between different groups.

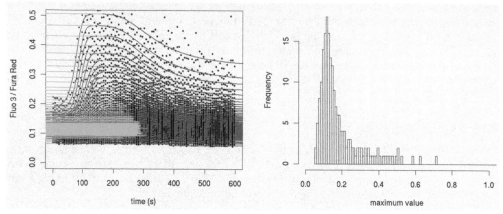

Fig. 5. Derivation of the distribution of parameter $y1$ (maximum value) of the *dlogist+* function. The 600 s long kinetic flow cytometric measurement time-frame was divided into 100 time intervals of equal length and 201 quantiles were calculated in each interval (black dots). *Dlogist+* function was fitted to each quantile (red curves). The maximum parameters (green lines) of each were collected and form a distribution (histogram on the right)

4. Experimental settings

Additionally to cytoplasmic Ca^{2+} signal (June et al., 1990) the development of a number of fluorescent probes (Johnson, 2001) provided an opportunity to measure the kinetics of other intracellular parameters such as membrane potential, mitochondrial Ca^{2+} levels and superoxide generation (Mészáros et al., 2011).

The authors of this chapter are mainly interested in investigating the activation process of T-lymphocytes. The experimental procedures described below are useful when examining the first short period of T-lymphocyte activation. We optimized incubation times, temperature and dye concentrations for Jurkat cells. The precise experimental settings can be found in (Mészáros et al., 2011). It should be emphasized that these experimental conditions may be different for other cell types.

4.1 Variables to be standardized during the measurements

Cells: although human peripheral blood samples are probably the most common specimen submitted to flow cytometry laboratory, to optimize an assay for experimental purposes, the most adequate way is to use different cell lines.

How specimens are obtained, handled, and stored is intimately related to the reason for obtaining them (for example, to analyse human peripherial blood lymphocytes, samples should be taken in a heparinized tubes and stored at room temperature). In most instances it is also crucial to assess the viability of separated cells. There are several methods available for this purpose (Stoddart, 2011). Additionally, viability should be determined prior to separation procedures.

The number of cells required per assay depends on the number of experiments planned and on the expected proportion of responding cells. Experiments with cells having a homogeneous response need ~106 cells per 10-min assay to quantify the response of a major population.

Solvents: the choice of cell culture medium can be influenced by two requirements: i) the metabolic requirements of the cells and ii) the solving properties of the fluorescent dyes used. For example cyanine and oxonol dyes are hydrophobic and will tend to plate out of aqueous solutions onto the surfaces of glass and plastic test tubes and onto the tubing in flow cytometers. Most of these compounds are highly soluble in DMSO (dimethylsulfoxide) and DMF (dimethyl formamide), which, because it does not readily evaporate, is a convenient vehicle for preparation of stock and working solutions.

In our protocols given below, and generally as well, the amount of DMSO added to cell suspensions is sufficiently small to keep the overall concentration at under 1% (v/v) (Brayton, 1986).

Dyes: evaluation of live cells using flow cytometry presents some difficult challenges. One of these is the need for maintaining the stability of cells. Therefore fluourescent dyes should not be toxic. During the staining procedure the cells and the reagents must be kept at appropriate temperature and should be carefully checked for pH (7.4), since the cells and some dyes are very sensitive to temperature and pH. The staining procedure should be performed under dimmed light and incubation should be in the dark, because of the light sensitivity of fluorochroms.

Dyes and ionophores for intracellular measurements may stick to cytometer tubing. Therefore, before measurement it is important to preequilibrate the tubing to prevent baseline shift. Likewise, tubing should be thoroughly cleaned after use by flushing the instrument with bleach. In case this is not done, residual dye may migrate from the tubing into cell samples that are introduced later into the instrument, and this may produce the appearance of immunofluorescent staining in unlabeled cells. In some cases ionophore in the sample may stick to the cytometer tubing and subsequently bind to cells in a later sample, producing a real change in the signal. When there is a possibility that this could happen, it may be prudent to flush with bleach between samples.

One has to be aware of a specific problem that may occur when several dyes are used in combination, i.e., their fluorescence emission spectra may overlap. In case quantitative information is required (i.e., the amount of fluorescence has to reflect a biochemical quantity), samples stained with a single dye should be included for use in setting fluorescence compensation (Roederer, 2002).

In the last years we established some flow cytometry methods that enable the users to monitor intracellular processes (membrane potential, mitochondrial Ca^{2+} levels and superoxide generation). We optimized incubation times, temperature and dye concentrations for T-lymphocytes. The dyes used are listed in Table 2. For more cellular function probes see (Johnson, 2001).

4.2 Considerations on timing

For many kinetic experiments with flow cytometry, the cell suspensions are analyzed for a period of 5-20 min. Timing is particularly critical when the change of the intracellular parameter to be observed lasts for a few minutes or less, in which case kinetic measurements incorporating time as a measurement parameter represent the only realistic approach to obtain consistent data. Even when stimuli produce effects lasting longer, it is important to maintain a relatively constant duration of cell incubation with dyes and stimuli from sample to sample.

4.3 Control measurements

Every experiment should include a baseline measurement which runs for 1-2 minutes with no additions to the sample. This provides a view of the degree of homogenity of individual cells in the population. Ideally the population distribution is very tight, but in case of presence of different cell types in the sample, the detected fluorescent signal intensity could be very diverse. In many cases, there is marked heterogenity in the changes that occur, sometimes even in populations of cells that were previously thought to be homogeneous. A limitation of flow cytometry, however, is that it does not permit kinetic resolution of certain complex kinetic responses such as cellular oscillatory responses. This requires video microscopy with digital image analysis, a technique that is complementary to flow cytometry for the study of various parameters of cell activation (Botvinick et al., 2007).

Name of the dye	Excitation max, (nm)	Emission max, (nm)	Excitation laser line (nm)	Use
Calcium sensitive ratiometric dyes				
Fluo-3-AM	506	526	488	**Cytomplasma Ca^{2+}-level**
Fura Red	436	655	488	
Indo-1	330-361	405-475	365	
Membrane potential sensitive dyes				
DiBAC4(3)	494	516	488	**Membrane potential**
DiBAC4(5)	590	616	488	
Mitochondrial probes				
JC-1	520	530, 590	488	**Mitochondrial potential**
TMRM	543	567	488	
Rhod-2/AM	552	581	488	**Mitochondrial Ca^{2+} level**
ROS sensitive dyes				
DAF-FM diacetate	495	515	488	**NO measurement**
HE	518	605	488	**O$_2^-$ generation measurement**

Table 2. Fluorescent probes for kinetic measurements

"Pseudo mixing test" is an important control in kinetic flow cytometry measurements. It determines whether the cell population is susceptible to mechanical shear forces like mixing and sample pressurization. These effects can activate pressure sensitive ion flux mechanisms present in some cell lines and can result in changes of the signal detection.

The fact that some of the fluorescent probes can be compartmentalized in cells forces us to use different control experiments. In order to test the validity of assay systems the use of specific activators and/or inhibitors is mandatory. At the end of the measurements to test whether the cells were properly loaded with reporter dye and whether the flow cytometry system was set up properly, the addition of positive or negative control chemicals is needed.

5. Applications of the FacsKin method

FacsKin provides a tool for the evaluation of kinetic flow cytometry data. This section presents some experiments that benefited from opportunities provided by FacsKin.

5.1 The investigation of calcium influx kinetics in Th1 and Th2 cells

Upon antigen presentation, a signal transduction pathway leads to a transient, biphasic elevation of $[Ca^{2+}]_{cyt}$ in lymphocytes (Figure 6). The first phase of the biphasic calcium signal is directly linked to the generation of IP_3, and calcium release from the endoplasmic reticulum (ER) upon the binding of IP_3 to its designated receptor (Lewis, 2001). The second phase is due to the activation of calcium release activated calcium (CRAC) channels in the cell membrane. The calcium signal converges to the activation of transcription factors leading to cytokine production and further factors needed for the development of an adequate lymphocyte response. The actual distribution of $[Ca^{2+}]_{cyt}$ depends on finely tuned interactions of mechanisms responsible for its elevation and decrease. Besides ER calcium release and calcium entry through the CRAC channels, mitochondria also contribute to the elevation of $[Ca^{2+}]_{cyt}$ via the regulation of CRAC channel functionality. However, in a later phase of lymphocyte activation, they may also take up and store large amounts of calcium via the mitochondrial calcium uniporter (MCU) and, therefore, decrease $[Ca2+]_{cyt}$ (Duchen, 2000). Other mechanisms that specifically contribute to the clearance of elevated $[Ca2+]_{cyt}$ in lymphocytes are the sarco/endoplasmic reticulum calcium ATPase (SERCA) (Feske, 2007) and the plasma membrane calcium ATPase (PMCA) (Di Leva et al., 2008).

The steps of activation begin with identical stimuli in the two major arms of T helper lymphocytes, Th1 and Th2 cells. However, they produce a different set of cytokines, and exert distinct effects on the inflammatory balance. While Th1 cells account for the development of a pro-inflammatory response, Th2 cells produce anti-inflammatory cytokines. Since the expression of cytokine genes is influenced by the characteristics of calcium influx kinetics, the differences in calcium handling of the Th1 and Th2 subset may remarkably contribute to variations in cytokine production (Dolmetsch et al., 1998). Therefore, we investigated the differences of calcium handling between Th1 and Th2 lymphocytes. We tested the contribution of the ER calcium release, the CRAC channel, the MCU, the SERCA pump and the PMCA pump to the regulation of $[Ca^{2+}]_{cyt}$ during the early period of T lymphocyte activation.

First, we compared the kinetics of calcium response in the Th1 and Th2 lymphocyte subsets following lymphocyte activation (Toldi et al., 2011a). AUC, Slope and Max values were lower in the Th2 subset. However, the t_{max} value did not significantly differ in the two subsets. The higher activity of the SERCA pump, along with the lower activity of mitochondrial calcium reuptake, and therefore of the CRAC channels account for the notion that Th2 cells go through a lower level of lymphocyte activation compared with Th1 cells upon identical activating stimuli.

In contrast to the SERCA pump, which functions from the beginning of calcium influx, the PMCA pump in Th1 cells has a role in the shaping of $[Ca^{2+}]_{cyt}$ kinetics from a stage when $[Ca^{2+}]_{cyt}$ is already elevated, thus ensuring the reconstitution of the original $[Ca^{2+}]_{cyt}$ level. This is represented in our results by the fact that the inhibition of this mechanism affects the Max values but not the Slope value in Th1 cells. The main regulator of the PMCA pump is the elevated level of $[Ca^{2+}]_{cyt}$ (Di Leva et al., 2008). Therefore, the initial calcium uptake of the ER has an important regulatory effect on the function of the PMCA pump. Since calcium reuptake by ER is more active in Th2 cells, $[Ca^{2+}]_{cyt}$ will not increase to the extent observed in Th1 cells. Therefore, it seems that $[Ca^{2+}]_{cyt}$ will not be elevated sufficiently in the Th2 subset to activate the PMCA pump during the initial phase of activation. In contrast, in Th1 cells, due to the lower extent of ER reuptake, $[Ca^{2+}]_{cyt}$ will increase sufficiently to activate the PMCA pump. Differential calcium signaling and distinct kinetics of the alterations of $[Ca^{2+}]_{cyt}$ may have an important contributing role to the production of dissimilar cytokines by Th1 and Th2 cells.

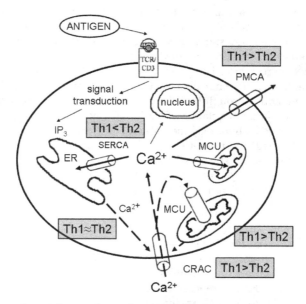

Fig. 6. The regulation of cytoplasmic free calcium level in lymphocytes. Interrupted arrows represent mechanisms responsible for the elevation, whereas bold arrows represent those responsible for the decrease of the cytoplasmic free calcium level. The differential activity of these mechanisms in Th1 and Th2 cells is marked in the gray boxes. TCR: T cell receptor, IP_3: inositol trisphosphate, ER: endoplasmic reticulum, CRAC: calcium release activated calcium channel, MCU: mitochondrial calcium uniporter, SERCA: sarco/endoplasmic reticulum calcium ATPase, PMCA: plasma membrane calcium ATPase

5.2 The role of lymphocyte potassium channels in the regulation of calcium influx

In order to maintain the electrochemical driving force for calcium entry from the extracellular space, depolarizing calcium influx needs to be counterbalanced by the efflux of cations, predominantly potassium. Therefore, lymphocyte activation is closely linked to and regulated by the function of lymphocyte potassium channels. There are two major types of potassium channels in T lymphocytes: the voltage-gated Kv1.3 and the calcium-activated IKCa1 channels. In our investigations, we studied the role of the Kv1.3 and IKCa1 channels in the process of lymphocyte activation not only in healthy individuals, but also in neonates, pregnancy and immune-mediated disorders, such as preeclampsia, multiple sclerosis and type 1 diabetes.

In healthy individuals, a triarylmethane compound (TRAM), the specific inhibitor of the IKCa1 channel decreased calcium influx in Th2 cells to a lower extent than in Th1 cells (Toldi et al., 2011b). This finding supports previous data from Fanger et al. (2000) obtained by patch clamp, showing that IKCa1 currents are smaller in Th2 cells when compared with the Th1 subset. A possible contributing element to this phenomenon might be that $[Ca^{2+}]_{cyt}$ which needs to reach a threshold to activate the IKCa1 channels, increases more rapidly in Th1 than in Th2 cells, due to the distinct function of the SERCA pump, as detailed above.

In contrast with IKCa1, the inhibition of Kv1.3 channels by a specific blocker, margatoxin (MGTX) results in a somewhat larger decrease of calcium in Th2 than in Th1 cells. Calcium influx in Th1 cells is less sensitive to the inhibition of the Kv1.3 channel. All in all, a larger amount of compensatory potassium leaves the cells through the Kv1.3 and IKCa1 channels upon lymphocyte activation in the Th1 subset, enabling a larger amount of calcium to enter from the extracellular space compared with the Th2 subset.

In healthy individuals, CD8 cells were more sensitive to the inhibition of the IKCa1 channels than the CD4 subset, responding with a higher level of decrease of the AUC value upon the application of TRAM. However, almost no difference was observed between the sensitivity of CD4 and CD8 cells to the inhibition of the Kv1.3 channel.

5.3 Lymphocyte activation and potassium channels in healthy pregnancy and preeclampsia

Both in pregnancy and in the neonatal period, the immune response and the kinetics of lymphocyte activation are altered compared with the adult, non-pregnant state. Healthy pregnancy is characterized by the development of an immune tolerance specific for the antigens presented by the developing fetus. The impairment of this tolerance and the development of an abnormal immune response play a major role in adverse pregnancy outcomes, including preeclampsia (Saito et al., 2007). This disorder is characterized by hypertension, proteinuria, edema and endothelial dysfunction usually evolving in the third trimester of pregnancy. The development of a maternal systemic inflammatory response has also been described in PE (Saito et al., 2007). An important feature of this disorder is the absence of Th2 skewness and thus the predominance of pro-inflammatory cytokines, as shown by a number of investigations (Darmochwal-Kolarz et al., 2002; Rein et al., 2002; Saito et al., 1999).

Our results indicate marked differences of calcium influx kinetics and sensitivity to Kv1.3 and IKCa1 channel inhibition (by margatoxin (MGTX) and a triarylmethane compound

(TRAM), respectively) in major lymphocyte subsets (i.e. Th1, Th2, CD4 and CD8 cells) between non-pregnant, healthy pregnant and preeclamptic lymphocytes (Toldi et al., 2010a). These properties in preeclampsia are more comparable to the non-pregnant state than to healthy pregnancy, suggesting that there is a characteristic pattern of calcium influx in healthy pregnant women that is missing in preeclamptic patients. This raises the notion that lymphocyte calcium handling upon activation may have a role in the characteristic immune status of healthy pregnancy.

AUC values of the calcium response are lower in healthy pregnancy in the Th1, CD4 and CD8 lymphocyte subsets. On contrary to Th1 cells, the activation induced calcium response of the Th2 subset is not decreased compared with the non-pregnant state. The decreased activation of the Th1 subset and the lack of decrease in Th2 cells may partly be responsible for the well established Th2 skewness in healthy pregnancy (Darmochwal-Kolarz et al., 2002; Rein et al., 2002; Saito et al., 1999). Unlike in healthy pregnancy, we could not detect a difference in the AUC values of calcium influx of Th1 and CD8 cells in preeclampsia compared to non-pregnant women. The maintained activation properties of Th1 lymphocytes in preeclamptic patients may contribute to the lack of Th2 dominance associated with normal pregnancy. It is of particular interest that calcium influx of Th2 lymphocytes in healthy pregnancy was insensitive to potassium channel inhibition, while calcium influx decreased significantly in non-pregnant samples upon treatment with the specific channel blockers. Of note, Th2 lymphocytes in preeclampsia presented with non-pregnant-like characteristics, and were also sensitive to MGTX and TRAM treatment. Since the regulatory function of Kv1.3 and IKCa1 channels on calcium influx appears to be limited in healthy pregnant samples, this may be an element contributing to the Th2 shift present in healthy pregnancy, but absent in preeclampsia. While calcium influx in CD8 and Th1 lymphocytes was resistant to potassium channel inhibition in preeclamptic, that of healthy pregnant lymphocytes was sensitive. Similarly to Th2 cells, while it is unclear whether the resistance of Th1 lymphocytes to potassium channel inhibition is reflected in their function, the insensitivity of the Th1 subset to the inhibition of regulatory lymphocyte potassium channels in preeclampsia may be linked to the Th1 skewness.

5.4 Lymphocyte activation and potassium channels in the newborn

Decreased functionality of neonatal T cells is a widely recognized experimental and clinical phenomenon. Reduced functioning is well characterized by a lower level of cytokine production compared to adult T cells (Cohen et al., 1999; García Vela et al., 2000). Several factors might be responsible for the decreased cytokine expression compared to adult lymphocytes. We hypothesized that short-term T lymphocyte activation properties are different in neonates compared to adults. We aimed to characterize the calcium influx kinetics upon activation in major T lymphocyte subsets (i.e. Th1, Th2, CD4 and CD8 cells) in the neonate, and its sensitivity to the specific inhibition of Kv1.3 and IKCa1 lymphocyte potassium channels (Toldi et al., 2010b).

Lower AUC and Max values in most of the investigated subsets suggest that short-term activation and associated calcium influx are decreased in neonatal lymphocytes, in line with the fact that newborns mount lower immune responses to distinct stimuli. Upon treatment of lymphocytes with selective inhibitors of the Kv1.3 and IKCa1 channels (MGTX and TRAM, respectively), calcium influx decreases in most investigated lymphocyte subsets isolated from adults. However, with the exception of the CD8 subset, such a reduction was not demonstrated in neonatal lymphocytes. These findings may partly explain why neonatal lymphocytes are less responsive to activating stimuli and, hence, exert a lower intensity of immune response. Our results improve the understanding of the mechanisms that prevent neonatal T cells from adequate activation upon activating stimuli, and partially elucidate previous experimental data indicating that a greater amount of stimulation is needed in neonatal lymphocytes compared with adults to achieve a similar immune response (Adkins, 1999; Cohen et al., 1999). The functional impairment of lymphocyte potassium channels may be of importance in those mechanisms.

5.5 Lymphocyte activation and potassium channels in multiple sclerosis

We also investigated calcium influx kinetics in multiple sclerosis (MS) patients without and with interferon (IFN) beta treatment compared to healthy individuals (Toldi et al., 2011b). We aimed to describe the effects of Kv1.3 and IKCa1 channel inhibitors on calcium influx, and to assess whether these inhibitors could potentially contribute to the treatment of MS via the selective reduction of lymphocyte activation as suggested before (Wulff et al., 2003). We demonstrated that the reactivity of lymphocytes isolated from MS patients receiving no IFN beta treatment is increased compared to healthy individuals, reflected by lower t_{max} and elevated Slope values in the CD4, Th1 and Th2 subsets (i.e. the peak of the calcium influx is reached more rapidly). Interestingly, these alterations were not present in the CD8 subset. Of note, while the kinetics of calcium influx is altered, comparable AUC and Max values indicate that the amount of calcium entering the lymphocytes in MS is not different from that in healthy individuals.

In healthy individuals, we found prominent differences between Th1 and Th2 cells with regard to their sensitivity to IKCa1 channel inhibition, as described above. In contrast with healthy subjects, the investigated lymphocyte subsets show altered sensitivity to the inhibition of the Kv1.3 and IKCa1 channels in MS patients without IFN beta. In this group, the inhibition of the IKCa1 channel results in a similar decrease of calcium influx in all investigated subsets. However, MGTX, the specific blocker of the Kv1.3 channel decreased the AUC value to a higher extent in CD8 cells than in CD4 cells. Therefore, the contribution of Kv1.3 channels to the activation of CD8 lymphocytes is increased in MS patients receiving no IFN beta. However, the specificity of Kv1.3 inhibition is limited to CD8 cells since other cell types, including Th1 and Th2 cells, behave in a similar manner in MS patients without IFN beta upon Kv1.3 channel inhibition. Therefore, on contrary to previous suggestions (Wulff et al., 2003), the inhibition of this channel does not seem to be specific enough, since it also affects anti-inflammatory cytokine producing Th2 cells. This would probably result in a setback of therapeutic efforts in MS. Therefore, the effects of administration of Kv1.3 channel inhibitors need to be further investigated and characterized in MS.

IFN beta therapy induces compensatory changes in calcium influx kinetics and lymphocyte potassium channel function in MS, shaping these properties more similar to those of healthy individuals. However, the suppressive effect of IFN beta treatment on lymphocyte activation is seen in Th1 cells selectively, while Th2 function is less affected. These observations might indicate a novel mechanism through which IFN beta exerts beneficial therapeutic effects on immune functionality in MS.

5.6 Lymphocyte activation and potassium channels in type 1 diabetes

In samples taken from type 1 diabetes mellitus (T1DM) patients, the activation characteristics of lymphocytes show an evident alteration compared to healthy samples (Toldi et al., 2010c). First, we noticed that the peak of calcium influx in the overall lymphocyte population and the Th1 subset is reached more rapidly in T1DM (i.e. t_{max} values were decreased), while AUC and Max values were comparable. Similarly to MS, this finding raised the notion of an increased reactivity of lymphocytes in T1DM.

In lymphocytes of healthy subjects both Kv1.3 and IKCa1 channels contribute to the maintenance of calcium influx upon activation. On the contrary, the sensitivity of T1DM lymphocytes to the inhibition of Kv1.3 channels is increased, probably due to the increased expression of Kv1.3 channels (Toldi et al., 2010c). The altered activation kinetics of T1DM lymphocytes may at least partly be attributed to the increased significance of Kv1.3 channels.

Based on promising animal data, the specific inhibition of Kv1.3 channels is under extensive investigation as a possible measure to prevent the development of the autoimmune response against pancreatic beta cells (Chandy et al., 2004). Our data indicate that by specific inhibition of Kv1.3 channels, lymphocyte activation can be modulated in T1DM. However, our findings provide clear evidence for Kv1.3 channels to have an important role in each lymphocyte subset in T1DM, including Th2 lymphocytes acting as counterbalancing factors in the development of T1DM through the production of anti-inflammatory cytokines (Yoon and Jun, 2001). Therefore, administration of Kv1.3 channel inhibitors would not have an exclusive effect on cells responsible for the autoimmune response in T1DM, but may have an impact on the activation characteristics of immune cells in general. The finding that increased significance of Kv1.3 channels in lymphocyte activation is not exclusive for a specific subset, but is characteristic for most of major lymphocyte subsets in T1DM, alerts us that the overall immunomodulatory effect upon inhibition of Kv1.3 channels in T1DM needs to be further characterized.

6. References

Adkins, B. (1999). T-cell function in newborn mice and humans. *Immunol Today*, Vol.20, No.7, (July 1999), pp. 330-335

Bailey, S., Macardle, P.J. (2006). A flow cytometric comparison of Indo-1 to fluo-3 and Fura Red excited with low power lasers for detecting Ca(2+) flux. *J Immunol Methods*, Vol.311, No.1-2, (April 2006), pp. 220-225

Bates, D.M., Watts, D.G. (1988). Nonlinear Regression Analysis and its Applications. Wiley Book, New York. ISBN 0471-816434

Botvinick, E.L. Shah, J.V. (2007). Laser-based measurements in cell biology. *Methods Cell Biol*, Vol.82, Part I, (2007), pp. 81-109.

Brayton, C.F. (1986). Dimethyl sulfoxide (DMSO): a review. *Cornell Vet*, Vol.76, No.1, (January 1986), pp. 61-90.

Chandy, K.G., Wulff, H., Beeton, C., Pennington, M., Gutman, G.A., Cahalan, M.D. (2004). K+ channels as targets for specific immunomodulation. *Trends Pharmacol Sci*, Vol.25, No.5, (May 2004), pp. 280–289

Cleveland, W.S. (1979). Robust Locally Weighed Regression and Smoothing Scatterplots. *Journal of the American Statistical Association*, Vol.74, No.368 (December 1979), pp. 829-836

Cohen, S.B., Perez-Cruz, I., Fallen, P., Gluckman, E., Madrigal, J.A. (1999). Analysis of the cytokine production by cord and adult blood. *Hum Immunol*, Vol.60, No.4, (April 1999), pp. 331-336

Darmochwal-Kolarz, D., Rolinski, J., Leszczynska-Goarzelak, B., Oleszczuk, J. (2002). The expressions of intracellular cytokines in the lymphocytes of preeclamptic patients. *Am J Reprod Immunol*, Vol.48, No.6, (December 2002), pp. 381-386

Demkow, U., Winklewski, P., Potapinska, O., Popko, K., Lipinska, A., Wasik, M. (2009). Kinetics of calcium ion concentration accompanying transduction of signals into neutrophils from diabetic patients and its modification by insulin. *J Physiol Pharmacol*, Vol.60, No.5, (November 2009), pp. 37-40

Di Leva, F., Domi, T., Fedrizzi, L., Lim, D., Carafoli, E. (2008). The plasma membrane Ca2+ ATPase of animal cells: structure, function and regulation. *Arch Biochem Biophys*, Vol.476, No.1, (August 2008), pp. 65–74

do Céu Monteiro, M., Sansonetty, F., Gonçalves, M.J., O'Connor, J.E.. (1999). Flow cytometric kinetic assay of calcium mobilization in whole blood platelets using Fluo-3 and CD41. *Cytometry*, Vol.35, No.4, (April 1999), pp. 302-310

Dolmetsch, R.E., Xu, K., Lewis, R.S. (1998). Calcium oscillations increase the efficiency and specificity of gene expression. *Nature*, Vol.392, No.6679, (April 1998), pp. 933–936

Duchen, M.R. (2000). Mitochondria and calcium: from cell signalling to cell death. *J Physiol*, Vol.529, Pt.1, (November 2000), pp. 57–68

Fanger, C.M., Neben, A.L., Cahalan, M.D. (2000). Differential Ca2+ influx, KCa channel activity, and Ca2+ clearance distinguish Th1 and Th2 lymphocytes. *J Immunol*, Vol.164, No.3, (February 2000), pp. 1153–1160

Feske, S. (2007). Calcium signalling in lymphocyte activation and disease. *Nat Rev Immunol*, Vol.7, No.9, (September 2007), pp. 690–702

García Vela, J.A., Delgado, I., Bornstein, R., Alvarez, B., Auray, M.C., Martin, I., Oña F., Gilsanz F. (2000). Comparative intracellular cytokine production by in vitro stimulated T lymphocytes from human umbilical cord blood (HUCB) and adult peripheral blood (APB). *Anal Cell Pathol*, Vol.20, No.2-3, (2000), pp. 93-98

Hawkins, D.M. (2004). The Problem of Overfitting. *J Chem Inf Comput Sci*, Vol.44, No.1 (January 2004), pp 1–12

Jakubczak, B., Wasik, M., Popko, K., Demkow, U. (2006). Kinetics of calcium ion concentration accompanying signal transduction in neutrophils from children with increased susceptibility to infections. *J Physiol Pharmacol*, Vol.57, No.4, (September 2006), pp. 131-137

Johnson, D.I. (2001). Cellular function probes. *Current Protocols in Cytometry*, Chapter4, Unit4.4, (May 2001).

June, C.H., Ledbetter, J.A., Rabinovitch, P.S., Martin, P.J., Beatty, P.G., Hansen, J.A. (1986). Distinct patterns of transmembrane calcium flux and intracellular calcium mobilization after differentiation antigen cluster 2 (E rosette receptor) or 3 (T3) stimulation of human lymphocytes. *J Clin Invest*, Vol.77, No.4, (April 1986), pp. 1224-1232

June, C.H., Rabinovitch, P.S. (1990). Flow cytometric measurement of intracellular ionized calcium in single cells with indo-1 and fluo-3. *Methods Cell Biol*, Vol.33, Chapter5, (1990), pp.37-58.

Kaposi, A.S., Veress, G., Vásárhelyi, B., Macardle, P., Bailey, S., Tulassay, T., Treszl, A. (2008). Cytometry-acquired calcium-flux data analysis in activated lymphocytes. *Cytometry A*, Vol.73, No.3, (March 2008), pp. 246-253

Koenker, R., Hallock, K.F. (2001). Quantile Regression. *Journal of Economic Perspectives*, Vol.15, No.4 (Fall 2001), pp. 143-156

Lewis, RS. (2001). Calcium signaling mechanisms in T lymphocytes. *Annu Rev Immunol*, Vol.19, (April 2001), pp. 497-521

Lund-Johansen, F, Olweus, J. (1992). Signal transduction in monocytes and granulocytes measured by multiparameter flow cytometry. *Cytometry*, Vol.13, No.7, (September 1992), pp. 693-702

Mészáros, G., Szalay, B., Toldi, G., Kaposi, A., Vásárhelyi, B., Treszl, A. (2011). Kinetic measurements using flow cytometry: new methods for monitoring intracellular processes. *Assay Drug Dev Technol*, [Epub ahead of print], (September 2011).

Motulsky, H.J., Ransnas, L.A. (1987). Fitting curves to data using nonlinear regression: a practical and nonmathematical review. *FASEB J*, Vol.1, No.5, (November 1987), pp. 365-374

Motulsky, H., Christopoulos, A. (2004). Fitting models to biological data using linear and nonlinear regression: a practical guide to curve fitting. Oxford University Press, USA

Nelder, J.A., Mead, R. (1965). A Simplex Method for Function Minimization. *Computer Journal*, Vol.7, No.4, (January 1965), pp. 308-313

Norgauer, J., Dobos, G., Kownatzki, E., Dahinden, C., Burger, R., Kupper, R., Gierschik, P. (1993). Complement fragment C3a stimulates Ca2+ influx in neutrophils via a pertussis-toxin-sensitive G protein. *Eur J Biochem*, Vol.217, No.1, (October 1993), pp. 289-94.

Omann, G.M., Harter, J.M. (1991). Pertussis toxin effects on chemoattractant-induced response heterogeneity in human PMNs utilizing Fluo-3 and flow cytometry. *Cytometry*, Vol.12, No.3, (March 1991), pp. 252-259

Picard R.R., Cook, R.D. (1984). Cross-Validation of Regression Models. *Journal of the American Statistical Association*, Vol.79, No.387 (September 1984), pp. 575-583

Politi, A., Gaspers, L.D., Thomas, A.P., Höfer, T. (2006). Models of IP3 and Ca2+ oscillations: frequency encoding and identification of underlying feedbacks. *Biophys J*, Vol.90, No.9, (May 2006), pp. 3120-3133

R Development Core Team. (2006). R: A Language and Environment for Statistical Computing. Vienna, Austria: R Foundation for Statistical Computing. ISBN 3-900051-07-0

Rein, D.T., Schondorf, T., Gohring, U.J., Kurbacher, C.M., Pinto, I., Breidenbach, M., Mallmann, P., Kolhagen, H., Engel, H. (2002). Cytokine expression in peripheral blood lymphocytes indicates a switch to T(HELPER) cells in patients with preeclampsia. *J Reprod Immunol*, Vol.54, No.1-2, (March 2002), pp. 133-142

Rijkers, G.T., Griffioen, A.W. (1993). Changes in free cytoplasmic magnesium following activation of human lymphocytes. *Biochem J*, Vol.289, No.2, (January 1993), pp. 373-377

Roederer, M., Treister, A., Moore, W., Herzenberg, L.A. (2001). Probability binning comparison: a metric for quantitating univariate distribution differences. *Cytometry*, Vol.45, No.1, (September 2001), pp. 37-46

Roederer, M. (2002). Compensation in flow cytometry. *Current Protocols in Cytometry*, Chapter 1, Unit 1.14, (December 2002).

Saito, S., Umekage, H., Sakamoto, Y., Sakai, M., Tanebe, K., Sasaki, Y., Morikawa, H. Increased T-helper-1-type immunity and decreased T-helper-2-type immunity in patients with preeclampsia. *Am J Reprod Immunol*, Vol.41, No.5, (May 1999), pp. 297-306

Saito, S., Shiozaki, A., Nakashima, A., Sakai, M., Sasaki, Y. (2007). The role of the immune system in preeclampsia. Mol. Aspects Med., Vol.28, No.2, (April 2007), pp. 192-209

Schepers, E., Glorieux, G., Dhondt, A., Leybaert, L., Vanholder, R. (2009). Flow cytometric calcium flux assay: evaluation of cytoplasmic calcium kinetics in whole blood leukocytes. *J Immunol Methods*, Vol.348, No.1-2, (August 2009), pp. 74-82

Stoddart, M.J. (2011). Cell viability assays: introduction. *Methods Mol Biol*, Vol.740, Mammalian Cell Viability, (2011), pp.1-6.

Stork, B., Neumann, K., Goldbeck, I., Alers, S., Kähne, T., Naumann, M., Engelke, M., Wienands, J. (2007). Subcellular localization of Grb2 by the adaptor protein Dok-3 restricts the intensity of Ca2+ signaling in B cells. *EMBO J*, Vol.26, No.4, (February 2007), pp. 1140-1149

Szalay, B., Mészáros, G., Cseh, A., Acs, L., Deák, M., Kovács, L., Vásárhelyi, B., Balog, A. (2012). Adaptive Immunity in Ankylosing Spondylitis: Phenotype and Functional Alterations of T-Cells before and during Infliximab Therapy. *Clin Dev Immunol*, Vol.2012, (Epub September 2011), 808724

Tang, Y., Stephenson, J.L., Othmer, H.G. (1996). Simplification and analysis of models of calcium dynamics based on IP3-sensitive calcium channel kinetics. *Biophys J*. Vol.70, No.1, (January 1996), pp. 246-263

Toldi, G., Stenczer, B., Treszl, A., Kollar, S., Molvarec, A., Tulassay, T., Rigo, J.Jr, Vasarhelyi, B. (2010a). Lymphocyte calcium influx characteristics and their modulation by Kv1.3 and IKCa1 channel inhibitors in healthy pregnancy and preeclampsia. *Am J Reprod Immunol*. Vol.65, No.2, (February 2011), pp. 154-163

Toldi, G., Treszl, A., Pongor, V., Gyarmati, B., Tulassay, T., Vasarhelyi, B. (2010b). T-lymphocyte calcium influx characteristics and their modulation by Kv1.3 and IKCa1 channel inhibitors in the neonate. *Int Immunol*, Vol.22, No.9, (September 2010), pp. 769-774

Toldi, G., Vasarhelyi, B., Kaposi, A.S., Meszaros, G., Panczel, P., Hosszufalusi, N., Tulassay, T., Treszl, A. (2010c). Lymphocyte activation in type 1 diabetes mellitus: the increased significance of Kv1.3 potassium channels. Immunol Lett. Vol.133, No.1, (September 2010), pp. 35-41

Toldi, G., Kaposi, A., Zsembery, Á., Treszl, A., Tulassay, T., Vásárhelyi, B. (2011a). Human Th1 and Th2 lymphocytes are distinguished by calcium flux regulation during the first ten minutes of lymphocyte activation. *Immunobiology*, [Epub ahead of print], (August 2011)

Toldi, G., Folyovich, A., Simon, Z., Zsiga, K., Kaposi, A., Mészáros, G., Tulassay, T., Vasarhelyi, B. (2011b). Lymphocyte calcium influx kinetics in multiple sclerosis treated without or with interferon beta. *J Neuroimmunol*, Vol.237, No.1-2, (August 2011), pp. 80-86

Wulff, H., Calabresi, P.A., Allie, R., Yun, S., Pennington, M., Beeton, C., Chandy, K.G. (2003). The voltage-gated Kv1.3 K+channel in effector memory T cells as new target for MS. *J Clin Invest*. Vol.111, No.11, (June 2003), pp. 1703–1713

Yoon, J.W., Jun, H.S. (2001). Cellular and molecular pathogenic mechanisms of insulin-dependent diabetes mellitus. *Ann N Y Acad Sci*, Vol.928, (April 2001), pp. 200–211

3

Flow Cytometric Sorting of Cells from Solid Tissue – Reagent Development and Application

P. S. Canaday and C. Dorrell
Oregon Health & Science University
USA

1. Introduction

The isolation and study of cells from solid tissues is one of the great current challenges for flow cytometry. The vast majority of mammalian cells reside within solid organs, but the application of flow cytometric techniques to such cells remains limited. The study of hematopoietic biology has been revolutionized by the development of standardized monoclonal antibodies recognizing defined antigens that permit flow cytometric isolation, definition, and characterization of blood cell types (Bendall et al., 2011; Fleming et al., 1993; Hoffman et al., 1980; Nakeff et al., 1979). Antibodies have been used to better identify and understand the role different blood cells play in the immune system (Springer et al., 1979). They have also been used to help zero in on the stem cell populations and to learn more about the processes of hematopoietic differentiation (Ikuta et al., 1992; Spangrude et al., 1988). Their use has impacted virtually every facet of hematopoietic study.

This same approach is likely to bring similar benefits to the study of solid tissue. Unfortunately, the reagents and techniques needed for studying cells from solid tissues are woefully limited. Most available surface-reactive antibodies have been defined by their activity against hematopoietic antigens. The lack of understanding about the surfaces of tissue-resident cells is one of the reasons why the isolation of adult stem cells and culture of primary cells from solid tissue remains problematic. In order to advance the study and characterization of these cells, new reagents and techniques must be developed. The development of monoclonal antibodies against solid-tissue specific antigens will help researchers sort and study subsets of cells from other organs. It will allow for comparison between subsets of cells from the same tissue and similar subsets from different tissues. It is likely that, as with the hematopoietic field, new subsets and classes of cells will be discovered. Hopefully, the development of more monoclonal antibodies will not only move research forward but will also aid in the development of therapeutic technologies.

This chapter will include information about the history of monoclonal antibodies with a focus on how they impacted the field of flow cytometry. The development of novel monoclonal antibodies and examples of their successful application to the field of solid tissue biology will be described, and suggestions will be offered for the advancement of flow cytometric research beyond the hematopoietic field.

2. A brief overview of the history of hematopoietic antibody production

In order to better understand how antibodies will benefit the study of solid tissues, it is helpful to examine how they have benefited the study of hematopoietic tissue. The study of hematopoietic tissue has been immensely successful and there is much benefit to be gained by understanding and learning from what has been done before.

2.1 Hybridomas

The advent of monoclonal antibody-producing hybridomas revolutionized the field of cytometry. Kohler and Milstein originally described the creation of immortal cells that produced monoclonal antibodies against a predefined target (Kohler and Milstein, 1975). They did this by fusing cells from a plasmacytoma line with splenocytes from a mouse that had been immunized with sheep erythrocytes. The resulting clonal cell line possessed the useful attributes of each parent; they could be cultured indefinitely and also continued to produce the desired immunoglobulin.

Prior to the development of hybridomas, polyclonal serum was used for cell or protein recognition. While there are applications where the use of polyclonal serum is advantageous, it has the disadvantages of being a finite resource and that the antigens are not fully characterized.

The innovation of monoclonal antibodies revolutionized cell analysis by allowing precise and accurate reproducibility both between experiments and between researchers. In the beginning of the prolific development of monoclonal antibodies, individual labs would make their own antibody clones (Lemke et al., 1978; Solter and Knowles, 1978). Very quickly, it became clear that a more standardized approach would be useful; it was not always clear, for example, whether antibodies that appeared to label the same cells were binding to the same antigens. During the first HLDA (Human Leucocyte Differentiation Antigens) workshop in 1992, 15 antibodies with known targets were standardized as CD (cluster of differentiation) markers (Zola and Swart, 2005). As was reflected in the name, the HLDA was concerned primarily with blood cells, which were readily accessible and a logical place to start.

2.1.1 Uses

Flow cytometry is not the only use for monoclonal antibodies; their application in biomedical and basic research is extensive. Although immunoassays were in use prior to the development of monoclonal antibodies, the increased availability and consistency of this new reagent dramatically increased the number of potential applications. Radioimmunotherapy would not be possible without the superb specificity of monoclonal antibodies. Pathogen classification used to require hours, or even days; there are now many cases where it can be done in a matter of minutes. Microscopy has taken advantage of monoclonal antibodies to produce exquisite images and reveal the internal structure of cells in ways not previously imagined. Many other examples exist, but a comprehensive list of monoclonal antibody applications is beyond the scope of this chapter.

2.2 Use of antibodies for flow cytometry

Even before Kohler and Milstein devised monoclonal-producing hybridomas, antibodies were being used for flow cytometry. The Herzenberg group, in 1971, became the first to employ antibody labeling as a cell identification tool in flow cytometry (Herzenberg, 1971). Polyclonal anti-sera were employed in subsequent studies, such as Epstein's work on lymphocyte biology (Epstein et al., 1974). In 1977, the first paper using fluorescently-labeled monoclonal antibodies was published, and the field has never looked back (Williams et al., 1977). Since then, thousands of publications have included the use of monoclonal antibodies in flow cytometry. They have become an indispensible tool for biomedical research.

2.3 Advances in flow cytometry

Advancements in the technical aspects of flow cytometry were driven largely by the rapid generation and characterization of new monoclonal antibodies. The demand for more fluorophores, more detectors, and more lasers in order to assess multiple properties on the same cell was the impetus for significant advancements in the field. Multicolored flow cytometry allowed for the discovery of multiple subsets of cells. Three color detection using a dual laser system was developed by 1983 (Parks et al., 1984). In 1985, BD utilized microcomputing in the new FACScan, increasing cell throughput and simplifying all aspects of the cell isolation process. By the turn of the millennium, eight simultaneous 'colors' (fluorochromes) were used in flow cytometry experiments (Roederer et al., 1997). Any time a new instrument becomes available, researchers immediately begin pushing the limits of the new technology. Advances in photomultiplier tube technology and the development of better dichroic filters also helped in the development of bigger and better flow cytometers.

The rise in technology necessitated new standardized ways of analyzing the data. The FCS (Flow Cytometry Standard) was proposed in 1984 by Murphy and Chused and was summarily adopted by the scientific community at large (Murphy and Chused, 1984). This facilitated data sharing between labs.

3. Antibody production targeting cells from solid tissue

The rapid development of antibodies and cell sorting technology to meet the needs of the hematopoietic research community has laid much of the ground work required for studies of other cell types. There is an ever-growing need to be able to identify subpopulations of cells from solid tissues. Diseases such as diabetes, liver disease, and a variety of cancers have benefited, and will continue to benefit from, an expanding pool of monoclonal antibodies that can be used to further investigate them (Jarpe et al., 1990; Larsen et al., 1986; Muraro et al., 1989). Using the same approach pioneered by Milstein and Kohler, research into all systems of the body has begun to follow in the footsteps of the hematopoietic field. While there are other ways of making hybridomas, the immunization of a rat or mouse with whole cells continues to be valuable because antigens are presented in their native configuration. Thus, the resulting antibodies are likely to recognize the epitopes available in a live, intact cell. In order to use these antibodies to identify and isolate cells for growth and/or functional studies, this is an essential property.

3.1 Immunization

Modern monoclonal antibody production uses a procedure that has changed little from the methods of Köhler and Milstein (Kohler and Milstein, 1975). In brief, the mouse or rat is immunized on multiple occasions using cells obtained from the tissue of interest. Four days after the final immunization the animal is sacrificed and the spleen harvested. Splenocytes are fused with myeloma cells using polyethylene glycol and cultured on semi-solid HAT (hypoxanthine/aminopterin/thymidine) media to prevent the growth of un-fused myeloma cells. Hybridoma clones are isolated and grown individually to produce antibody for testing, and interesting clones are expanded to larger cultures and cryopreserved for further study. Of these steps, the selection of the immunogen and the treatments used to isolate it from solid tissue prior to immunization are the most critical to the success of the procedure. One can immunize with partially dissociated tissue, fully dispersed cells in suspension or a FACS-purified cell subpopulation. Each of these has advantages and disadvantages.

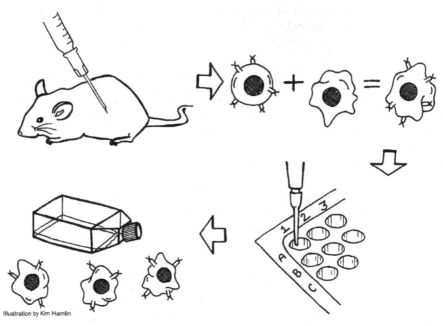

Illustration by Kim Hamlin

Fig. 1. Making Hybridomas. A mouse is injected with an immunogen, antibody producing splenocytes are fused with myeloma cells to produce hybridoma lines, clones are grown in 96-well plates using drug selection to exclude un-fused myeloma cells and selected clones are scaled up

Immunization using minimally-dispersed fragments of tissue is simple, quick and requires limited (if any) proteolytic enzyme treatment. Thus, this approach is most likely to preserve sensitive antigens. Because it does not exclude indigestible material (which can include important tissue substructures), this method also offers the most complete representation of all antigens in the tissue. The disadvantage is that this method includes antigenic material that gives rise to immune responses against unwanted antigens that are present in debris.

If tissue is instead enzymatically dissociated to a single-cell suspension, much of the intercellular debris can be eliminated and the cell-surface antigens are more available. Although native epitopes might be altered by proteolysis, it is important to consider that this same process will be required to recover cells for future analysis; raising antibodies against a proteolysis-sensitive antigen is of limited utility. Unfortunately, it is not always clear what the optimal digestion protocol will turn out to be for a given tissue type. If it is possible to perfuse the organ with a collagenase, this can be a very productive approach (Klaunig et al., 1981). Regardless of whether perfusion is performed, dispersal to a single-cell state usually requires three elements: Mechanical dissociation (by mincing and/or passage through a narrow aperture such as a needle), collagenase treatment (to digest extracellular matrix) and a broad-specificity protease (e.g. trypsin) to disperse tightly-associated cells. Specific protocols successfully used for tissues such as mouse liver (Dorrell et al., 2008b), and human pancreas (Dorrell et al., 2008a) are described in the next section.

Immunization with a subset of a single-cell suspension allows for the generation of antibodies targeted against a subset of cells. At a minimum, it should be possible to exclude defined "contaminating" populations such as blood and endothelium. The cell subset may be isolated using flow sorting, immunomagnetic separation, or physical properties such as density or size. This approach has the same downside as tissue digestion because the researcher must know how to digest the tissue in order to liberate the cells of interest without damaging them. In addition, the number of recovered cells tends to be smaller, meaning that the immunization might be less efficient.

A useful method for increasing the likelihood of obtaining antibodies targeted to a particular cell type is subtractive immunization (Williams et al., 1992). In this method, the first immunization uses tissue, cells and/or debris containing antigens which are irrelevant or undesirable. Subsequent treatment with cyclophosphamide kills the lymphocytes which have responded to any of these antigens, so that after subsequent immunizations with the tissue/cells of interest, only the new and relevant antigens cause an immune response. This approach was used successfully to raise antibodies against human pancreatic islet cells after pre-immunization with trypsin and calf serum, which might otherwise have been confounding antigens required for islet dispersal and cell storage (Dorrell et al., 2008a). Subtractive immunization is compatible with any of the immunogen sources described above.

3.1.1 Proven immunization strategies for mouse liver and human islets

This chapter will focus on two different tissue preparations from two dissimilar tissues. Understanding why they would be treated differently will help the researcher determine the best course of action for future tissue digestion.

Preparing single-cell solution from a mouse liver requires several stages. The first is a two-step portal vein perfusion with calcium and magnesium-free Hanks salt solution followed by a solution containing collagenase. After about twenty minutes, when the liver is visibly degraded it is removed, placed in a dish with media, and teased apart with forceps to release as many cells as possible. The resulting slurry should be allowed to flow through a 40 µm strainer. The filtered cells can be spun for two minutes at 50g in order to pellet the hepatocytes; other cell types will remain in suspension after this treatment. Tissue that does not pass through the filter should be subjected to an ex vivo digest of about 20 minutes in 2.5 mg/mL collagenase D at 37°C before collection through a 40 µm strainer. Any remaining solid tissue is then digested with 0.05% trypsin and mechanically dissociated with a pipetter. Following all of these treatments little or no undigested material should remain. At each stage, it is important to collect dissociated cells by passing them through a 40µm strainer and to store them on ice without further exposure to enzyme (Dorrell et al., 2008b; Klaunig et al., 1981).

We receive intact human islets from the Islet Cell Resource Center network. They are essentially small aggregates of cells obtained as the end product of a whole-organ perfusion protocol, and therefore require only a simple enzymatic digest to reduce to a single cell suspension. Islets are washed with calcium- and magnesium-free DPBS (Dulbecco's Phosphate-Buffered Saline) and then incubated in 0.05% trypsin at 37°C. Gentle pipetting is done every three minutes in order to aid the dissociation until the islets are visibly dispersed into a single-cell state.

3.2 Hybridoma screening

Regardless of the immunization strategy and target cell population, most of the resulting hybridomas will not produce desirable antibodies. It is therefore essential to have a strategy for evaluating many hybridomas to obtain as many useful ones as possible. Labeling of tissue sections with hybridoma supernatants followed by microscopic evaluation can be labor-intensive, but it provides invaluable information regarding cell type-specificity. A detailed anatomical knowledge of the target tissue is not generally required by the screener; the important requirement at this stage is the identification of clones that selectively label populations of cells found in common structures (e.g. ducts, blood vessels, islets, etc.).

For the screening of large numbers of hybridomas for reactivity, cryosections of OCT (Optimal Cutting Temperature) cryomatrix-embedded tissue is recommended. To save time, hybridoma supernatants can be pooled and evaluated in batches. Batches exhibiting a promising tissue labeling pattern are then reevaluated, clone-by-clone. Figure 2 illustrates immunofluorescent labeling using an antibody with an interesting labeling pattern that emerged after an immunization series using human islet cells; this one made the cut. "HIC1-8G12" labels both the ducts (as shown by co-labeling with KRT19) and a subset of cells within the islet.

Further examination revealed that these islet cells were alpha cells (as shown by co-labeling with transthyretin [TTR], which is strongly expressed in this population (Dorrell et al., 2011b)). Using "interesting labeling pattern" as a criterion for antibody selection is subjective, of course, but the study of markers like HIC1-8G12 can lead to the identification of previously unexpected relationships between cell types. Flow cytometric screening of this antibody (described in the next section) showed that this particular antibody recognizes a cell surface antigen and can be used for live cell flow cytometry studies.

4. Flow cytometry

Protocols describing analysis and sorting by flow cytometry have been developed with primary blood cells and/or transformed cell lines in mind. For solid-tissue derived cells, some modifications and special considerations are needed.

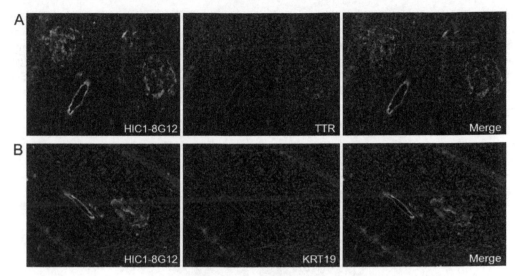

Fig. 2. Human pancreatic tissue labeling by HIC1-8G12. Co-labeling with TTR (A) or KRT19 (B) shows that this antibody binds an antigen that is present on both ducts and alpha cells

4.1 Selecting antibodies of interest by flow cytometric screening

After the initial screening on slides, it is important to determine which antibodies have cell surface reactivity. Flow cytometry is ideal for this purpose. Because the goal is to create antibodies against surface molecules, if an antibody with an interesting tissue labeling pattern does not label live cells with sufficient intensity and specificity to be distinguished by flow cytometry, it is not worth pursuing.

As discussed previously, there are many ways of getting cells into a single-cell suspension. It is recommended that if an enzymatic dispersal was used for immunization then the same method should be used to prepare cells for flow cytometry. If no enzymatic dispersal was used, it is worth taking the time to do the screening using a variety of digestion methods.

Many people worry that the dispersal method will influence the preservation of epitopes. They are right to worry. Some methods will alter or remove antigens of interest rendering them unrecognizable. We have found, for example, that the antigen recognized by one of our in-house favorite antibodies (MIC1-1C3) is unusually sensitive to digestion with Pronase, a mixture of strong proteases (Dorrell et al., 2011a). It is very important to try to optimize your digestion protocol but bear in mind that, at the end of the day, if you can't get the cells into single-cell suspension then they can't be sorted and studied in isolation.

With precious human samples, it can be much more difficult to do the experiments and learn which digestion method will work best but it is worth optimizing in the beginning for higher reproducibility later on.

As with screening on tissue sections, hybridoma supernatants can be pooled for screening by flow cytometry. This may not be necessary if the tissue section screen reduced the candidate list to a reasonable number, but will be necessary if tissue section screening was not performed or if few cells are available. During screening by flow cytometry, the antibody isotype can also be conveniently determined. Cells incubated in the supernatant from a single clone are detected using a combination of isotype-specific secondary antibodies selective for mouse or rat IgG or IgM. Figure 3 is a dot plot showing cells labeled with a mixture of primary antibodies, some IgG and some IgM. Because these were detected with both PE-conjugated anti-mouse IgM and APC-conjugated anti-mouse IgG, primary antibodies of these two isotypes can be used together.

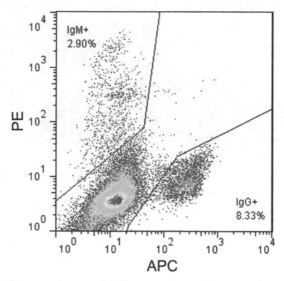

Fig. 3. Isotype specific secondary antibodies permit simultaneous detection of IgG and IgM-labeled cells

4.2 Multi-step antibody labeling to isolate populations of interest

Using unconjugated monoclonal antibodies has advantages and disadvantages. On one hand, it takes longer to label the cells because a secondary antibody is required for detection, necessitating additional steps. On the other hand, the researcher is not limited by pre-conjugated fluorochromes and can change "colors" to suit the experiment or instrument. Although pre-conjugated monoclonal antibodies marking known lineages can be an important tool in the study of solid tissue cells, such reagents are limited and most experiments will require the use of novel, unconjugated antibodies as well. Thus, despite its added complexity, a multi-step labeling protocol incorporating both fluorochrome-conjugated and unconjugated antibodies is ideal.

Generally speaking, monoclonal antibodies made against mouse antigens are made in rats and monoclonal antibodies targeting human antigens are made in mice. It is possible to do combination labeling using both conjugated and unconjugated antibodies raised in the same host species, but special care is required; a detailed protocol is provided later in this section. Polyclonal antibodies are produced by immunization of a variety of species, and are often easy to combine with monoclonal antibodies for co-detection. It is vital to know the host species of each antibody, its isotype and whether or not each secondary antibody is adsorbed to prevent cross-reactivity against immunoglobulins from another species.

Once an antibody has been found with reactivity against a cell population of interest, it is important to do a simple titration in order to ensure that the optimal concentration is being utilized. This helps to optimize the intensity of labeling by minimizing non-specific "background" labeling and also avoids waste. A proper titration involves labeling a fixed number of cells with antibody a range of different concentrations. When the antibody concentration is unknown (which is often the case for hybridoma supernatants), it is convenient to test a series of dilutions (e.g. undiluted, 1:10, 1:100, 1:500:, 1:1000). We prefer to use the highest concentration that does not result in a detectable elevation in non-specific labeling of "negative"cells. It is important to have an excess of antibody to ensure that all target antigens are saturated. The staining will be much easier to reproduce if the researcher takes the time to do one titration experiment at the beginning.

Once the sample is dissociated into single cells, it can be resuspended in holding buffer at a concentration of between $1X10^6$ and $1X10^7$ per ml. This has proven to be a good working concentration for the labeling of solid-tissue derived cells. A recommended general-purpose holding buffer is DMEM with 1% FBS and 0.1 mg/ml DNase I; this mixture provides nutrients, "blocks" cells against non-specific antibody binding by exposure to bovine immunoglobulins and reduces cell aggregation by digesting nuclear material released by dead cells.

It can be advantageous to use a combination of newly developed and commercially available antibodies. This allows for the exclusion (by electronic gating) of unwanted cell types. If multiple antibody-defined cell types are undesirable, the experiment can be simplified by using a common fluorochrome to label them for combined exclusion. This is sometimes referred to as a "dump channel". For example, if both blood and endothelium are unwanted, anti-CD45-FITC and anti-CD31-FITC can be employed together.

We have found it useful to employ a labeling technique incorporating two different unconjugated primary antibodies and three conjugated antibodies in a five color sort. This allowed us to learn whether the new antibodies were reactive against blood or endothelium in addition to epithelial subpopulations.

A sample labeling protocol follows:

Step 1. Using 5ml round-bottom tubes (polystyrene or polypropylene, of the specific type required by the instrument), incubate the appropriate samples with unconjugated primary antibodies on ice for 20 minutes. Primary antibodies that can be distinguished by isotype or host species can be combined at this stage. If the cells are prone to settling, resuspend them periodically by vortexing.

Step 2. Wash by increasing the volume and centrifuging at the speed appropriate for your cells. Aspirate all but about 50 µl of supernatant.

Step 3. Resuspend the pellet in the small volume that remains by gently flicking the tube until there is no longer a visible pellet. Do this before adding appropriate volume of holding buffer. (Failure to do this may result in clumping)

Step 4. Add all secondary antibodies and incubate for 20 minutes on ice.

Step 5. Wash as described above and resuspend in blocking buffer. This solution is holding buffer supplemented with serum/sera of the species in which the conjugated antibodies to be used in Step 6 were raised. The immunoglobulins present in such sera will block unoccupied sites on the divalent secondary antibodies added in Step 4. Allow 20 minutes for blocking to proceed, on ice.

Step 6. Add conjugated primary antibodies and incubate for 20 minutes on ice.

Step 7. Wash and resuspend in holding buffer containing a dead cell-marking agent such as propidium iodide. The optimal concentration for sorting tends to be between $1X10^6$ and $1X10^7$ depending on the amount of debris present after cell isolation. At this point, it is important to visually inspect each tube for the presence of cell clumps. If these cannot be dispersed by vortexing they must be excluded by cell straining to avoid clogging the sorter's nozzle. BD makes tubes with strainer caps that allow convenient filtration of samples immediately prior to sorting.

4.3 Sorting

There are several instrument-related considerations for sorting tissue-derived cells. The selection of optimal nozzle diameter, fluidic pressure, forward and side scatter voltage settings are all vital and will be different depending on the organ and tissue type. Furthermore, because tissue-derived cells are being stored in an unnatural environment - a single cell suspension – care must be taken to avoid cell aggregation and to minimize ongoing cell death. These are issues that seldom arise in the world of hematopoietic sorting.

Cells from solid tissues tend to be larger than hematopoietic cells so it is often necessary to use a larger nozzle in order to avoid damaging the cells or clogging the nozzle. Cells from solid tissues are also often more fragile and are potentially more likely to be damaged by the shear forces when they pass through a smaller nozzle. The correct nozzle diameter must be determined empirically, but when in doubt one should err towards using a larger nozzle and accepting reduced cell throughput.

For sorting cells from human pancreatic islets or nonparenchymal cells isolated from a mouse liver, we use a 100 μm nozzle at a pressure of 15 PSI. Under these conditions, cells were sorted at between 2,000 and 4,000 events per second with over 98% purity (Dorrell et al., 2008b). When sorting mouse hepatocytes, which are large and fragile, the optimal nozzle for highest post-sort viability is the 150 μm nozzle at 4.5 PSI (Duncan et al., 2010).

There are a few tricks to keeping cells in a single cell suspension. If a visible pellet of settled cells is observed in the source tube, the sorter's tube vortexer or manual flicking/vortexing should be employed. As discussed previously, minimizing the amount of calcium can reduce cell aggregation; a low $[Ca^{2+}]$ holding buffer may be helpful. Keeping the sample cold using a chilling apparatus will also inhibit clumping.

Size-standard beads (e.g. 10 μm diameter) should be used to establish consistent forward and side scatter voltages. This facilitates the direct comparison of the size and granularity of cell samples collected on different days and even on different instruments. Sample-to-sample variability of solid tissue derived cells can be much greater than that of hematopoietic cells, so this sort of calibration is very important.

It is common practice when using flow cytometry with hematopoietic cells to set the voltages for the fluorescent detectors so that negative population falls below 10^1. This approach is not recommended for the more heterogeneous populations found in isolated tissue-derived cells, however. It is important to set the voltage on the PMTs so that very few cells are against the axis; if this means that some of the "negative" cells exhibit a significant degree of non-specific labeling, so be it. Note that this will be less of an issue when using software that permits visualization using a logical scale. Figure 5 illustrates how it becomes very difficult to identify distinct populations when detector voltages are set so that not all cells are on scale.

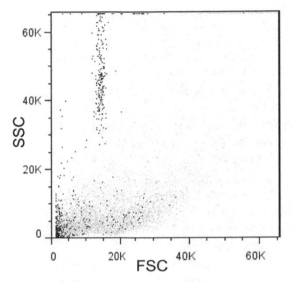

Fig. 4. 10 micron diameter beads and cells both on scale

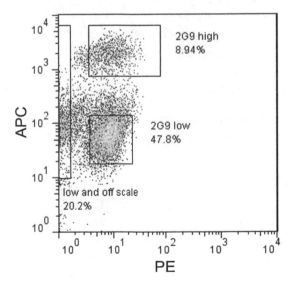

Fig. 5. Some cells are "off-scale low"

The use of "back-scattering" to evaluate the size/granularity of antibody-defined cell subpopulations requires the ability to gate on distinct populations. Since these properties are hard to predict on solid tissue-derived cells, back-scattering can reveal where the cells appear on a FSC/SSC plot and thus insure that both the optimal voltages and gating are employed.

5. Post-sort analysis

While flow cytometry data is valuable, it is the ability to isolate cell populations and do follow-up experiments that it such an overwhelmingly powerful tool. Cell populations can be analyzed for RNA content, they can be cultured, reprogrammed, transplanted, or even simply cytospun and stained for other markers.

5.1 Microarray and RT-PCR

Microarray and QRTPCR are arguably two of the most powerful tools for more comprehensive understanding of each cell type. When post-sort RNA analysis is the goal, there are a few things to keep in mind in order to acquire the highest quality RNA.

A very high purity sorted sample will result in less contamination of RNA from other cell types. A typical sort will result in a purity of 95% or higher. Unless the available cell numbers are very limited, a higher purity level is probably desirable. Re-sorting can be advantageous because it can allow you to make the initial sort faster, not aborting any events.

There are, of course, advantages and disadvantages of sorting and then resorting. The obvious advantage is that the resorted sample is very close to 100% pure; there are very few contaminating cells that might negatively affect results. A potential disadvantage is that the cells are being put through the sorter again. This increases the likelihood of them being damaged. The question of whether the RNA expression might change as the cells are subject to shear forces multiple times is a valid but hypothetical concern.

5.2 Cell culture

Culturing cells that have never been isolated before poses exciting challenges. How does one optimize culture conditions for cells that have not been thoroughly characterized? It requires trial and experimentation but at least if the cells can be analyzed first to see what genes they are expressing, the researcher has information that will aid with making educated guesses.

5.2.1 Stem cell assays

If the goal of sorting is to isolate stem/progenitor cells, there are a number of things that can be done in order to determine whether the cells of interest can be accurately described as stem cells.

One good example of this is the work done by Sato and colleagues, where LGR5 positive cells were isolated from the mouse intestine and used to initiate three dimensional cultures in Matrigel (Sato et al., 2009). In this case, previously gained knowledge, about what the stem cells were expressing and what the requirements were to maintain them, were used in order to cultivate the intestinal stem cell and maintain them indefinitely.

The most rigorous stem cell assay is transplantation. If the unique population is sorted and found to be able to engraft a recipient animal and give rise to multiple types of daughter cells it is a strong indicator that the population contains a stem cell. It is fortunate that many human stem cells are transplantable into immune-deficient mice and give rise to cells that can replace the equivalent mouse cell (Grompe, 2001; Kamel-Reid and Dick, 1988; Wang et al., 2001).

5.2.2 Cell reprogramming

There are several ways that new antibodies might contribute to the success of cell reprogramming. First, if tissue-resident progenitors can be identified and isolated, their inherent plasticity would make them obvious targets for reprogramming. Additionally, if antibodies are created against cells in various stages of development, it will aid in the understanding of cell expression and signaling during development, and the replication of these events in adult cells. Knowing what cells express at various stages of development also allows for quick assays to determine whether the researcher is on the right path.

5.2.3 Culture for research and for transplant

Currently, it is very difficult to get cells for experimentation from cadaveric human donors. Researchers need sources of tissue and unfortunately the cell lines that are available are almost always tumor-derived. This limits their utility with regard to learning about the normal biology of certain cell types. Great strides have already been made in the field of solid tissue culture with the goal of future transplantation. Tissue culture has been shown to be remarkably successful at producing skin (Ueda, 2011). There has also been some success culturing hepatocytes with the hope of creating an external liver support device (Niu et al., 2009). As our understanding becomes more comprehensive, so too will our ability to maintain these cells in culture. Eventually, tissue banks might augment or even supplant the need for cadaveric organ donation.

6. Conclusion

The more tools we have to work with, the more quickly and thoroughly we can address issues of disease and expand our understanding of healthy tissue. Being able to physically isolate different cell subtypes is crucial to that understanding if we want to utilize powerful tools, such as microarray and RNA sequencing, which require large numbers of highly purified populations of cells. Experiments using up to 17 fluorochromes and 31 metal-conjugated antibodies are already being performed on blood and bone marrow cells (Bendall et al., 2011; Perfetto et al., 2004). In our lab, we have produced a number of antibodies against solid tissue and used these to isolate and analyze defined subpopulations that were not previously accessible by flow cytometry (Dorrell et al., 2011b; Dorrell et al., 2011c). As the need for these antibodies continues, it is hoped that more researchers will contribute to the array of tools that can be used in solid-tissue research.

The HLDA has expanded their mission to include characterization of surface epitopes found on all tissues relevant to the immune system. This expansion was reinforced with a name change; they are now the HCDM (Human Cell Differentiation Molecules) (Zola et al., 2007) Unfortunately there is not the same sort of all-inclusive drive toward a comprehensive characterization and evaluation of antibodies against tissue-bound cells. We feel that a comparable initiative directed towards solid organs, using the strategies and techniques described in this document, would yield many additional tools and reagents for the study of normal and pathological tissues.

7. Acknowledgements

The authors would like to thank Kim Hamlin for illustrating figure 1.

8. References

Bendall, S.C., Simonds, E.F., Qiu, P., Amir el, A.D., Krutzik, P.O., Finck, R., Bruggner, R.V., Melamed, R., Trejo, A., Ornatsky, O.I., *et al.* (2011). Single-cell mass cytometry of differential immune and drug responses across a human hematopoietic continuum. Science *332*, 687-696.

Dorrell, C., Abraham, S.L., Lanxon-Cookson, K.M., Canaday, P.S., Streeter, P.R., and Grompe, M. (2008a). Isolation of major pancreatic cell types and long-term culture-initiating cells using novel human surface markers. Stem Cell Res *1*, 183-194.

Dorrell, C., Erker, L., Lanxon-Cookson, K.M., Abraham, S.L., Victoroff, T., Ro, S., Canaday, P.S., Streeter, P.R., and Grompe, M. (2008b). Surface markers for the murine oval cell response. Hepatology *48*, 1282-1291.

Dorrell, C., Erker, L., Schug, J., Kopp, J.L., Canaday, P.S., Fox, A.J., Smirnova, O., Duncan, A.W., Finegold, M.J., Sander, M., *et al.* (2011a). Prospective isolation of a bipotential clonogenic liver progenitor cell in adult mice. Genes Dev *25*, 1193-1203.

Dorrell, C., Grompe, M.T., Pan, F.C., Zhong, Y., Canaday, P.S., Shultz, L.D., Greiner, D.L., Wright, C.V., Streeter, P.R., and Grompe, M. (2011b). Isolation of mouse pancreatic alpha, beta, duct and acinar populations with cell surface markers. Mol Cell Endocrinol *339*, 144-150.

Dorrell, C., Schug, J., Lin, C.F., Canaday, P.S., Fox, A.J., Smirnova, O., Bonnah, R., Streeter, P.R., Stoeckert, C.J., Jr., Kaestner, K.H., *et al.* (2011c). Transcriptomes of the major human pancreatic cell types. Diabetologia *54*, 2832-2844.

Duncan, A.W., Taylor, M.H., Hickey, R.D., Hanlon Newell, A.E., Lenzi, M.L., Olson, S.B., Finegold, M.J., and Grompe, M. (2010). The ploidy conveyor of mature hepatocytes as a source of genetic variation. Nature *467*, 707-710.

Epstein, L.B., Kreth, H.W., and Herzenberg, L.A. (1974). Fluorescence-activated cell sorting of human T and B lymphocytes. II. Identification of the cell type responsible for interferon production and cell proliferation in response to mitogens. Cell Immunol *12*, 407-421.

Fleming, W.H., Alpern, E.J., Uchida, N., Ikuta, K., and Weissman, I.L. (1993). Steel factor influences the distribution and activity of murine hematopoietic stem cells in vivo. Proc Natl Acad Sci U S A *90*, 3760-3764.

Grompe, M. (2001). Mouse liver goes human: a new tool in experimental hepatology. Hepatology *33*, 1005-1006.

Herzenberg, L.A. (1971). II Selective depletion of lymphocytes. In Immunologic Intervention (Academic Press), pp. 96-104, 145-148, 281, 289.

Hoffman, R.A., Kung, P.C., Hansen, W.P., and Goldstein, G. (1980). Simple and rapid measurement of human T lymphocytes and their subclasses in peripheral blood. Proc Natl Acad Sci U S A 77, 4914-4917.

Ikuta, K., Uchida, N., Friedman, J., and Weissman, I.L. (1992). Lymphocyte development from stem cells. Annu Rev Immunol 10, 759-783.

Jarpe, A.J., Hickman, M.R., Anderson, J.T., Winter, W.E., and Peck, A.B. (1990). Flow cytometric enumeration of mononuclear cell populations infiltrating the islets of Langerhans in prediabetic NOD mice: development of a model of autoimmune insulitis for type I diabetes. Reg Immunol 3, 305-317.

Kamel-Reid, S., and Dick, J.E. (1988). Engraftment of immune-deficient mice with human hematopoietic stem cells. Science 242, 1706-1709.

Klaunig, J.E., Goldblatt, P.J., Hinton, D.E., Lipsky, M.M., Chacko, J., and Trump, B.F. (1981). Mouse liver cell culture. I. Hepatocyte isolation. In Vitro 17, 913-925.

Kohler, G., and Milstein, C. (1975). Continuous cultures of fused cells secreting antibody of predefined specificity. Nature 256, 495-497.

Larsen, J.K., Munch-Petersen, B., Christiansen, J., and Jorgensen, K. (1986). Flow cytometric discrimination of mitotic cells: resolution of M, as well as G1, S, and G2 phase nuclei with mithramycin, propidium iodide, and ethidium bromide after fixation with formaldehyde. Cytometry 7, 54-63.

Lemke, H., Hammerling, G.J., Hohmann, C., and Rajewsky, K. (1978). Hybrid cell lines secreting monoclonal antibody specific for major histocompatibility antigens of the mouse. Nature 271, 249-251.

Muraro, R., Nuti, M., Natali, P.G., Bigotti, A., Simpson, J.F., Primus, F.J., Colcher, D., Greiner, J.W., and Schlom, J. (1989). A monoclonal antibody (D612) with selective reactivity for malignant and normal gastro-intestinal epithelium. Int J Cancer 43, 598-607.

Murphy, R.F., and Chused, T.M. (1984). A proposal for a flow cytometric data file standard. Cytometry 5, 553-555.

Nakeff, A., Valeriote, F., Gray, J.W., and Grabske, R.J. (1979). Application of flow cytometry and cell sorting to megakaryocytopoiesis. Blood 53, 732-745.

Niu, M., Hammond, P., 2nd, and Coger, R.N. (2009). The effectiveness of a novel cartridge-based bioreactor design in supporting liver cells. Tissue Eng Part A 15, 2903-2916.

Parks, D.R., Hardy, R.R., and Herzenberg, L.A. (1984). Three-color immunofluorescence analysis of mouse B-lymphocyte subpopulations. Cytometry 5, 159-168.

Perfetto, S.P., Chattopadhyay, P.K., and Roederer, M. (2004). Seventeen-colour flow cytometry: unravelling the immune system. Nat Rev Immunol 4, 648-655.

Roederer, M., De Rosa, S., Gerstein, R., Anderson, M., Bigos, M., Stovel, R., Nozaki, T., Parks, D., and Herzenberg, L. (1997). 8 color, 10-parameter flow cytometry to elucidate complex leukocyte heterogeneity. Cytometry 29, 328-339.

Sato, T., Vries, R.G., Snippert, H.J., van de Wetering, M., Barker, N., Stange, D.E., van Es, J.H., Abo, A., Kujala, P., Peters, P.J., et al. (2009). Single Lgr5 stem cells build crypt-villus structures in vitro without a mesenchymal niche. Nature 459, 262-265.

Solter, D., and Knowles, B.B. (1978). Monoclonal antibody defining a stage-specific mouse embryonic antigen (SSEA-1). Proc Natl Acad Sci U S A 75, 5565-5569.

Spangrude, G.J., Muller-Sieburg, C.E., Heimfeld, S., and Weissman, I.L. (1988). Two rare populations of mouse Thy-1lo bone marrow cells repopulate the thymus. J Exp Med 167, 1671-1683.

Springer, T., Galfre, G., Secher, D.S., and Milstein, C. (1979). Mac-1: a macrophage differentiation antigen identified by monoclonal antibody. Eur J Immunol 9, 301-306.

Ueda, M. (2011). Skin, Applied Tissue Engineering (InTech).

Wang, X., Al-Dhalimy, M., Lagasse, E., Finegold, M., and Grompe, M. (2001). Liver repopulation and correction of metabolic liver disease by transplanted adult mouse pancreatic cells. Am J Pathol 158, 571-579.

Williams, A.F., Galfre, G., and Milstein, C. (1977). Analysis of cell surfaces by xenogeneic myeloma-hybrid antibodies: differentiation antigens of rat lymphocytes. Cell 12, 663-673.

Williams, C.V., Stechmann, C.L., and McLoon, S.C. (1992). Subtractive immunization techniques for the production of monoclonal antibodies to rare antigens. Biotechniques 12, 842-847.

Zola, H., and Swart, B. (2005). The human leucocyte differentiation antigens (HLDA) workshops: the evolving role of antibodies in research, diagnosis and therapy. Cell Res 15, 691-694.

Zola, H., Swart, B., Banham, A., Barry, S., Beare, A., Bensussan, A., Boumsell, L., C, D.B., Buhring, H.J., Clark, G., et al. (2007). CD molecules 2006--human cell differentiation molecules. J Immunol Methods 319, 1-5.

4

Analysis of Cellular Signaling Events by Flow Cytometry

Jacques A. Nunès, Guylène Firaguay and Emilie Coppin
Institut National de la Santé et de la Recherche Médicale, Unité 1068,
Centre de Recherche en Cancérologie de Marseille, Institut Paoli-Calmettes,
Aix-Marseille Univ., Marseille,
France

1. Introduction

All the cells are engaged on a cell communication by sending and receiving signals. Cells having the correct receptors on their surfaces will encode the plasma membrane signal recognition to intracellular signals corresponding to cellular signaling events. This process corresponds to a huge intracellular network of protein-protein or lipid-protein interactions. To avoid to be lost in this kind of maze, some general signaling pathways are most of the time present in many cell systems such as MAP kinases or PI-3 kinases pathways. These signaling pathways induce some protein phosphorylation events that can be identified by specific antibodies directed against these phosphosites (anti-phospho antibodies). Thus, these crossroads can be determined by using antibodies. A widely used method to identify these signaling events is the Western blotting or protein immunoblotting from cell extracts. However, several research groups are now developing new analytical techniques based also on the use of anti-phospho antibodies, but allowing them to detect signaling events at the single cell level. These experimental approaches are using the principles of the flow cytometry (FCM). Phosphoflow technology will provide rapid, quantitative and also multi-parameter analyses on single cells.

2. Cell signaling

In biology, cell signaling corresponds to the mechanisms of communication at the cellular level (Gomperts et al., 2009). The molecules involved in these exchanges provide three functions: transport of information via chemical signals, decoding the messages carried by these signals through receptors (intercellular communication itself), and finally transfer the orders contained in these messages to the intracellular machinery (intracellular communication).

Cellular communication may be endocrine (exchange of information remotely using hormones), paracrine (local exchanges between adjacent cells, such as neurotransmission) or autocrine (messages sent and received by the same cell to self-regulate). The molecules that carry information (hormones, mediators) can be compared with "keys" (called ligands) adapted to the "locks" represented by the receivers. In this metaphor, the intracellular signaling is comparable to a "bolt" activated by movement of the key in the lock. Despite their diversity, the signaling mechanisms are guided by common characteristics.

2.1 General laws involved in cell signaling

All the reactions based on a system [key-lock] and downstream events follow the law of mass action where molecules are engaged each other and then disengaged. These states are corresponding respectively to the reactions of association or dissociation. The affinity of one molecule for another one is defined by the ratio between association and dissociation constants. Thus, the affinity is high when the time of association is longer than the time of dissociation. These parameters are dependent of the conformational changes of the molecules. These alterations open some sites at the surface of the molecule that can be accessible, for instance, to enzymes such as the protein kinases. Thus, these opened sites become substrates for the kinases and a phosphate ion (PO_4^{3-}) is transferred. Thus, cell signaling is in part a large equilibrium between phosphorylation and dephosphorylation states for proteins and but also lipids. Edmond H. Fischer and Edwin G. Krebs were able to show that reversible protein phosphorylation affects the structure, shape, function and activity of proteins that are responsible for the regulation of nearly all aspects of cellular life (Krebs & Fischer, 1989). Fischer and Krebs received the Nobel Prize in Physiology or Medicine in 1992 for the discovery of reversible protein phosphorylation and its importance as a biological regulatory mechanism.

2.2 Intercellular signaling

Intercellular signaling is governed by ligand-receptor interactions. The families of ligands or signals often differ only by tiny molecular substitutions. The addition of a radical hydroxyl (-OH) or methyl (-CH3) is sufficient to differentiate the action of soluble mediators such as dopamine, norepinephrine and epinephrine. In the family of morphine (short sequences of amino acids called peptides and acting in a manner similar to morphine, especially by intervening in the modulation of pain), the substitution of one amino acid can alter the recognition by its ligand receiver called receptor. The study of immunological recognition provides many opportunities to demonstrate that the replacement of a few amino acids in a protein could markedly change binding properties. The antigen receptors such as the T-cell receptors (TCRs) adjust to different agonist, partial agonist and antagonist peptides by subtle conformational changes in their complementarity-determining regions (CDRs), in induced-fit mechanisms of antibody/antigen recognition (Rudolph & Wilson, 2002). This knowledge has been largely improved by high resolution of ligand-receptor co-crystal structures. Thus, various receptors recognize their ligands with different affinities. These differences are important because they help to refine considerably the transmission of information: each peptide acts optimally on its specific receptor, but is also able to affect other receptor family with a lower affinity. This heterogeneity of recognition generates an affinity gradient into the ligand-receptor interactions that in part, will explain selectivity in the intracellular encoding of the signals delivered by the receptors.

2.3 Intracellular signaling

The receptors are made up of successive sequences with the recognition motif of ligand (exposed to the outside of the cell), a transmembrane segment that provides the binding of the receptor to the plasma membrane, and finally a sequence on intracellular enzyme activity. They fall into three groups that differ in the mode of intracellular as a result of the binding of the signal. In the simplest family (insulin receptor, for example), both ends of the same molecule provide signal reception and the enzyme activation. In the two other groups, receptor and enzyme activation functions are not contained in the same protein sequence such as i) a receptor directly associated with an independent cytoplasmic enzyme (for example, the CD4 molecule associated with the Src protein tyrosine kinase family member, Lck), ii) or a receptor linked to an independent cytoplasmic enzyme by a third partner named an adaptor protein (for example, the G protein-coupled receptor (GPCR) β adrenergic receptor coupled to the adenylate cyclase via heterotrimeric G proteins).

Activation of the receptor is itself relayed by families of specialized molecules that provide signal transduction, that is to say their relays, the membrane to all intracellular processes (Fig. 1). Some of them, such as JAK (activating protein kinases called Janus, by reference to the Latin god of thresholds and crossings, to the extent that they "open" signaling downstream of receptors) (Li et al., 2008) or MAPK (mitogen-activated protein kinases) (Mendoza et al., 2011), integrate information from different receptors expressed at the cell membrane. The cell can respond appropriately, for example by phosphorylating its own receptors and thereby adjusting at all times the sensitivity of the cell to signals from its environment (the process of sensitization and desensitization). Several other main signaling pathways have been described such as the phosphatidylinositol 3-kinase-mammalian target of rapamycin (PI3K-mTOR) pathway. The signaling pathways are the major routes to control, for instance, cell differentiation, proliferation in response to extracellular stimuli. Moreover, the major pathways are playing as a concert where these pathways have common regulators and might intersect each other to co-regulate downstream events (Mendoza et al., 2011).

Figure 1 illustrates a simplest view of the major signaling pathways as vertical boxes where always lipid or protein kinase activities are involved. Upon extracellular signals induced by ligand-receptor interactions, these different boxes are turned-on by some adaptor molecules that are connecting receptors to the first steps of these signaling cascades.

Several websites are dedicated to describe these signaling pathways (for example, the Database of Cell Signaling from Science Signaling, a journal published by the American Association for the Advancement of Science (http://stke.sciencemag.org/) or the UCSD Signaling Gateway, a comprehensive resource in signal transduction (http://www.signaling-gateway.org/)). Some nuclear factor involved in gene transcription can be involved very rapidly in these signaling cascades and located transiently at the inner face of the plasma membrane such as the Nuclear Factor of kappa light chain gene enhancer in B cells (NFκB) or the Signal Transducers and Activators of Transcription (STATs). As already mentioned (Mendoza et al., 2011), the signaling pathways are inter-connected. For instance, the phosphoinositide-specific phospholipase C (PI-PLC) clives the phosphatidylinositol-4,5-bisphosphate to generate inositol-1,4,5-trisphosphate (IP3, involved in ionized Ca^{2+}) and diacylglycerol (DAG). This increase of DAG will participate to i) the activation of some serine/threonine protein kinase C (PKC) isoforms that are able to activate NFκB signaling by phosphorylating a repressor of these transcription factors (first box in Figure 1) and ii) the activation of the RAS / Extracellular signal-regulated kinases 1 & 2 (ERK-1/2) pathway via the involvement of a DAG-dependent RAS guanyl releasing protein (RasGRP) (third box in Figure 1).

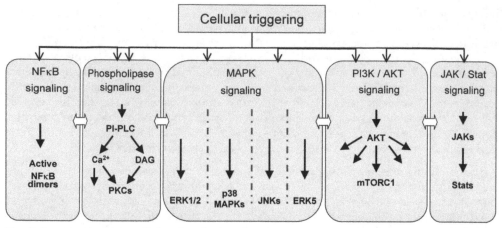

Fig. 1. Some major intracellular signaling pathways activated under an exogenous triggering are here illustrated such as Nuclear Factor of kappa light chain gene enhancer in B cells (NFκB), phosphoinositide-specific phospholipase C (PI-PLC), mitogen-activated protein kinases (MAPK), phosphatidylinositol 3-kinase (PI3K)/Akt and Janus kinase (JAK)/ Signal Transducers and Activators of Transcription (STATs). These pathways are regulated by phosphorylation steps and inter-connected (see text). These signaling cascades are activating transcription factors involved in gene expression

Thus, the intracellular signaling acts as a primary target for control of the expression of specific genes in the cell based on varying combinations of signals it receives. The intracellular signaling mechanisms involved in initiating motor responses (in muscle) or secretory (in the exocrine glands or the brain), but their main role is to control gene expression. They are thus responsible for the process of cell differentiation and a referral to a cell cycle proliferation or, conversely, apoptosis (programmed cell death, which led to the elimination of atypical cells); hence their importance in the process of tumor formation. The majority of cancers results from a dysfunction of intracellular signaling molecules, which in turn causes an imbalance between uncontrolled proliferation and loss of the ability of tumor cells to be self-destructed by apoptosis.

3. Methods for the detection of signaling events

There are several methods involving chemistry, biochemistry, microscopy and physics that are used to detect and analyze intracellular signals. Here, we briefly describe a general method used by who is interested on signal transduction as the Western blot and then we develop the experimental approaches using flow cytometers.

3.1 Western blot (also called protein immunoblot): An universal method for the detection of signaling events

The method is based on the use of electrophoresis to transfer proteins from a gel to a membrane (Towbin et al., 1979). The aim of SDS-PAGE (PolyAcrylamide Gel Electrophoresis) is to separate proteins according to their size, and no other physical feature. Western blots allow investigators to determine the molecular weight of a protein and to measure relative amounts of the protein present in different samples. Prior to SDS-PAGE, the proteins should be extracted from isolated cells or tissues. The proteins are then transferred to a membrane (for example, nitrocellulose), where they are probed using specific antibodies.

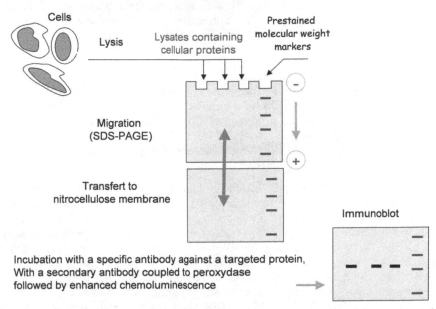

Fig. 2. Western blot method needs to extract the proteins from a reasonable number of cells (but not from a single cell)

The signaling events are governed by phosphorylation/dephosphorylation steps. When a protein is phosphorylated, its apparent molecular weight will be higher (for instance, the molecular weight of the protein Ser/Thr kinase ERK-2 will shift from 42 to 44 kDa upon phosphorylation). The ERK-2 molecular weight mobility shift detected by Western blot correlates with an increase of ERK-2 Ser/Thr kinase activity detected by *in vitro* kinase assays (Nunès et al., 1994). However, the detection of kinase activity based on slower mobility of activated kinases upon SDS-PAGE is not always validated because it does always correlate with the real enzymatic activity (Yao et al, 2000). Thus, some specific antibodies have been developed against the phosphorylated peptide sequences of activated kinases (Yung et al., 1997). Now, many anti-phosphosite antibodies against many phosphorylated proteins such as protein kinases or protein kinase substrates are commercially available. Furthermore, the quality of these antibodies employed is of great importance for the success of the detection of a specific phosphorylation step. The validation of these antibodies will be discussed later on in the text.

The conventional Western blot results are semi-quantitative and require a large number of cells (between 10^2 to 10^6 cells per point). New approaches using similar immuno-detection as Western blot, have been developed to reduce the number of cells per point. High-resolution capillary isoelectric focusing (IEF) allows isoforms and individual phosphorylated forms to be resolved using around 25 cells per point (O'Neill et al., 2006). This capillary–based nano-immunoassay is able evaluate the different ratio of unphosphorylated to mono or bi-phosphorylated ERK1/2 using anti-ERK1/2 antibodies as a probe. This system is now commercialized as the NanoPro 1000 system (ProteinSimple, Santa Clara, CA, USA). Despite all efforts in this field of the Western blot, it is virtually impossible to reach the detection of signaling events at the single cell level. By definition, the flow cytometry is able to detect events at the single cell level using specific antibodies, which is essential in a heterogeneous cell population.

3.2 Phosphoflow analysis

Flow cytometry (FCM) has been used for a long time to detect specific markers at the cell surface using antibodies labeled with fluorescent dyes. Then, similar approaches have been performed to detect intracellular proteins such as cytokine production in hematopoietic cells. Intracellular cytokine detection by FCM opens the door to analyze other intracellular proteins such as phosphorylated signaling molecules (Chow et al., 2001; Krutzik & Nolan, 2003).

Detection of phospho-proteins by FCM requires that the protein is stable and accessible to the antibody. Cells are usually stimulated and fixed with formaldehyde or paraformaldehyde to cross-link the phospho-proteins and stabilize them for analysis. The fixed cells must then be permeabilized to allow for entry of phospho-specific antibodies into the cells. Different permeabilization techniques are often useful for various subcellular locations. A mild detergent will allow for detection of cytoplasmic proteins, while alcohol may be required for antibody access to nuclear proteins. Alcohol permeabilization may also enhance phospho-protein detection using peptide specific antibodies due to the denaturing property of alcohol (Krutzik et al., 2004). This approach has been improved to visualize many parameters at the single cell level (multidimensional molecular profiles of signaling) to define cell network phenotypes in many cell types (Irish et al., 2004; Irish et al., 2006).

The consideration of signaling networks as dynamic systems is crucial for a full understanding, and this requires methods applicable to analyze individual living cells (Johnson & Hunter, 2005). A protocol has been developed to determine by FCM in a dynamic system and in single-cell, the phospho-protein activation status (Firaguay & Nunès, 2009). For this approach, a pleitropic activator of the signaling pathway such as pervanadate, has been used. In this protocol, the phospho-protein staining corresponds to a sandwich labelling to increase the brightness (for instance, a rabbit monoclonal antibody against anti-phosphoSer473 AKT followed by biotinylated anti-rabbit immunoglobulins (Ig) and then streptavidin conjuguated with the Phycoerythrin (PE).

All the specific antibodies used for phosphoflow analysis should be well optimized for this technology. For instance, it will be easier to use monoclonal antibodies than polyclonal antibodies. These antibodies should be validated in Western blot by showing that they are able to detect the targeted phospho-protein at the right molecular weight (without detecting other protein, meaning other bands on the immunoblot). Then for a proper validation in FCM, the phospho-protein staining should be sensitive to the cell treatment with some specific inhibitors of signaling pathways (for anti-phosphoSer473 AKT staining in activated cells should be decreased by a cell treatment with a PI3K inhibitor). Finally, the phospho-protein staining should be decreased or abolished in cells derived from mice invalidated for the gene encoding the targeted protein (KO mice) or in cellular models using a RNA interference.

Using validated anti-phospho-protein antibodies, this phosphoflow analysis is able to define signaling events upon receptor triggering (multi intracellular phospho-protein staining) at the single cell level (multi extracellular surface marker staining). This purpose is illustrated in Figure 3.

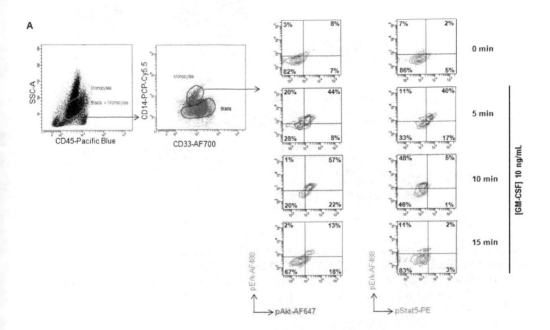

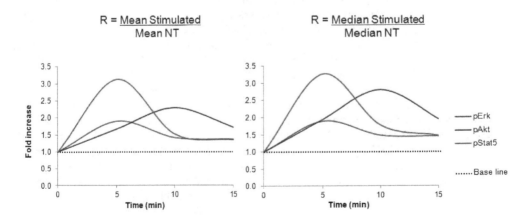

c

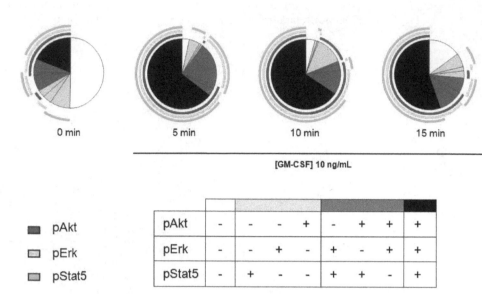

pAkt	-	-	-	+	-	+	+	+
pErk	-	-	+	-	+	-	+	+
pStat5	-	+	-	-	+	+	-	+

■ pAkt
□ pErk
▨ pStat5

Fig. 3. Peripheral blood mononuclear cells (PBMC) from bone marrow sample were obtained by Ficoll (AbCyss LymphoPrep, #1114545) gradient (centrifugation at 2500 rpm for 20 min without brake) followed by two washes (centrifugation at 900 rpm for 15 min) in RPMI-1640 medium (Gibco, #21875) added to 2% Fetal Calf Serum (FCS) (Eurobio, #CVFSV F00-01). Cells number was determined by Trypan blue staining. Whole PBMCs were maintained for 16 hrs in RPMI 2% FCS at 2.10^5 cells/mL

For GM-CSF stimulation, 5.10^5 cells were pre incubated 1 min at 37°C, and incubated with 10 ng/mL human GM-CSF (Peprotech # 300-03) in RPMI-1640-2% FCS at 37°C for 5, 10 and 15 min. Cells were fixed by 1,6% paraformaldehyde (PFA) (Sigma Aldrich, # 158127) at room temperature for 10 min. Cells were washed (centrifugation at 1800 rpm for 5 min) twice with phosphate-buffered saline (PBS) (Life Technologies, #70013065). Cells were stained 30 min at 4°C for extracellular lineage markers: CD45 - Pacific Blue (BioLegend, #304029), CD14 - PerCP-Cy5.5 (BD Pharmingen, #550787) and CD33 - AlexaFluor 700 (BD Pharmingen, #561160). After washes in FCM buffer (PBS, 2%, 1mM EDTA) viability was evaluated by Live/Dead Staining (Invitrogen, LIVE/DEAD Fixable Aqua Dead Cell Stain, #L34957) for 15 min at 4°C. Then for intracellular staining, cells were washed twice in FCM buffer and were permeabilized and fixed with Cytofix/Cytoperm (BD Biosciences, #554722) for 10 min at 37°C followed by washes in Perm/Wash buffer (BD Biosciences 10X Perm/Wash Buffer, #554723). Intracellular staining for phosphoflow analysis was performed as previously described (Firaguay & Nunès, 2009). Cells were incubated 30 min at 4°C by adding: i) directly coupled antibodies: anti-phospho-Akt S473 Alexa Fluor 647 (Beckman Coulter, #A88915), anti-phospho-ERK-1/2 T202/Y204 Alexa Fluor 488 (Beckman Coulter, #A88928) and ii) Uncoupled antibodies: anti-phospho-STAT5 (Cell Signaling Technology, #9314). Cells were wash twice in Perm/Wash buffer (BD Biosciences), incubated 20 min at 4°C with biotinylated secondary antibodies (Jackson ImmunoResearch product, Beckman

Coulter, Biotin-SP-AffiniPure F(ab')2 Frag Donkey AntiRabbit IgG (H+L), #711-066-152). After two washes, cells were incubated 15 min at 4°C with Streptavidin conjugated with phycoerythrin (Streptavidin-PE) (Beckman Coulter, #IM3325). Cells were washed twice, re-suspended in PBS and processed on a LSR II SORP 4 lasers flow cytometer (BD Becton Dickinson). Data were collected and analyzed using DIVA software (BD Biosciences) and FlowJo software (Tree Star).

Panel A: Monocyte gating strategy. Monocytes were gated on i) intermediate Side scatter cells and CD45 positive cells and then ii) on both CD14 and CD33 positive cells. Then phospho-Erk, phospho-Akt and phospho-Stat5 were visualized on the monocyte population in resting condition and under 10 ng/mL of human GM-CSF stimulation for different times (5, 10 and 15 min). For this illustration, dot plots were analyzed using DIVA software (BD Biosciences).

Panel B: AKT, ERK1/2 and Stat5 phosphorylation kinetics in monocytes under GM-CSF stimulation. Kinetics Ratio (stimulated / non treated NT) Mean (left) and Median (right) of fluorescence in response to 10 ng/mL human GM-CSF stimulation. As represented, kinetics are similar on both Mean and Median Fluorescence intensity parameters: phospho-Stat5 and phospho-ERK1/2 increase are early: at 5 min of stimulation and they decreased to a constant "sub-basal" level at 10 min, whereas phospho-AKT is increased later: at 10 min of stimulation and decreased at 15 min. For this representation statistical data were extracted using FlowJo software (Tree Star).

Panel C: Statistical representation of phospho-AKT, ERK1/2 and Stat5 negative and positive monocytes populations under GM-CSF stimulation. Color gradation represents percentages of monocytes negative for phospho-ERK1/2, phospho-AKT and phospho-Stat5 (in white), positive for one and two of these parameters (respectively in light and dark grey) and positive for all of them (in black). Colored curves surrounding circles shows positive phospho-proteins represented in each portion of the circle: phospho-ERK1/2, phospho-AKT and phospho-Stat5 are respectively in yellow, red and green. As represented, phopho-Stat5 alone and associated to phospho-ERK alone or with phospho-AKT is the major signaling response of monocytes to GM-CSF stimulation, in particular at 5 min. Phospho-AKT response to stimulation appears to be later since at 10 min the percentage of phospho-AKT positive monocytes is distinctly increased, in particular for proportion only phospho-AKT positive cells at this time of stimulation. For this representation, statistical data were extracted using FlowJo software (Tree Star) and represented with SPICE v5.2 software (Apple Mac U.C. Berkeley, CA, USA).

4. Concluding remarks and recent investigations in phosphoflow analysis

As illustrated by data shown in Figure 3, the phosphoflow analysis permits to use one or multiple signal transduction markers in association with cell surface markers to dissect complex biological processes in heterogeneous populations. This technique is quite rapid, quantitative and a major point: working at the single cell level. However, the investigators must be careful with the buffers used for cell fixation and permeabilization that could affect the binding of the antibodies to the phosphosites (Krutzik et al., 2004). Another major point is a perfect validation of the antibodies used for phosphoflow analysis (see text, paragraph 3.2). In some case, the brightness of a signaling marker will be enhanced by using a sandwich labeling method (Firaguay & Nunès, 2009).

The ability to analyze multiple single-cell parameters is critical for understanding cellular heterogeneity. In the context of oncology, cancer cells harboring a hyperactivation of signaling pathways such PI3K/AKT, Ras/ERK or JAK/Stats pathways can be treated with specific inhibitors of these pathways (Vivanco & Sawyers, 2002; Irish et al., 2006). However, such inhibitors will target the cancer cells but can also affect other cell types that will be involved in the tumor environment (Le Tourneau et al., 2008). Thus, these chemotherapy protocols could affect the homeostasis of the normal cell compartment that could be evaluated by analyzing multiple single-cell parameters. This balance is illustrated by a picture into the Figure 4 (see below).

Fig. 4. By targeting Ras/ERK or PI3K/AKT pathways in cancer cells, these treatments could affect for instance immune cells such as Natural Killer (NK) cells or different T cell subsets such as regulatory T cells (Tregs). This cell homeostasis will be affected as illustrated by this picture as a Mikado game where sticks could be a signaling parameter or a cell type. The artwork of this picture has been performed by the STUDIO HULKETTE (*http://www.hulkette.com*)

To analyze a large number of parameters at the single cell level in a heterogeneous population, new apparatus has been developed by Scott Tanner's group, the single-cell mass cytometer (Bandura et al., 2009). In this study, single cell 20 antigen expression assay was performed on cell lines and leukemia patient samples immuno-labeled with lanthanide-tagged antibodies. Recent studies from Gary Nolan's group push this technology to analyze simultaneously 34 cellular parameters, where they compare the consequences of different drug treatments on the immune cells and hematopoiesis (Bendall et al., 2011). The beauty of these new technologies in the field of cell signaling opens a new area on the signal monitoring of treatments in many diseases such as cancers, infectious or autoimmune diseases. However, a brake to the development of these approaches will be the huge number of data that will be generated and how to analyze these data. Some clues of computational approaches were already built such as spanning-tree progression analysis of density-normalized events (SPADE) (Qiu et al., 2011).

Phosphoflow analysis will become an universal technology to investigate cell signaling events at the single cell level. By using validated antibodies for FCM and right protocols of fixation/permeabilization, the phosphoflow analysis will be a standard method in many labs across the world to analyze few parameters at the same time. Moreover, the new technologies are already under the market to analyze more than 30 parameters at the same time.

5. Acknowledgements

We would like to thank Dr. Françoise Gondois-Rey and Dr. Christine Arnoulet for their help on the FCM settings for cell surface markers and le Studio Hulkette for the artwork. This work was supported by grants from Institut National de la Santé et de la Recherche Médicale and the Institut National du Cancer (# PL-06026 and # INCa/DHOS 2009) (to J.A. Nunès). J.A. Nunès is a recipient of a Contrat d'Interface Clinique with the Department of Hematology (Institut Paoli Calmettes). G. Firaguay was supported by fellowships from the Institut National du Cancer. E. Coppin is supported by a fellowship from the Région Provence Alpes Côte d'Azur (PACA) – Innate Pharma.

6. References

Bandura, D.R., Baranov, V.I., Ornatsky, O.I., Antonov, A., Kinach, R., Lou, X., Pavlov, S., Vorobiev, S., Dick, J.E., & Tanner, S.D. (2009). Mass cytometry: technique for real time single cell multitarget immunoassay based on inductively coupled plasma time-of-flight mass spectrometry. *Anal Chem*, 81, pp. 6813-6822, ISSN 1520-6882.

Bendall, S.C., Simonds, E.F., Qiu, P., Amir el, A.D., Krutzik, P.O., Finck, R., Bruggner, R.V., Melamed, R., Trejo, A., Ornatsky, O.I., Balderas, R.S., Plevritis, S.K., Sachs, K., Pe'er, D., Tanner, S.D., & Nolan, G.P. (2011). Single-cell mass cytometry of differential immune and drug responses across a human hematopoietic continuum. *Science*, 332, pp. 687-696, ISSN 1095-9203.

Chow, S., Patel, H., & Hedley, D.W. (2001). Measurement of MAP kinase activation by flow cytometry using phospho-specific antibodies to MEK and ERK: potential for pharmacodynamic monitoring of signal transduction inhibitors. *Cytometry*, 46, pp. 72-78, ISSN 0196-4763.

Firaguay, G., & and Nunes, J.A. (2009). Analysis of signaling events by dynamic phosphoflow cytometry. *Sci Signal*, 2, pl3, ISSN 1937-9145.

Gomperts B.D., Kramer I.M., Tatham P.E.R. (Ed(s).). 2009. *Signal Transduction, 2nd Edition*, Academic Press, ISBN: 978-0-12-369441-6, Maryland Heights, MO, USA.

Irish, J.M., Hovland, R., Krutzik, P.O., Perez, O.D., Bruserud, O., Gjertsen, B.T., & Nolan, G.P. (2004). Single cell profiling of potentiated phospho-protein networks in cancer cells. *Cell*, 118, pp. 217-228, ISSN 0092-8674.

Irish, J.M., Kotecha, N., & Nolan, G.P. (2006). Mapping normal and cancer cell signalling networks: towards single-cell proteomics. *Nat Rev Cancer*, 6, pp. 146-155, ISSN 1474-175X.

Johnson, S.A., & Hunter, T. (2005). Kinomics: methods for deciphering the kinome. Nat Methods 2, 17-25, ISSN 1548-7091.

Krebs, E.G., & Fischer, E.H. (1989). The phosphorylase b to a converting enzyme of rabbit skeletal muscle. 1956. *Biochim Biophys Acta*, 1000, pp. 302-309, ISSN 0006-3002.

Krutzik, P.O., and Nolan, G.P. (2003). Intracellular phospho-protein staining techniques for flow cytometry: monitoring single cell signaling events. *Cytometry A*, 55, pp. 61-70, ISSN 1552-4922.

Krutzik, P.O., Irish, J.M., Nolan, G.P., & Perez, O.D. (2004). Analysis of protein phosphorylation and cellular signaling events by flow cytometry: techniques and clinical applications. *Clin Immunol*, 110, pp. 206-221, ISSN 1521-6616.

Le Tourneau, C., Faivre, S., Serova, M., & Raymond, E. (2008). mTORC1 inhibitors: is temsirolimus in renal cancer telling us how they really work? *Br J Cancer*, 99, pp. 1197-1203, ISSN 1532-1827.

Li, W.X. (2008). Canonical and non-canonical JAK-STAT signaling. Trends Cell Biol, 18, pp. 545-551, ISSN 1879-3088.

Mendoza, M.C., Er, E.E., & Blenis, J. (2011). The Ras-ERK and PI3K-mTOR pathways: cross-talk and compensation. Trends Biochem Sci, 36, pp. 320-328, ISSN 0968-0004.

Nunes, J.A., Collette, Y., Truneh, A., Olive, D., & Cantrell, D.A. (1994). The role of p21ras in CD28 signal transduction: triggering of CD28 with antibodies, but not the ligand B7-1, activates p21ras. *J Exp Med*, 180, 1067-1076, ISSN 0022-1007.

O'Neill, R.A., Bhamidipati, A., Bi, X., Deb-Basu, D., Cahill, L., Ferrante, J., Gentalen, E., Glazer, M., Gossett, J., Hacker, K., Kirby, C., Knittle, J., Loder, R., Mastroieni, C., Maclaren, M., Mills, T., Nguyen, U., Parker, N., Rice, A., Roach, D., Suich, D., Voehringer, D., Voss, K., Yang, J., Yang, T., & Vander Horn, P.B. (2006). Isoelectric focusing technology quantifies protein signaling in 25 cells. *Proc Natl Acad Sci U S A*, 103, pp. 16153-16158, ISSN 0027-8424.

Qiu, P., Simonds, E.F., Bendall, S.C., Gibbs, K.D., Jr., Bruggner, R.V., Linderman, M.D., Sachs, K., Nolan, G.P., & Plevritis, S.K. (2011). Extracting a cellular hierarchy from high-dimensional cytometry data with SPADE. *Nat Biotechnol*, 29, pp. 886-891, ISSN 1546-1696.

Rudolph, M.G., & Wilson, I.A. (2002). The specificity of TCR/pMHC interaction. *Curr Opin Immunol*, 14, pp. 52-65, ISSN 0952-7915.

Towbin, H., Staehelin, T., & Gordon, J. (1979). Electrophoretic transfer of proteins from polyacrylamide gels to nitrocellulose sheets: procedure and some applications. *Proc Natl Acad Sci U S A*, 76, pp. 4350-4354, ISSN 0027-8424.

Vivanco, I., & Sawyers, C.L. (2002). The phosphatidylinositol 3-Kinase AKT pathway in human cancer. *Nat Rev Cancer*, 2, pp. 489-501, ISSN 1474-175X.

Yao, Z., Dolginov, Y., Hanoch, T., Yung, Y., Ridner, G., Lando, Z., Zharhary, D., & Seger, R. (2000). Detection of partially phosphorylated forms of ERK by monoclonal antibodies reveals spatial regulation of ERK activity by phosphatases. *FEBS Lett*, 468, pp. 37-42, ISSN 0014-5793.

Yung, Y., Dolginov, Y., Yao, Z., Rubinfeld, H., Michael, D., Hanoch, T., Roubini, E., Lando, Z., Zharhary, D., & Seger, R. (1997). Detection of ERK activation by a novel monoclonal antibody. *FEBS Lett*, 408, pp. 292-296, ISSN 0014-5793.

5

Gamma Radiation Induces p53-Mediated Cell Cycle Arrest in Bone Marrow Cells

Andrea A. F. S. Moraes et al.*
Universidade Federal de São Paulo – Unifesp,
Universidade Nove de Julho – Uninove,
Universidade Estadual de Santa Cruz – Uesc
Brazil

1. Introduction

The hematopoietic system is organized in a hierarchical manner in which rare hematopoietic stem cells initiate the hierarchy and have the ability to self-renew, proliferate and differentiate into different lineages of peripheral blood cells as well as to intermediate hematopoietic progenitor cells. Most hematopoietic stem cells are quiescent under steady-state conditions and function as a stock population to protect the hematopoietic system from exhaustion due to various stressful conditions. In contrast, hematopoietic progenitor cells are rapidly proliferating cells with limited self-renewal ability. The proliferation and differentiation of hematopoietic progenitor cells fulfills the requirements of normal hematopoiesis allowing the hematopoietic system to react promptly and effectively to meet the demand for increasing the output of mature cells during hematopoietic crisis such as loss of blood, hemolysis, infection, the depletion of HPCs by chemotherapy and/or radiotherapy (Reya, 2003; Weissman *et al.*, 2001; Walkley *et al.*, 2005).

Reactive oxygen species (ROS) are produced in organisms due to radiation, biotransformation of dietary chemicals, some diet components, transient metal ions, inflammatory reactions and during normal cellular metabolism.

The effect of gamma radiation ionization affects the main components of biological material such as carbon, hydrogen, oxygen, nitrogen. Radiobiology is described as the action of ionizing radiation on living things. On the molecular and cellular levels direct ionizing radiation can affect molecules in cells, especially DNA, lipids and proteins, promoting breakage and / or modifications or indirectly acting on water molecules and generating excitation ionization products of water radiolysis which include free radicals present and reactive oxygen species (ROS). Both direct and indirect effects of radiation on cells and tissues have biological effects in the short or long term. Such effects can be mitigated or

* Lucimar P. França, Vanina M. Tucci-Viegas, Fernanda Lasakosvitsch, Silvana Gaiba, Fernanda L. A. Azevedo, Amanda P. Nogueira, Helena R. C. Segreto, Alice T. Ferreira and Jerônimo P. França
Universidade Federal de São Paulo – Unifesp, Brazil
Universidade Nove de Julho – Uninove, Brazil
Universidade Estadual de Santa Cruz – Uesc, Brazil

eliminated by the antioxidant system and cell repair system that work against oxidizable stress and / or cellular stress (Figure 1).

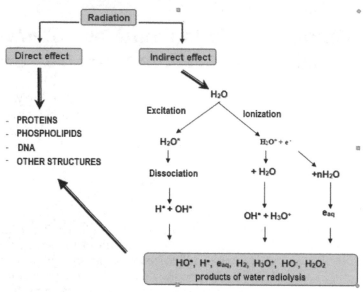

Fig. 1. Direct and indirect effect of radiation produced by free radicals of water radiolysis. Modified scheme of Stark, 1991

The pro-oxidant/antioxidant balance leads to a disturbance which, in turn, results in a condition of oxidative stress subsequently oxidizing cell components, activating cytoplasmic and/or nuclear signal transduction pathways, modulating gene and protein expression and changing both DNA and RNA polymerases activities. Normal cellular metabolism seems to be a primary source for endogenous ROS (such as the participation of oxidatively damaged DNA and repair in aging or cancer development). Oxidative damage to cellular DNA often causes mutagenesis as well as programmed cell death. While mutagenesis might result in carcinogenesis, programmed cell death often causes degenerative disorders (Nakabeppu *et al.*, 2007). Hydroxyl radicals generated from ionizing radiation attack DNA resulting in single strand breaks and oxidative damage to sugar and base residues. Hydroxyl radicals cause ionization of DNA bases as well as of other cellular components. Unsaturated fatty acids play an important role, since lipid peroxidation yields a plethora of stable derivates, which add to nucleic acids forming exocyclic DNA adducts of high miscoding potential, as well as DNA-DNA and DNA-protein cross-links (Bartsch *et al.*, 2004).

Damage of cells by ionizing radiation includes mainly modifications of DNA molecules, such as single and double-strand breaks. While single-strand breaks are quickly repaired in a process that requires poly-(ADP-ribose)-polymerase (PARP), double-strand breaks represent potentially lethal damage and their repair is complicated. Imperfect DNA repair causes mutations and contributes to genome instability. This is mostly manifested as chromosome aberrations, interchromosomal and intrachromosomal rearrangements (dicentric aberrations, translocations, or inversions). Detection of chromosomal aberrations in peripheral lymphocytes is an important indicator of obtained dose of radiation (Kozubek, 2000).

When the DNA is damaged an interconnected network of signaling is activated, resulting in damage repair, temporary or permanent cell cycle arrest or cell death. Cell cycle arrest allows time for DNA damage repair. If the repair is unsuccessful, the cells are removed by apoptosis, necrosis or their proliferation is permanently suppressed by initiation of stress-induced premature senescence (SIPS). Cmielová et al., 2011) reported that it is possible that the major mechanism of response of these tissues to irradiation is not apoptosis, but induction of SIPS. Both pathways often work together to induce replicative and premature senescence. In general, activation of p53 and upregulation of p21 in cells undergoing senescence are transient (Toussaint et al., 2000; Robles & Adami, 1998).

Increased p53 activity and p21 expression usually occur during the onset of senescence and then subside when the expression of p16 starts to rise. Before p16 upregulation, inactivation of p53 can prevent senescence induction in some cells. However, once p16 is highly expressed cell cycle arrest becomes irreversible simply by downregulation of p53 (Campisi et al., 2005; Beausejour et al., 2004; Narita et al., 2004). This suggests that while both p53 and p21 play an important role in the initiation of senescence, only p16 is required for the maintenance of senescence. In agreement with this suggestion, we found that IR-induced activation of p53 and upregulation of p21 occurred prior to the increased expression of p16 and p19 in murine BM HSCs (Meng et al., 2003; Neben et al., 1993; Wang et al., 2006)

Recent studies showed that a majority of murine BM hematopoietic cells including HSCs died by apoptosis after exposure to a moderate dose of IR in vitro. However, a subset of these cells survived IR damage up to 35 days in a long-term BM-cell culture although having lost their clonogenic function. These surviving cells exhibited an increased AS β gal activity, a biomarker for senescent cells, and expressed elevated levels of the proteins (p16Ink4a and p19Arf) encoded by the Ink4a-Arf locus, whose expression has been implicated in the establishment and maintenance of senescence by direct inhibition of various cyclin-dependent kinases (CDKs), (Meng et al., 2003; Dimri et al., 1995; Lowe et al., 2003; Sharpless et al., 1999).

The main function of mitochondria is ATP production, which occurs during mitochondrial oxidative phosphorylation (ox-phos). In several cell types, mitochondria also act as a very efficient $Ca2+$ buffer, taking up substantial amounts of cytosolic $Ca2+$ at the expense of mitochondrial membrane potential ($\Delta\Psi m$). The pathways of $Ca2+$ entry into mitochondrial matrix are known as the mitochondrial calcium uniporter (MCU), the "rapid mode" mechanism, and the mitochondrial ryanodine receptor. The main role of mitochondrial $Ca2+$ is the stimulation of the ox-phos enzymes. In addition to ox-phos, mitochondria are central players in cellular $Ca2+$ signaling by shaping and buffering cellular $Ca2+$ signals. As a consequence of $Ca2+$ uptake, mitochondria can suffer $Ca2+$ overload, triggering the opening of the permeability transition pore (PTP) which is associated with apoptosis via the mitochondrial pathway or necrosis due to mitochondrial damage. PTPs have been shown to be promoted by thiol oxidation and inhibited by antioxidants, supporting a role of ROS in pore opening. In addition, it has been demonstrated that mitochondrial $Ca2+$ uptake can lead to free radical production. From a thermodynamic point of view, however, it has been noted that $Ca2+$ uptake occurring at the expense of membrane potential should result in a decrease in ROS production (Crosstalk signaling between mitochondrial $Ca2+$ and ROS) (Brookes et al., 2004). Mitochondrial PTP is formed from a number of proteins within the matrix, and mitochondrial inner and outer membranes (Ishas & Mazat, 1998; Brookes *et al.*, 2004). One of the processes through which mitochondria contribute to cell death is through PTP opening (Bratton & Cohen, 2001). The PTP precise composition remains unclear, but it is evident that this is a multi-subunit protein channel that spans the mitochondrial inner and outer membrane. Critical components appear to include the mitochondrial VDAC (voltage-dependent anion channel), the ANT (adenine nucleotide translocase), and cyclophilin D (Ishas & Mazat, 1998). It has also been shown that pre-treatment of isolated mitochondria with pro-oxidants can lower the threshold at which the PTP opening occurs (Brookes & Darley-Usmar, 2004). It seems that the opening of this Ca^{2+}-dependent channel plays an important role in controlling the commitment of the cell to death through apoptotic or necrotic mechanisms (Kokoszka *et al.*, 2004; Baines *et al.*, 2005).

2. Objective

The purpose of this work is to evaluate ionizing radiation-induced apoptosis in mice bone marrow cells and the role of p53, p21 and Ca++ in this process, cell cycle alterations and indirect determination of reactive oxygen specimens (ROS).

3. Methods

3.1 Animals

Mice C57BL/10 (3 months) were provided by the Instituto Nacional de Farmacologia (located at Rua Três de Maio 100, Vila Clementino, SP, Brazil). The animals were maintained on standard mouse feed and water *ad libitum*.

3.2 Irradiation

Gamma irradiation was carried out with an Alcyon II [60]CO teletherapy unit with the mice at a distance of 80 cm from the source. The dose rate was 1.35 Gy/min. Animals in batches of ten were placed in a well–ventilated acrylic box with an individual cell for each mouse and exposed whole – body to 7Gy (Segreto *et al.*, 1999).

3.3 Preparation procedure of bone marrow cells

The mice were killed by cervical dislocation 4 hours after gamma irradiation, and both femurs were removed from each mouse. After cutting of the proximal and distal ends of the femurs, the bone marrow cells were gently flushed out with 5ml suspension using phosphate buffered saline (PBS) (Segreto *et al.*, 1999).

3.4 Intracellular reactive oxygen species

Intracellular peroxides were determined by incubating $2X10^6$ cells/ml in medium (defined above) with 5 nM 2',7'-dichlorodihydrofluorescein diacetate (DCFH-DA) (Molecular Probes) for 30 min at 37°C, then analyzed using a Becton Dickinson (BD) Bioscience Flow Cytometer - model FACScalibur (San Jose, CA) equipped with an argon laser emitting at 488 nm (Jagetia & Venkatesh, 2007). BD CellQuest Pro software was used for fast reliable acquisition, analysis and presentation of information.

3.5 Measurement of intracellular Ca++

Calcium was measured after incubation of the bone marrow cells with the fluorescence indicator Fura- 2/AM in the form of acetoxymethyl ester (AM). The bone marrow cells at concentration of 10^6 cells/ml were suspended in 2.5 ml of Tyrode's solution (137 mM NaCl, 2.68 KCl, 1.36 mM CaCl2_2H2O, 0.49 mM MgCl2_6H2O, 12 mM NaHCO3, 0.36 mM NaH2PO4, and 5.5 mM D-glucose) for 30 min. The suspension was then centrifuged at 100 g for 4 min, the supernatant was aspirated, and the result pellet was suspended in 2.5 ml of Tyrode and transferred to the quartz cuvette of a SPEX fluorometer (AR CM System) for fluorescence determination. Measurements were made at alternated 340- and 380-nm excitation wavelengths, with emission at 505 nm. The autofluorescence ratio was less than 10% and therefore was not subtracted from the fluorescence measurements before calculation of the fluorescence ratio. The cells were incubated with 0.01% pluronic 127 detergent and 2M Fura-2/AM, and the cuvette was transferred to a PerkinElmer Life Sciences spectrofluorometer (LS 5B, Buckinghamshire, UK) to determine the fluorescence spectrum of the indicator in the excitation range of 300 to 400 nm, with emission at 520 nm. In the esterified form, maximum fluorescence was observed at 390 nm. As the indicator Fura- 2/AM was transformed into the acid form, the fluorescence peak shifted to 350 nm within an average period of 2 h, thus indicating the maximum amount of indicator incorporation into the cell suspension. At that time the cell suspension was washed with 15 ml of Tyrode and centrifuged at $100g$ for 4 min. The supernatant was discarded, and the pellet was suspended in 2.5 ml of Tyrode and transferred to a SPEX fluorometer programmed for excitation at two wavelengths (340 and 380 nm) with emission at 505 nm, under constant stirring at 37 °C. The first reading of this phase corresponded to basal calcium. At the end of each experiment, a control was performed using 50 mM digitonin, 1mM MnCl2, and 2 mM EGTA. The results are calculated by the relative ratio of 340 and 380 nm, considering the reading for digitonin to be 100% (Rmax) and 0% for EGTA (Rmin). Using the ratio Rmax/Rmin, calcium concentration was estimated by the formula of (Grynkiewicz et al., 1985).

3.6 Apoptosis assay

The bone marrow cells were labeled with annexin V-FITC (Roche), binding to phosphatidylserine at the cell surface of apoptotic cells and propidium iodide (PI) provided by Sigma, is used as a marker of cell membrane permeability following manufacturer's instructions. Samples were examined by fluorescence-activated cell sorter (FACS) analysis, and the results were analyzed using Cell-Quest software (Becton Dickinson Model: Facscalibur, San Jose, CA) (Vermes et al., 1995).

3.7 Cell cycle analysis

For cell cycle analysis the cells were washed with cold PBS and fixed with 70% ethanol. For detection of low-molecular-weight fragments of DNA, the cells were incubated for 5 min at room temperature in a buffer (1.92 ml: 0.2 mol/L Na_2HPO_4 + 8 ml: 0. 1 mol/L citric acid, pH 7.8) and then stained with PI in Vindelov's solution for 40 min at 37 °C. The measurements were performed in a Becton Dickinson Flow Cytometer model FACScalibur (San Jose, CA); data were analyzed using Cell Quest software (Vindelov & Christensen, 1990)

3.8 Flow cytometric analysis of p53 and p21

For analysis of p53 and p21 expression and activation, cells were washed once in phosphate-buffered saline (PBS) containing 0.1% sodium azide (Sigma), centrifuged at 1.4G, fixed in 2% paraformaldehyde for 10 minutes at 4°C, washed 3 times in PBS, and washed twice in PBS with 50 mmol/L NH_4Cl. Cells were then permeabilized with 0.1% saponin in PBS containing 10% normal bovine serum for 30 minutes at 22°C. The first primary antibody incubation was performed in PBS containing 10% normal bovine serum and 0.1% saponin. Aliquots were then incubated for 60 minutes with anti-p53 and anti-p21 antibodies (Santa Cruz Biotechnology, Santa Cruz, CA), final dilution 1:800, or rabbit IgG as a control, followed by washing in PBS containing 0.1% saponin 3 times for 5 minutes at 22°C.   Cells were then incubated with the first fluorochrome-conjugated secondary antibody Alexa 488 and 594, final dilution 1:1600, for 40 minutes at 37°C in the dark (Danova *et al.*, 1990).

3.9 Immunocytochemistry

After cutting off the proximal and distal ends of the femurs, the bone marrow cells were gently flushed out with 5ml suspension using phosphate buffered saline (PBS) (Segreto *et al.*, 1999). Bone marrow cells were washed 3 times with phosphate-buffered saline (PBS), cytofuged onto glass slides, fixed in 4% paraformaldehyde for 10 minutes at 4°C, washed 3 times in PBS, and washed twice in PBS with 50 mmol/l NH_4Cl. Cells were permeabilized with 0.1% saponin in PBS containing 10% normal bovine serum for 30 minutes at 22°C. The first primary antibody incubation was performed in PBS containing 10% normal bovine serum and 0.1% saponin. Slides were incubated with anti-p53 and anti-p21 antibodies at the dilution 1:100 (Santa Cruz Biotechnology, Santa Cruz, CA) for 2 hours, followed by the nuclear staining dye DAPI (4_6-diamidino-2-phenylindole) 1:10000 (catalog #D1036; Molecular Probes, Invitrogen, Carlsbad, CA) which was diluted in the preparatory solution A of Slowfade Antifade kit (Molecular Probes, Eugene, OR) for a final dilution 1:100 and the fluorescent secondary antibody Alexa 594 and 647, final dilution 1:500 (Invitrogen). Neither one used lectin conjugated with Alexa fluor 488 for the cellular membrane. After an initial primary and secondary antibody staining, the procedure was repeated for the second primary and secondary antibody staining, and the slides were then mounted in the preparatory solution A containing Dapi. Each fluorochrome was analyzed individually by means of an inverted confocal laser scanning Fluorescence Microscope (LSM; Zeiss, Germany).

3.10 Ethics

The present study was performed in accordance with the ethical standards laid down in the updated version of the 1964 Declaration of Helsinki and was approved by the Research Ethics Committee of the Federal University of São Paulo, Brazil (CEP N° 1248/01).

3.11 Statistical analysis

The results obtained were analyzed using a one-way analysis of variance (ANOVA) followed by the Student – Newman – Keuls Multiple Range Test.

4. Results and discussion

Cell exposure to DNA-damaging agents can result in timely repair of the damage and maintenance of genetic fidelity in daughter cells, cell death, or the development of heritable genetic changes in viable daughter cells through replication of damaged DNA or segregation of damaged chromosomes before repair (Nakabeppu *et al.*, 2007; Christine *et al.*, 1995; Szymczyk *et al.*, 2004). The last one could generate the genetic changes that contribute to the development of a malignant phenotype. As a result, the gene products controlling timely repair and appropriate cell death after DNA damage are expected to be critical determinants of neoplastic evolution. Moreover, as most antineoplastic agents are DNA-damaging agents, these same gene products probably influence response and cure rates during the treatment of human tumors (Christine *et al.*, 1995).

The number of cells in apoptosis was determined using Annexin-V and cell viability by propidium iodide using the technique of flow cytometry, Figure 2. These results indicated the loss of transport function or structural integrity of the cytoplasmic membrane, which is crucial to distinguish viable from unviable cells, given the wide variety of cell viability assays using different cationic markers such as (PI, 7 -AAD) or (Hoechst 333420). These fluorochromes determine the transport characteristics of the plasma membrane (Ormerod *et al.*, 1992, 1993; Poot *et al.*, 1997). In addition, flow cytometry has identified the viability of bone marrow cells (Figure 2) i.e. the number of non-viable cells that characterize death by late apoptosis (Vermes *et al.*, 1995). Contour diagram of FITC-Annexin V/PI flow cytometry of bone marrow cells after 7 Gy irradiation. The lower left quadrants of each panels show the viable cells, which exclude PI and are negative for FITC-Annexin V binding. The upper right quadrants (7.0 ± 1.2) % contain the non-viable, necrotic cells, positive for FITC-Annexin V binding and for PI uptake. The lower right quadrants (50.1 ± 1.2) % represent the apoptotic cells, FITC-Annexin V positive and PI negative demonstrating cytoplasmic membrane integrity.

Ionizing radiation induces G2/M cell cycle arrest and triggers a p53-dependent signaling pathway that, in turn, may induce apoptotic cell death (Szymczyk *et al.*, 2004). Nonetheless, this response may vary and be cell type-dependent because p53 response causes cell cycle arrest in untransformed cells (Bates *et al.*, 1999) and apoptosis in transformed cells (Bates & Vousden, 1999). Cell cycle distribution was determined by propidium iodide staining for both the control group and irradiated group, the latter with cells exposed to 7 Gy radiation (GI). An increase in the percentage of apoptotic cells was observed in the irradiated group 4h after irradiation (Figure 3B) when compared to the control group (Figure 3A). Cells in G2 are more susceptible to ionizing radiation, as this is the moment when the cell must be repaired prior to entering mitosis. It is possible that the increased activity of protein kinase ATM (ataxiatelangiectasia) modifies the p53 levels thereby altering the cell cycle distribution (Figure 3).

Cell cycle progression and cell fate are determined by the dynamic balance between different p53 downstream effectors, including p21 (Waf1/Cip1) (Waldman *et al.*, 1996). DNA damage-induced cell cycle arrest is regulated by the tumor suppressor p53 by direct stimulation of p21[WAF1/CIP1] expression, an inhibitor of cyclin-dependent kinases (Cdks). Working with the cyclin proteins, Cdks ensure, for example, that DNA replication in the S phase follows from the G_1 resting phase. p21[WAF1/CIP1] then, inhibits both the G_1/S and the

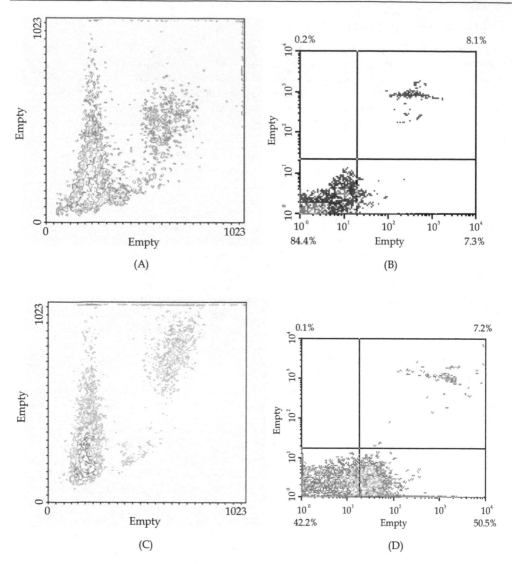

Fig. 2. Flow Cytometric Analysis of control (A and B) and irradiated (C and D) groups to demonstrate the basis for the gating of viable, apoptotic and necrotic cells. The mice were killed by cervical dislocation 4 hours after gamma irradiation. The exposure time was 5.18 minutes in order to achieve 7 Gy of radiation. On Figure 2, A and C, axis X refers to forward scatter and axis Y refers to side scatter. According to the control samples, the following quadrants of the cytograms were defined: - the lower left quadrant showing viable cells; the lower right quadrant representing early apoptotic cells, binding annexin V, but still retaining their cytoplasmic integrity; the upper right quadrant representing nonviable, late apoptotic/necrotic cells, positive for annexin V and the upper left quadrant showing nonviable necrotic cells/nuclear fragments with no annexin V

G_2/mitosis (M) transitions, by exerting its negative effects on various Cdks. For example, according to current theories, by inhibiting cyclin B/Cdc2 (Cdk1) p21[WAF1/CIP1] induces G_2/M phase arrest. Even without being required for initiating G_2/M cell cycle arrest, p53 and p21[WAF1/CIP1] seem to be critical for sustaining it as, for instance, cells lacking p53 and p21[WAF1/CIP1] leave the G_2/M cell cycle arrest prematurely and either enter mitosis or reinitiate DNA replication, causing genomic instability and possibly leading to accumulation of oncogenic mutations (Bork *et al.*, 2009). Changes in the cell cycle in response to ionizing radiation were evaluated using the p53 protein (phosphorylated - serina15). The p53 protein activation is the major route of cellular response after DNA damage in many cell types, as described by Yu & Zhang, 2005. We demonstrated that this machinery is activated when apoptosis is induced by radiation. The signaling pathway after p53 activation leads to an increased expression of p21, which consequently affects the profile of the cell cycle and also leads to cell death by apoptosis. Phosphorylation of serine 15 is a modification that is due to DNA damage induced by radiation. This was evaluated 4h after irradiation, using specific antibodies against p53 and p21 (Figures 4 and 5). We observed the increased expression of p53 and p21 by confocal fluorescence microscopy. These data corroborate the method presented by flow cytometry, which showed an increase in fluorescence intensity of

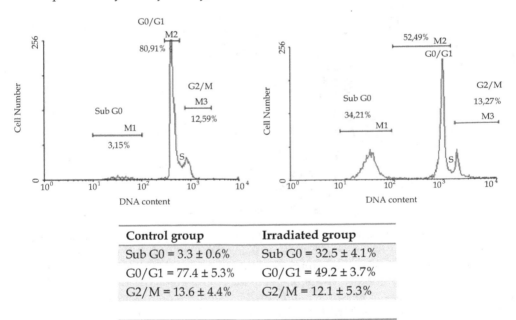

Control group	Irradiated group
Sub G0 = 3.3 ± 0.6%	Sub G0 = 32.5 ± 4.1%
G0/G1 = 77.4 ± 5.3%	G0/G1 = 49.2 ± 3.7%
G2/M = 13.6 ± 4.4%	G2/M = 12.1 ± 5.3%

Fig. 3. Analysis of cell cycle distribution: A) Control group and B) Irradiated group. Progression through the cell cycle was monitored 4 h after ionizing radiation. Bone marrow cells were analyzed for DNA content by FACS using PI staining. The histogram was generated using CellQuest software (Alam *et al.*, 2004). Representative histograms are shown from $n=4$ independent experiments. Table show cell percentages quantified by the mean (±SD) of four different peak analyses. Statistical analysis: Anova - Newman Keuls. **p < 0.05**

these proteins when compared to the control group (Figures 4 and 5). Overlapped images for immunocytochemistry of p53 and p21 proteins and cellular structures (Nuclei: blue-fluorescent DNA and cell membrane: green-fluorescent Lectin). In the irradiated group there is an increase in fluorescence intensity and a change in the distribution of p53 and p21 when compared to the control group (Figures 6 and 7).

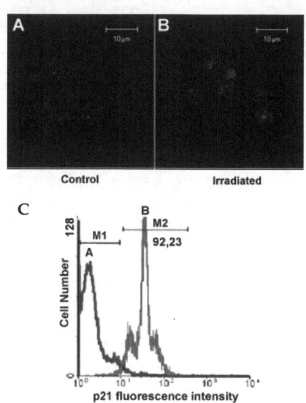

Fig. 4. (above – A and B) Confocal microscopy analysis: bone marrow cells of mice were immunostained in red-fluorescent. Specific staining using antibody p21 (red-fluorescent Alexa Fluor® 594). The irradiated group (B) shows an increase in fluorescence intensity and a change in p21 distribution when compared to the control group (A). (graph below-C) Flow cytometric analysis of p21 protein in paraformaldehyde-fixed cells harvested from mice femurs. Control group A (left) and irradiated group B (right). The bone marrow cells were analyzed 4h after radiation. The cells were pooled, washed, permeabilized with saponin, and stained with anti-p21 antibody and then analyzed by flow cytometry. Histograms were generated and analyzed using CellQuest cell cycle software. Percentage of cells showing fluorescence intensity. Histogram related to gated p21 positive bone marrow cells of GC and GI (89.8 ± 5.3). Statistical analysis: Anova - Newman Keuls. **p < 0.05, N = 4.**

Cells with not completely repaired DNA damage entering SIPS instead of being removed by apoptosis cause the selection of a resistant population, creating a risk of developing a population of cells with dangerously damaged genome. Stem cells are currently exhaustively studied due to their self-renewal capability, ability to produce a broad spectrum of cell types and high proliferation potential and they can be used in bone marrow transplantation and peripheral blood stem cell transplantation for patients who had radiation treatment or high dose of chemotherapy in treatment of cancer. Nonetheless, their reaction to DNA damage as a response to genotoxic stress is not widely known, such as ionizing radiation used in the treatment of many cancer types. Hematopoietic stem cells react to irradiation mostly by apoptosis induction, and sometimes by senescence induction. (Vávrová *et al.*, 2002; Meng *et al.*, 2003). Even though mesenchymal stem cells isolated from bone marrow do not die by apoptosis after irradiation, they are notable to proliferate anymore. The irradiation with the dose 2.5-15 Gy do not destruct the cells, but induces telomere shortening, stops cell division and increases the activity of senescence-associated β-galactosidase increases (Serakinci *et al.*,

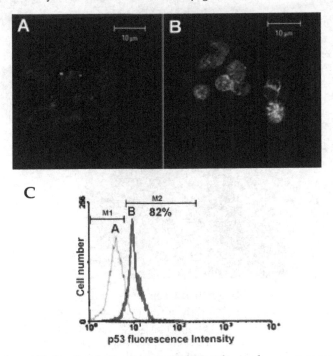

Fig. 5. (above – A and B) Confocal microscopy analysis of mice bone marrow cells. Specific staining using p53 antibody (blue-fluorescent Alexa Fluor® 647). The irradiated group (B) showed an increase in fluorescence intensity and a change in p21 distribution when compared to the control group (A). (graph below - C) Flow cytometric analysis of p53 protein in paraformaldehyde-fixed cells harvested from mice femurs. Control group A (left) and irradiated group B (right). The bone marrow cells were analyzed 4h after radiation. The cells were pooled, washed, permeabilized with saponin, and stained with anti-p53 antibody and then analyzed by flow cytometry. Histograms were generated and analyzed using CellQuest cell cycle software. Percentage of cells showing fluorescence intensity. Histogram related to gated p53 positive bone marrow cells of GC and GI (87.6 ± 5.3). Statistical analysis: Anova - Newman Keuls. **p < 0.05, N = 4**

2007). Induction of premature senescence have implicated two major pathways: the p53-p21Cip1/Waf1 or p19Arf-Mdm2-p53- p21Cip1/Waf1 pathway, triggered by DNA damage; and the p16Ink4a-Rb pathway, activated by the Ras-mitogen-activated protein kinase cascade (Livak & Schmittgen, 2001; Takano *et al.*, 2004; Cheng *et al.*, 2000). Activation of either pathway is sufficient to induce senescence, but they frequently work together causing premature senescence. Induction of apoptosis or premature senescence, or both, in HSCs and progenitors can result in inhibition of their hematopoietic function. (Wang *et al.*, 2005).

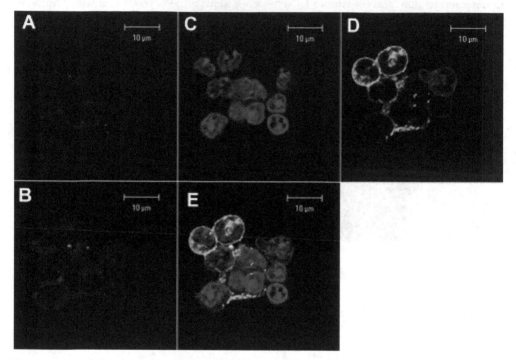

Fig. 6. Laser confocal images of bone marrow cells of the control group mice, were stained by immunofluorescence with combinations of the four fluorophores (red, green, blue and pseudo-colored gray): A) Red channel showing a specific staining using antibody anti-p21 (Alexa fluor 594). B) Pseudo-colored gray channel showing a specific staining using antibody anti-p53 (Alexa fluor 647). C) Blue channel showing nuclei counterstained with the blue-fluorescent DNA stain DAPI (diamidino-2-phenylindole) which is a blue fluorescent probe that fluoresces brightly upon selectively binding to the minor groove of double stranded DNA. D) Green channel showing a Lectin staining using the conjugated with green - fluorescent Alexa fluor 488. Lectin staining using the conjugated with green - fluorescent Alexa fluor 488. E) Specific combination of the four channels reveals overlapped images A, B, C and D. These findings suggest that the bone marrow is more highly susceptible to oxidative damage by radiation ionizing, it also induces alteration in intracellular calcium levels and cell death apoptosis. In summary, the present work demonstrates that radiation induced apoptosis is increase when p53 or p21 responsive G2 arrest being suggestive induction of premature senescence may represent a novel underlying mechanism for radiation

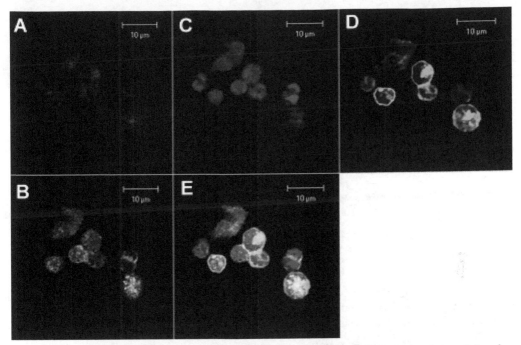

Fig. 7. Laser confocal images of bone marrow cells of the irradiated group mice, were stained by immunofluorescence with combinations of the four fluorophores (red, green, blue and pseudo-colored gray): A) Red channel showing specific staining using antibody anti-p21 (Alexa fluor 594). B) Pseudo-colored gray channel showing a specific staining using antibody anti-p53 (Alexa fluor 647). C) Blue channel showing nuclei counterstained with the blue-fluorescent DNA stain DAPI (diamidino-2-phenylindole) which is a blue fluorescent probe that fluoresces brightly upon selectively binding to the minor groove of double stranded DNA. D) Green channel showing a Lectin staining using the conjugated with green - fluorescent Alexa fluor 488. Lectin staining using the conjugated with green - fluorescent Alexa fluor 488. E) Specific combination of the four channels reveals overlapped images A, B, C and D. In the irradiated group there is an increase in fluorescence intensity and a change in the distribution of p53 and p21 when compared to the control group

In many circumstances, changes in cytosolic Ca2+ concentration may be observed. A surplus of cellular Ca2+ is dangerous for cell life, since it activates many proteases and phospholipases. This Ca2+ surplus can, thus, lead to necrotic cell death. Apoptotic cell death also relies on increased Ca2+ concentrations (Murgia *et al.*, 2009). Both deaths are mediated by endoplasmic reticulum (ER) calcium release and by capacitative Ca2+ influx through Ca2+ release-activated Ca2+ channels (Pinton & Rizzuto, 2006). In Figure 8, we observed the radiation-induced biological effects on cell membranes and the resulting damage which can raise the cytosolic calcium levels triggering cell death by apoptosis, as described by other authors (Starkov *et al.*, 2004). The elevation of cytosolic calcium results in activation of a variety of enzymes sensitive to this ion, such as phosphatases, caspases and calpains, which can be activated from the mitochondria by signaling molecules (Hajnóczky *et al.*, 2003). So it was relevant to determine calcium levels as an indicator of oxidative stress and smooth endoplasmic reticulum stress. Our results showed a significant increase of basal levels of calcium after irradiation (GI). The baseline values of calcium are associated with various cellular responses such as death by necrosis, apoptosis and cell proliferation (Hirota *et al.*, 1998; Anza *et al.*, 1995).

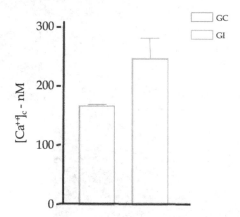

Fig. 8. Effect of γ radiation on basal intracellular Ca^{2+} concentration in mice granulocyte of the control (GC) and irradiated (GI) groups with doses of 700cCy – measured by fura-2 fluorophore. Statistical analysis: Newman Keuls. **p < 0.05, N = 4**

Either necrosis or apoptosis is activated depending on the range of the insult, and consequently on the amount of Ca2+ increase. Ca2+ can establish local concentrations within the cell, due to its low rate of diffusion, when compared to other second messengers and on the dynamic sequestration of this ion by several organelles (Rizzuto & Pozzan, 2006). This unique property allows the cell to decipher various signals, triggering different consequences through a single molecule. ER is crucial to regulate calcium concentration and the sensitivity to apoptosis. Ca2+ accumulated in the ER can be released upon apoptotic stimuli coupled to IP3, and being detected by mitochondria. This process is regulated by Bcl-2 family proteins, strategically located at the ER and mitochondria surfaces. Cells that overexpress Bcl-2 showed a considerable decrease in Ca2+ levels within the ER and the Golgi apparatus. Consequently, reduced Ca2+ concentration increases upon stimuli coupled to IP3 generation were detected both in the cytosol and in the mitochondria (Pinton *et al.*, 2000). In cells in which the pro-apoptotic members of Bcl-2 family, Bax and Bak, were deleted, the same effect was observed. On the other hand, Bax and Bak double knockout cells are protected against apoptotic stimuli (Scorrano *et al.*, 2003). In these cells, ER Ca2+ values were restored to control level through silencing of Bcl-2 (Danial & Korsmeyer, 2004). Stimuli that produce cytosolic Ca2+ increase bring out large Ca2+ influxes in the mitochondria matrix, regardless of the low affinity of mitochondrial Ca2+ carriers for this ion. This is explained by the existence of mitochondria-ER contacts, where microdomains of high Ca2+ concentrations are found and trigger fast accumulation of Ca2+ in the matrix (Hayashi *et al.*, 2009). A variety of responses is caused by mitochondria Ca2+ uptake, from stimulation of metabolism (ATP production) - when they are subject to a transient stimuli, to apoptosis - in case of a more persistent or excessive Ca2+ increase. Additionally, mitochondria Ca2+ accumulation triggered by apoptotic stimuli causes swelling and fragmentation, followed by cytochrome c release. The opening of permeability transition pores (PTP) was also observed upon ceramide-induced apoptosis, which sensitizes mitochondria to the otherwise physiological IP3-mediated Ca2+ signal (Szalai *et al.*, 1999). In which concerns ER calcium concentrations, Bcl-2 family members control this apoptotic pathway. In particular, upon apoptotic stimuli, Bax and Bak localize at the outer mitochondria membrane triggering its permeabilization and release of apoptotic factors in the cytosol (Szalai *et al.*, 1999). Apoptotic cell death may also be caused by mitochondrial malfunction due to ROS-induced mtDNA damage, lipid peroxidation or protein oxidation (Migliaccio *et al.*, 1999).

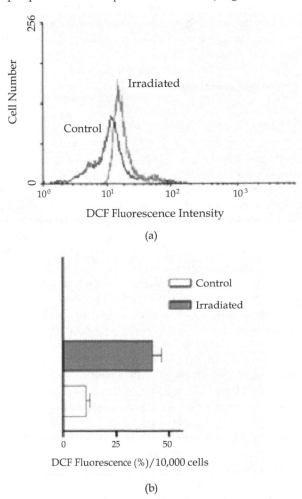

(a)

(b)

Fig. 9. Measurement of reactive oxygen species (ROS) induced by ionizing radiation in bone marrow cells. The cells were loaded with DCFH-DA (10μM) for 30 minutes. After conversion of the ester to DCFH by intracellular esterases, the number of cells exhibiting increased fluorescence of oxidized DCF was evaluated by flow cytometry. A- Histograms represent four independent experiments in bone marrow cells for groups control and exposed to ionizing radiation. B- Statistical analysis for groups control and irradiated: Anova - Newman Keuls. **p < 0.05, N = 4**

ROS cellular levels induced by ionizing radiation were evaluated by measuring DCFH-fluorescence. DCFH is permeable to membrane cell where it is rapidly hydrolyzed by esterases, thus the production of ROS leads to increased DCF fluorescence. We observed an increase of about 50% in DCF fluorescence induced by radiation in bone marrow cells (Figure 9A) in the irradiated group and with flow cytometry we were able to distinguish differences to this type of analysis (Figure 9B).

However, since the caspases are dependent on a reduced thiol group in their active site to cleave their substrates, apoptosis progression is inhibited by excessive ROS, or reactive oxygen species production (Hengartner, 2000). Once more, it should be pointed out, that a pro-oxidant state is induced in mitochondria by CaCl2 through indirect action, since this molecule did not present an intrinsic capacity to induce oxidative stress directly. As previously reported, the redox effects induced by CaCl2 is very likely to be consequence of CaCl2-dependent MPTP induction (Zamzami & Kroemer, 2001).

5. Conclusion

There is controversy in the literature regarding the use of antioxidants as inhibitor of apoptotic response. In some systems it is almost certain that free radicals and ROS are produced, and in these cases, antioxidants have been shown to reduce or delay apoptosis. Oxidative stress induced by gamma radiation can induce to clonogenic death or apoptosis. Although physical and chemical agents can promote cytotoxicity or genotoxicity action leading to apoptosis independent of production of free radicals, oxidative stress induced by radiation can, without any doubts, elicit cell death, and mild oxidative stress which in turn initiate apoptosis. ROS can cause oxidative stress even though not being essential for the apoptotic processes. A very interesting hypothesis is that perturbation of cellular redox homeostasis may control these key events. p53 protein is widely regarded as an important sensor of genotoxic damage in cells, and mutations in p53 are the most frequent observed in human cancers. After radiation induced damage, a protein expression such as p53 occur, which takes to p21 transcription, leading to cell cycle arrest in G0/G1 in order to repair the DNA damage, irreversible growth arrest, terminal differentiation, or apoptosis. Major processes resulting from oxidative stress include alteration of metabolic pathways, lipid peroxidation and the loss of intracellular calcium homeostasis. Thus, the lethality of gamma radiation is well known as much as cellular radiosensitivity. The syndrome of acute radiation may occur by failure of the bone marrow, gastrointestinal tract or central nervous system. The management of patients suffering from acute radiation syndromes still remains a major challenge. Survival of radiation induced bone marrow failure depends on the dose of radiation received. At radiation doses of 3 to 8 Gy, morbidity and lethality are primarily caused by hematopoietic injury. Our results suggest that bone marrow is highly susceptible to oxidative damage by the gamma radiation dose of 7 Gy, whose mechanism of action is directly related with changes in the intracellular calcium levels, increasing ROS and cell death apoptosis. It was observed that major processes resulting from oxidative stress include alteration of metabolic pathways, lipid peroxidation and the loss of intracellular calcium homeostasis. We demonstrated that this machinery is activated when apoptosis is induced by radiation. The signaling pathway after p53 activation leads to an increased expression of p21, which consequently affects the profile of the cell cycle and also leads to cell death by apoptosis. This confirms the hypothesis that perturbation of cellular redox homeostasis may control these key events in the process of cell death by apoptosis. In summary, our work demonstrates that radiation induced apoptosis is increased when p53 or p21 responsive G2 arrest occur, suggesting induction of premature senescence which may represent a novel underlying mechanism for radiation and may be able to contribute to the treatment of cancer by radiotherapy.

6. Acknowledgements

We would like to thank Unifesp and Uesc (collaborators) for their help in experiments with bone marrow cells mice. The authors gratefully acknowledge the financial support from Fapesp, Fapesb and CNPq grants.

7. References

Alam, S.; Sen, A.; Behie, L.A. & Kallos, M.S. (2004). Cell cycle kinetics of expanding populations of neural stem and progenitor cells in vitro. *Biotechnology and Bioengineering.* Vol.88,No.3,pp.332-347, ISSN 0006-3592

Baines, C.; Kaiser, R.; Purcell, N.; Blair, N.; Osinska, H.; Hambleton, M.; Brunskill, E.; Sayen, M.; Gottlieb, R.; Dorn, G. *et al.* (2005). Loss of cyclophilin D reveals a critical role for mitochondrial permeability transition in cell death. *Nature (London)* Vol.434, pp.658–662, ISSN 0028-0836

Bates, S. & Vousden, K. (1999). Mechanisms of p53-mediated apoptosis. *Cellular and Molecular Life Sciences*, Vol.55, No.1, pp.28–37, ISSN 1420-682X

Bates, S.; Hickman, E. & Vousden, K. (1999). Reversal of p53-induced cell-cycle arrest. *Molecular Carcinogenesis*, Vol.24, No.1, pp. 7–14, ISSN 0899-1987

Beausejour, C.; Krtolica, A.; Galimi, F.; Narita, M.; Lowe, S.; Yaswen, P. & Campisi, J. (2003). Reversal of human cellular senescence: roles of the p53 and p16 pathways. *The EMBO journal*, Vol.22, pp.4212-4222, ISSN 0261-4189

Bratton, S. & Cohen, G. (2001). Apoptotic death sensor: an organelle's alter ego? *Trends in Pharmacological Sciences*, Vol.22, pp.306–315, ISSN 0165-6147

Brookes, P. & Darley-Usmar, V. (2004). Role of calcium and superoxide dismutase in sensitizing mitochondria to peroxynitrite-induced permeability transition. *American Journal of Physiology. Heart and Circulatory Physiology*, Vol.286, pp.H39–H46, ISSN 0363-6135

Brookes, P. & Darley-Usmar, V. (2004). Role of calcium and superoxide dismutase in sensitizing mitochondria to peroxynitrite-induced permeability transition. *American Journal of Physiology. Heart and Circulatory Physiology*, Vol.286, pp.H39–H46, ISSN 0363-6135

Brookes, P.; Yoon, Y.; Robotham, J.; Anders, M. & Sheu, S. (2004). Calcium, ATP, and ROS: a mitochondrial love–hate triangle. *American Journal of Physiology. Cell physiology*, Vol.287, pp.C817–C833, ISNN 0363-6143

Campisi, J. (2005). Senescent cells, tumor suppression, and organismal aging: good citizens, bad neighbors. *Cell*, Vol.120, pp.513-522, ISSN 0092-8674

Canman, C.; Gilmer, T.; Coutts, S. & Kastan, M. (1995). Growth factor modulation of p53-mediated growth arrest versus apoptosis. *Genes & Development*, Vol.1, No.9, pp. 600-611, ISSN 0890-9369

Cmielová, J.; Havelek, R.; Jiroutová, A.; Kohlerová, R.; Seifrtová, M.; Muthná, D.; Vávrová, J. & Řezáčová, M. (2011). DNA Damage Caused by Ionizing Radiation in Embryonic Diploid Fibroblasts WI-38 Induces Both Apoptosis and Senescence. *Physiological Research*, Vol.60, pp.667-677, ISSN 0862-8408

Dimri, G.; Lee, X.; Basile, G. *et al*. A biomarker that identifies senescent human cells in culture and in aging skin in vivo. *Proceedings of the National Academy of Sciences of the United States of America*, Vol.92, pp.9363-9367, ISSN 0027-8424

Grynkiewicz, G.; Poenie, M.; Roger, Y. & Tsien, B. (1985). A New Generation of Ca2+ Indicators with Greatly Improved. Fluorescence Properties. *The journal of Biological Chemistry*, Vol.260, No.6, pp. 3440-3450, ISSN 1067-8816

Hengartner, M. (2000). The biochemistry of apoptosis. *Nature*, Vol.401, pp.770–6, ISSN 0028-0836

Ichas, F. & Mazat, J. (1998). From calcium signaling to cell death: two conformations for the mitochondrial permeability transition pore Switching from low- to highconductance state. *Biochimica et Biophysica Acta*, Vol.1366, pp. 33–50, ISSN 0006-3002

Vermesa, I.; Haanena, C.; Steffens-Nakkena, H. & Reutellingspergerb, C. (1995). A novel assay for apoptosis. Flow cytometric detection of phosphatidylserine expression on early apoptotic cells using fluorescein labelled Annexin V. *Journal of Immunological Methods*, Vol.184, No.1, pp.39-51, ISSN 0022-1759

Jagetia, G. & Venkatesh, P. (2007). .Inhibition of radiation-induced clastogenicity by Aegle marmelos (L.) correa in micebone marrow exposed to different doses of gamma-radiation. *Human & Experimental Toxicology*, Vol.26, No.2, pp.111-24, ISSN 0960-3271

Kokoszka, J.; Waymire, K.; Levy, S.; Sligh, J.; Cai, J.; Jones, D.; MacGregor, G. & Wallace, D. (2004). The ADP/ATP translocator is not essential for the mitochondrial permeability transition pore. *Nature (London)* Vol.427, pp.461–465, ISSN 0028-0836

Lowe, S. & Sherr, C. (2003). Tumor suppression by Ink4a-Arf: progress and puzzles. *Current Opinion in Genetics & Development*. Vol.13, pp.77-83, ISSN 0959-437X

Meng, A.; Wang, Y.; Van Zant, G. & Zhou, D. (2003). Ionizing radiation and busulfan induce premature senescence in murine bone marrow hematopoietic cells. *Cancer Research*, Vol.63, pp.5414-5419, ISSN 0008-5472

Nakabeppu, Y.; Tsuchimoto, D.; Yamaguchi, H. & Sakumi, K. (2007). Oxidative damage in nucleic acids and Parkinson's disease. *Journal of Neuroscience Research*, Vol.85, pp. 919–93, ISSN 1097-4547

Narita, M. & Lowe, S. (2004). Executing cell senescence. *Cell Cycle*, Vol.3, pp.244- 246, ISSN 1538-4101

Neben, S.; Hellman, S.; Montgomery, M.; Ferrara, J.; Mauch, P. & Hemman, S. (1993). Hematopoietic stem cell deficit of transplanted bone marrow previously exposed to cytotoxic agents. *Experimental hematology*, Vol.21, pp.156-162, ISSN 0301-472X

Nicoletti, I.; Migliorati, G.; Pagliaci, M.; Grignani, F. & Riccardi, C. (1991). A rapid and simple method for measuring thymocyte apoptosis by propidium iodide staining and flow cytometry. *Journal of Immunological Methods*, Vol.139, pp.271-279, ISSN 0022-1759

Paredes-Gamero, E.; Craveiro, R.; Pesquero, J.; França, J.; Oshiro, M. & Ferreira, A. (2006). Activation of P2Y1 receptor triggers two calcium signaling pathways in bone marrow erythroblasts. *European Journal of Pharmacology*, Vol.534, pp.30–38, ISSN 0014-2999

Poele, R.; Okorokov, A.; Jardine, L.; Cummings, J. & Joel, S. (2002). DNA damage is able to induce senescence in tumor cells in vitro and in vivo. *Cancer Research*, Vol.62, pp.1876-1883, ISSN 0008-5472

Rao, J.; Xu, D.; Zheng, F.; Long, Z.; Huang, S.; Wu, X.; Zhou, W.; Huang, R. & Liu, Q. (2011). Curcumin reduces expression of Bcl-2, leading to apoptosis in daunorubicin-insensitive CD34+ acute myeloid leukemia cell lines and primary sorted CD34+ acute myeloid leukemia cells. *Journal of Translational Medicine*, Vol.9, pp.71, ISSN 1479-5876

Reya, T. (2003). Regulation of hematopoietic stem cell self-renewal. *Recent Progress in Hormone Research*, Vol.58, pp.283-95, ISSN 0079-9963

Robles, S. & Adami, G. (2002). Agents that cause DNA double strand breaks lead to p16INK4a enrichment and the premature senescence of normal fibroblasts. *Oncogene*, Vol.16, pp.1113-1123, ISSN 0950-9232

Segreto, R.; Egami, M.; França, J.; Silva, M.; Ferreira, A. & Segreto, H. (1999). The bone marrow cells radioprotection by amifostine - nn/n ratio, apoptosis, ultrastructural and lipid matrix evaluation. *Interciencia*, Vol. 24, No.2, pp.127-134, ISSN 0378-1844

Sharpless, N. & DePinho, R. (1999). The INK4A/ARF lócus and its two gene products. *Current Opinion in Genetics & Development*. Vol.9, pp.22-30, ISSN 0959-437X

Szymczyk, K.; Shapiro, I. & Adams, C. (2004). Ionizing radiation sensitizes bone cells to apoptosis. *Bone*, Vol.34, No.1, pp. 148–156, ISSN 8756-3282

Toussaint, O.; Medrano, E. & von Zglinicki, T. (2000). Cellular and molecular mechanisms of stressinduced premature senescence (SIPS) of human diploid fibroblasts and melanocytes. *Experimental Gerontology*, Vol.35, pp.927-45, SSN 0531-5565

Vermes, V. I. , Haanen, C; Steffens-Nakken, H & Reutelingsperger, C. (1995). A novel assay for apoptosis - Flow cytometric detection of phosphatidylserine early apoptotic cells using fluorescein labeled expression on Annexin. *Journal of Immunological Methods*, Vol.184, pp. 39-51, ISSN 0022-1759

Vindeløv, L. & Christensen, I. (1990). A review of techniques and results obtained in one laboratory by an integrated system of methods designed for routine flow cytometric DNA analysis. *Cytometry*. Vol.11, pp.753-770, ISSN 1552-4922

Waldman, T.; Lengauer, C.; Kinzler, K. & Vogelstein, B. (1996). Uncoupling of S phase and mitosis induced by anticancer agents in cells lacking p21. *Nature*, Vol.381, No.6584, pp.713–716, ISSN 0028-0836

Walkley, C.; McArthur, G. & Purton, L. (2005). Cell division and hematopoietic stem cells: not always exhausting. *Cell Cycle*, Vol.4, pp.893-896, ISSN 1538-4101

Wang, Y.; Schulte, B.; LaRue, A.; Ogawa, M. & Zhou, D. (2006). Total body irradiation selectively induces murine hematopoietic stem cell senescence. *Blood*, Vol.107, pp.358-366, ISSN 0006-4971

Weissman, I.; Anderson, D. & Gage, F. (2001). Stem and progenitor cells: origins, phenotypes, lineage commitments, and transdifferentiations. *Annual Review of Cell and Developmental Biology*, Vol.17, pp.387-403, ISSN 1081-0706

Zamzami, N. & Kroemer, G. (2001). The mitochondrion in apoptosis: how Pandora's box opens. *Nature Reviews. Molecular Cell Biology*, Vol.2, pp. 67–71, ISSN 1471-0072

6

Early Events in Apoptosis Induction in Polymorphonuclear Leukocytes

Annelie Pichert, Denise Schlorke, Josefin Zschaler,
Jana Fleddermann, Maria Schönberg, Jörg Flemmig and Jürgen Arnhold
Institute for Medical Physics and Biophysics, University of Leipzig, Leipzig
Germany

1. Introduction

Different white blood cells are involved in immune responses to pathogens, environmental stress, alterations in energy and nutrition supply as well as traumata. To find an adequate answer to the myriad of exogenous and endogenous noxes is a high challenge for the human immune system. Chronic inflammatory diseases like rheumatoid arthritis, arteriosclerosis, inflammatory bowel disease, and many others are associated with a disturbed regulation of immune functions (Peng, 2006; Zernecke & Weber, 2010). Despite numerous worldwide investigations regulatory aspects are only scarcely understood in innate and acquired immunity.

Polymorphonuclear leukocytes (PMNs, also called neutrophils) are among the first cells that are rapidly recruited to infected and/or injured tissue. Their infiltration into inflammatory sites is highly regulated and supported by adhesion molecules, cytokines, and extracellular matrix components in both blood vessel wall and adjacent tissue (Muller, 2002; Taylor & Gallo, 2006). These cells, by releasing special proteins and generating reactive metabolites, contribute to pathogen defense, regulation of the inflammatory process, and to tissue injury.

Unperturbed or slightly activated PMNs die by apoptotic cell death (Walker et al., 2005; Erwig & Henson, 2007). The rapid clearance of apoptotic PMNs by macrophages is crucial for efficient resolution of inflammation including the activation of different anti-inflammatory mechanisms that stop the recruitment of novel immune cells, deactivate pro-inflammatory cytokines, depress pro-inflammatory and anti-apoptotic pathways, and promote tissue repair. Apoptosis is induced either by release of mitochondrial constituents or by signalling via death receptors. In these pathways, different initiator and executioner caspases are activated that induce the degradation of molecules of the cytoskeleton, DNA, and others. In the result, numerous apoptotic vesicles will be formed without the release of internal constituents.

Apoptosis induction is mainly counterregulated by signals from phosphoinositide 3-kinase and protein kinase B that suppress caspase activation (Simon, 2003). In PMNs, the anti-apoptotic pathway is activated when the cells phagocytose foreign microorganisms or become attached to other cells or materials. This pathway ensures the functional responsibility of PMNs at inflammatory sites.

Both apoptotic cells/vesicles or cells, in which the anti-apoptotic pathway is activated, can undergo a necrosis that represents a cell death accompanied by cell and organelle swelling, plasma membrane rupture, and release of cytoplasmic content. Besides uncontrolled necrosis due to physical stress, necrotic cell death may be induced and regulated by signalling pathways (Degterev & Yuan, 2008). The uptake of necrotic PMNs by macrophages is associated with the release of pro-inflammatory mediators that further promote the inflammatory process (Vandivier et al., 2006). Thus, the interplay between macrophages and apoptotic/necrotic cells, mostly with PMNs, considerably determines the fate of an inflammation. This scheme, although very simplified, is also supported by the fact that PMN apoptosis is highly delayed and dysregulated in patients with severe sepsis (Jiminez et al., 1997; Fanning et al., 1999).

Special products of apoptotic PMNs apparently contribute to induction of anti-inflammatory signaling pathways in macrophages. Among them, PMN-derived chloramines like taurine chloramine and monochloramine are known to dampen the activation of NFκB in pro-inflammatory cells (Kontny et al., 2003; Ogino et al., 2005). Moreover, the 5-lipoxygenase of PMNs is involved in the transcellular synthesis of lipoxins that are able to stop the invasion of unperturbed PMNs to inflammatory sites (Serhan et al., 2008). However, fine mechanisms of their involvement in regulation of inflammatory processes are only poorly understood.

Thus, the induction of apoptosis in PMNs is an important prerequisite for the successful resolution of immune responses. From this background, we give here an overview about flow cytometry approaches to study early events in activation of PMNs and in apoptosis induction in these cells. These methods enable determination of specific properties of a large number of individual cells in a very short time. Because it is impossible to consider flow cytometry analysis for all PMN constituents, we will focus here on the analysis of those components that are unique to PMNs and highly necessary for specific functions of this type of immune cells.

2. Early activation of PMNs

Recruitment of PMNs from peripheral blood is mediated by selectins and integrins (Springer, 1994). After firm adhesion to the endothelium, PMNs invade into the inflamed tissue. Several chemotactic factors form a gradient for the directed movement of PMNs to the inflammatory loci. The bacterial product fMet-Leu-Phe, soluble immune complexes, leukotriene B_4 and the cytokine interleukin 8 (IL-8) are important chemoattractant agents (Lin et al., 2004). PMNs express specific receptors to these chemotactic factors. Receptor activation by these agents causes local changes in the contractility of cytoskeleton components and favors a directed movement of the cells. As PMNs have to pave their way through the tightly packed vessel wall and adjacent regions filled with different extracellular matrix components, these cells express and release also special proteases helping to degrade tissue and matrix components. Among these proteases are collagenase, gelatinase, different other matrix metalloproteases and the specific PMN serine proteases proteinase 3, cathepsin G, and elastase.

Here we will focus on flow cytometry approaches for binding of IL-8 to PMNs and for the expression of PMN specific serine proteases on the surface of PMNs.

2.1 Binding of IL-8 to PMNs

The cytokine IL-8 (also called CXCL8 according to the chemokine nomenclature) is released at inflammatory sites from fibroblasts, monocytes, endothelial and epithelial cells. It is a strong chemoattractant to PMNs (Rajarathnam et al., 1994). Besides the receptor binding site (the ELR motif near the NH_2-terminus) on IL-8, there are positive charged epitopes on its surface involved in binding to sulfated glycosaminoglycans (Kuschert et al., 1999; Lortat-Jacob et al., 2002; Pichert et al., 2012). Around inflammatory loci, IL-8 is mostly fixed to sulfated extracellular matrix components forming, thus, a gradient for the invading PMNs.

Neutrophils have two kinds of G protein-coupled receptors for IL-8 called CXCR1 and CXCR2 (Stillie et al., 2009). Receptor activation causes phosphorylation of Akt, calcium influx, formation of F-actin, and cytoskeletal rearrangement. These events are important for chemotactic movement. Several factors are known to modulate the binding of IL-8 to its receptors. Alpha-1 antitrypsin and IL-8 form a complex that is unable to interact with CXCR1 (Bergin et al., 2010). Truncation of amino acids at the NH_2-terminus by matrix metalloproteases, CD13, cathepsin L, or proteinase 3 creates a series of natural isoforms of IL-8. Some of them have a higher biological activity than the original IL-8 protein with 77 amino acid residues (Padrines et al., 1994; Mortier et al., 2011).

Oligomerization of chemokines affects also their interaction with receptors. While both monomeric and dimeric IL-8 forms were capable of inducing cell recruitment, the dimeric form induced a stronger migration in a mouse lung model (Das et al., 2010). Binding to sulfated glycosaminoglycans promotes oligomerization of IL-8 (Hoogewerf et al., 1997).

The binding of recombinant IL-8 to PMNs is shown in Fig. 1.

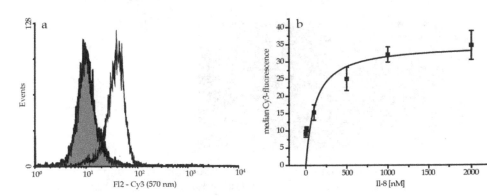

Fig. 1. Binding of IL-8 on the surface of freshly isolated PMNs. PMNs (2×10^6) were incubated with varying IL-8 concentrations for 60 min, followed by washing steps to remove unbound protein. After incubation of cells with the primary anti-human IL-8 rabbit antibody (4 µg/ml) for 1 h, PMNs were washed again and the secondary Cy3-conjugated goat anti-rabbit antibody (1.5 µg/ml) was added for 30 min in the dark. A representative example of fluorescence distribution in the presence of 1 µM IL-8 is shown in (a). The control is given in grey. Using median values from three independent measurements a binding curve was created with a K_d value of (112.7 ± 94.3) nM (b)

2.2 Detection of serine proteases on the surface of PMNs

In invading PMNs, the serine proteases proteinase 3, cathepsin G, and elastase are involved in microbicidal activity, penetration of cells through endothelium and adjacent connective tissue, and processing of various cytokines (Meyer-Hoffert, 2009; Kessenbrock et al., 2011). Although all three proteases have striking structural similarities (Korkmaz et al., 2008), there are clear differences in their functional response at inflammatory loci (Fleddermann et al., 2011). Unlike elastase, proteinase 3 and cathepsin G are already released from resting or slightly activated PMNs. While proteinase 3 binds heavily to cell surface epitopes, cathepsin G interacts preferentially with sulfated glycosaminoglycans. Proteinase 3 is apparently involved in both the infiltration of unperturbed PMNs into inflammatory sites and in cell necrosis, while cathepsin G plays most likely an important role in the degradation of specific components of the extracellular matrix during PMN invasion. In contrast, elastase probably contributes to shedding of surface molecules on macrophages helping to induce a pro-inflammatory feature in these cells (Pham, 2006). Examples for the incubation of resting PMNs with proteinase 3, cathepsin G, and elastase are given in Fig. 2.

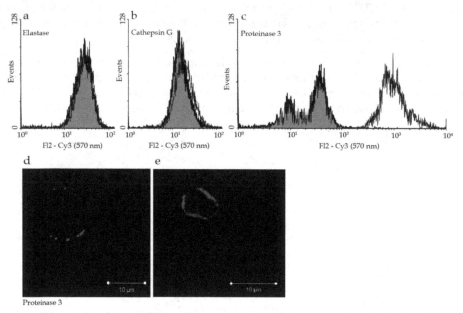

Fig. 2. Binding of externally added serine proteases on the surface of freshly isolated PMNs. Cells (2×10^6) were incubated with elastase (0.5 µM), cathepsin G (1.06 µM) and proteinase 3 (1.09 µM) for 1 h. After removal of unbound proteases by washing steps, PMNs were incubated with anti-neutrophil elastase rabbit antibody (134 µg/ml), anti-cathepsin G rabbit antibody (474 µg/ml) or mouse monoclonal proteinase 3 antibody (4 µg/ml) for 60 min. After washing and resuspension, Cy3-conjugated goat anti-mouse or goat anti-rabbit secondary antibodies (1.5 µg/ml) were added and incubated for 30 min in the dark (a-c). Flow cytometry distribution curves with externally added proteases are given in white. Controls are highlighted in grey. Representative confocal fluorescence microscopy images of freshly isolated PMNs treated with Cy3-labeled antibodies against proteinase 3 without (d) or after addition of proteinase 3 (e). Representative data from four measurements are given

As shown in Fig. 2, the difference in the amount of elastase and cathepsin G on the PMN surface is very small after protease addition. However, a drastic shift in fluorescence yield is observed in case of proteinase 3, indicating the existence of a high number of potential binding sites.

3. Alterations in PMNs during apoptosis induction

In contrast to other cell types, systems to resist an apoptosis induction such as glutathione-dependent antioxidant enzymes are expressed to a minor degree in PMNs (Kinnula et al., 2002). In non-affected tissue, PMNs are known to undergo a spontaneous apoptosis (Payne et al., 1994). This type of cell death can be simulated *in vitro* keeping the cells at 37 °C under sterile conditions for several days. Phorbol 12-myristate 13-acetate (PMA) or the calcium ionophore ionomycin accelerate this process and enable investigation of PMN apoptosis within few hours. PMNs activated by 50 nM PMA are characterized by enhanced hydrogen peroxide levels, reduced cell size, condensed nuclei, and enhanced DNA fragmentation (Lundqvist-Gustafsson & Bengtsson, 1999). A late event in apoptosis induction is the breakdown of the original cell into smaller apoptotic vesicles. Their rapid removal by macrophages prevents the release of toxic components due to secondary necrosis of these apoptotic bodies.

We compare sensitive flow cytometry approaches for the detection of changes in the vitality status of PMNs during spontaneous or PMA-mediated apoptosis. PMA was used at very low nanomolar concentrations in order to better visualize early changes during apoptosis induction. Here we present data about forward and sideward scattering of PMNs, binding of fluorescence-labeled annexin V to phosphatidylserine epitopes and the intercalation of propidium iodide into DNA as well as measurement of the integrity of mitochondria using the dye 5,5',6,6'-tetrachloro-1,1',3,3'-tetraethylbenzimidazolylcarbocyanine iodide (JC-1).

3.1 Forward and sideward scattering

While the value of the forward scatter depends on the cell size, the sideward scatter is a measure of the cell granularity. In flow cytometry analysis, the first parameter is usually plotted on a linear, the later one on a logarithmic scale.

The assessment of both kinds of scattering already provides a lot of information about the apoptotic process (Fig. 3). After apoptosis induction, a slight increase in the cell size takes place as visualized by a shift in the forward scattering values (Fig. 3b). The granularity remains unaffected. This alteration can be interpreted as a round-up of cells due to the progressive loss of linkages of the actin cytoskeleton with the plasma membrane. An early event in apoptosis induction in PMNs is the degradation of the corresponding link proteins (Kondo et al., 1997). Additionally, parts of the endoplasmic reticulum of the cells will be incorporated into the plasma membrane (Franz et al., 2007).

At later stages of apoptosis (Fig. 3c), enhanced forward scattering values vanished and smaller cell sizes became more prominent. At the same time a broader distribution of sideward scattering values was observed. This indicates the formation of polymorphic apoptotic bodies. The appearance of small apoptotic bodies (low forward and sideward scattering values) can also be seen during later apoptosis. Their appearance is a clear sign for the breakdown of apoptotic cells into smaller vesicles at later stages of apoptosis.

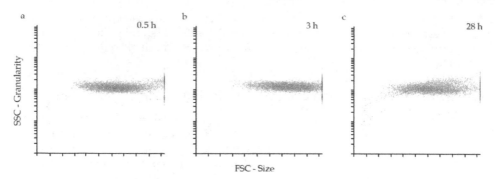

Fig. 3. Changes in cell size and granularity upon spontaneous apoptosis of PMNs. Cells (10^6) were cultivated at 37 °C for 0.5 h (a), 3 h (b), and 28 h (c), respectively. Changes in cell size and granularity were analyzed by flow cytometry using forward scattering (FSC) and sideward scattering (SSC). One representative example from four different cell preparations is shown

3.2 Binding of annexin V and uptake of propidium iodide

This approach is often used to assess the vitality status in a cell population. Vital cells do neither bind annexin V nor incorporate propidium iodide. Apoptotic cells are also unable to take up propidium iodide, but they express phosphatidylserine epitopes on the outer leaflet of the plasma membrane that are able to bind fluorescently labeled annexin V. In necrotic cells, propidium iodide permeates through the damaged plasma membrane and becomes highly fluorescent due to intercalation into DNA.

An example for analysis with both dyes in spontaneous and PMA-induced apoptosis in PMNs is given in Fig. 4.

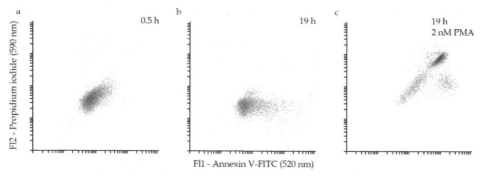

Fig. 4. Detection of apoptosis and necrosis in PMNs by annexin V and propidium iodide. PMNs (10^6) were cultivated at 37 °C for 0.5 h (a), or 19 h (b,c). In (c), cells were additionally activated by 2 nM PMA. Afterwards, PMNs were stained with fluorescently labeled annexin V (annexin V-FITC) and propidium iodide to detect apoptosis and necrosis, respectively. One representative example from four different experiments is shown

Freshly isolated PMNs did neither show any externalization of phosphatidylserine nor an uptake of propidium iodide (Fig. 4a). At prolonged incubation, part of the cells became annexin V-positive, but they did not incorporate propidium iodide (Fig. 4b). Thus, an apoptosis induction was observed. In contrast, the prolonged incubation of PMNs with PMA yielded double-positive cells with enhanced values in both channels (Fig. 4c). From these data necrotic processes in these PMNs can be assumed.

Although this method is very convenient, there are two main problems necessary to consider in its application. An accumulation of phosphatidylserine at the outer leaflet of the plasma membrane is observed in dying cells, where these epitopes serve as a matrix for a number of proteins like annexin 1, thrombospondin, and β_2-glycoprotein 1 and as a signal for cell clearance by macrophages (Lauber et al., 2004; Walker et al., 2005). In apoptotic cells, myeloperoxidase binds also to phosphatidylserine epitopes (Leßig et al., 2007; Flemmig et al., 2008). It is generally assumed that the phosphatidylserine exposure on the cell surface is associated with activation of scramblases and inhibition of translocases (Yoshida et al., 2005). In apoptotic PMNs, there are caspase-dependent and -independent mechanisms in the appearance of annexin V-positive epitopes (Blink et al., 2004; Chen et al., 2006). On the other hand, a transfer of phosphatidylserine from internal granule stores to the cell surface cannot be ruled out during apoptosis (Mirnikjoo et al., 2009).

The second problem concerns changes in the cell size and the breakdown of cells into smaller vesicles during apoptosis. These alterations are already illustrated in forward and sideward scattering (see the previous section). Thus at later stages of apoptosis, the appearance of smaller vesicles and cell debris would affect all fluorescence measurements. It cannot be excluded that these smaller vesicles, even though they are fluorescent, appear in the same gate where originally vital unperturbed cells were found.

3.3 Changes in mitochondria

Neutrophils possess only a few mitochondria that are functionally different in contrast to mitochondria of other cells. Their participation in ATP synthesis and cytochrome *c* content is limited, but they maintain a membrane potential (van Raam et al., 2006). Cellular stress is

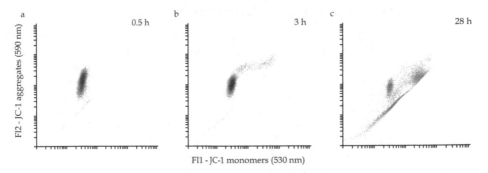

Fig. 5. Application of the dye JC-1 to detect early apoptotic events in PMNs. PMNs (10^6) were cultivated at 37 °C for 0.5 h (a), 3 h (b), or 28 h (c). Afterwards, cells were incubated with JC-1 (0.77 µM) for 10 min. After washing steps, the fluorescence of JC-1 monomers (channel 1) and aggregates (channel 2) was analyzed. Control experiments (not shown) were performed using the mitochondrial membrane ionophore valinomycin. One representative example from five different cell preparations is shown

sensitively reflected at the mitochondrial level leading to a rupture in processes maintaining the mitochondrial potential. In PMNs, pro-apoptotic proteins such as Smac/DIABLO, HtrA2/Omi and cytochrome c are released from mitochondria (Maianski et al., 2004). Early changes in mitochondrial integrity can be easily measured by the dye 5,5',6,6'-tetrachloro-1,1',3,3'-tetraethylbenzimidazolylcarbocyanine iodide (also called J-aggregate forming cationic dye, JC-1) (Cossarizza et al., 1993). This dye emits a red fluorescence when it accumulates in aggregated form in intact mitochondria. In apoptotic cells with damaged mitochondria, JC-1 is distributed over the whole cell cytoplasm and emits a green fluorescence.

We provide here an example for the application of this dye to PMNs undergoing a spontaneous apoptosis (Fig. 5). In freshly isolated vital PMNs with intact mitochondria, the dye is mainly incorporated in aggregated form with a dominating emission at 590 nm (Fig. 5a). The onset of apoptosis in PMNs was observable by increasing monomer fluorescence (Fig. 5b). At later times, nearly all cells showed a strong fluorescence of JC-1 monomers while the number of JC-1 aggregates was diminished (Fig. 5c).

3.4 Further systems for analysis of apoptosis induction

There are further flow cytometry approaches to detect changes in PMN vitality. As later stages of apoptosis are associated with progressive DNA fragmentation, the appearance of nicks in the DNA can be identified by the terminal deoxynucleotidyl transferase dUTP nick end labeling (TUNEL) assay. The investigation of PMN apoptosis in granulocyte preparations by the TUNEL assay can be affected by eosinophils that yield false positive results (Kern et al., 2000).

Active caspase-3 can be visualized in PMNs by enzyme inhibitors bearing a fluorescence label. This approach has been used to investigate effects of nicotinamide on PMN apoptosis (Fernandes et al., 2011).

4. Generation of oxidants during PMN apoptosis

PMNs are equipped with special enzymes for generation of a large amount of oxidants that are involved in antimicrobial activity, regulation of immune functions and apoptosis induction. These enzymes are NADPH oxidase (Shatwell & Segal, 1996) and myeloperoxidase (Klebanoff, 2005). While the first enzyme generates superoxide anion radicals that further dismutate to hydrogen peroxide, the heme protein myeloperoxidase uses H_2O_2 to oxidize (pseudo)halides to the corresponding (pseudo)hypohalous acids. Under physiological relevant conditions, the production of hypochlorous acid (HOCl) and hypothiocyanite ($^-$OSCN) is important (van Dalen et al., 1997). As myeloperoxidase is strongly expressed in PMNs (and to a lesser extent in monocytes) there is a great interest to understand specific functions of this enzyme during immune reactions of PMNs (Arnhold & Flemmig, 2010).

In PMNs, reactive oxygen species (ROS) are important for apoptosis induction. Apoptosis is delayed in NADPH oxidase deficiency (Lundqvist-Gustafsson & Bengtsson, 1999) and by scavenging of H_2O_2 by glutathione or catalase (Kasahara et al., 1997; Yamamoto et al., 2002). Interestingly, oxidants down-regulate phosphoinositide 3-kinase γ activity and inhibit actin polymerization during PMN apoptosis (Xu et al., 2010). Myeloperoxidase deficiency suppresses also PMN apoptosis (Tsurubuchi et al., 2001; Milla et al., 2004). Several myeloperoxidase products such as hypothiocyanite, monochloramine and taurine chloramine are potent inducers of apoptosis (Emerson et al., 2005; Lloyd et al., 2008; Ogino et al., 2009).

Here we focused our attention on two sensitive flow cytometry approaches for oxidant generation in close association with apoptosis induction in phorbol ester- and IL-8-activated PMNs. While dihydrorhodamine 123 detects very sensitively but non-specifically the oxidative activity in PMNs (Bizyukin et al., 1995), the combination of the dyes 2-[6-(4'-amino)phenoxy-3H-xanthen-3-on-9-yl]benzoic acid (APF) and 2-[6-(4'-hydroxy)phenoxy-3H-xanthen-3-on-9-yl]benzoic acid (HPF) allows to visualize specifically the generation of the myeloperoxidase product HOCl in PMNs (Setsukinai et al., 2003). Structures of these dyes are given in Fig. 6.

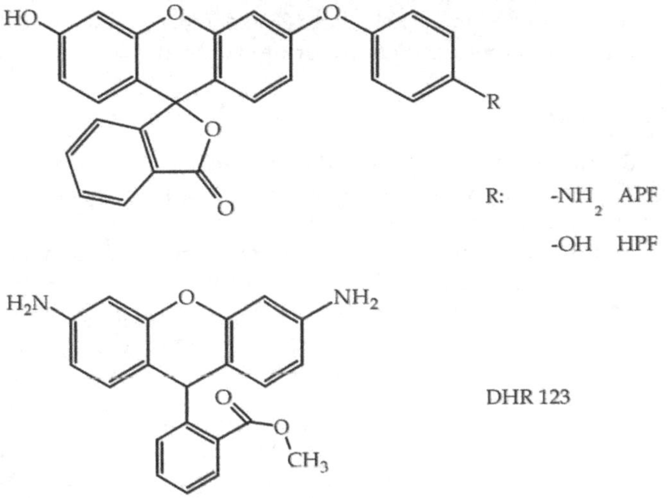

R: $-NH_2$ APF

     $-OH$ HPF

DHR 123

Fig. 6. Chemical structures of dyes applied for oxidant detection in PMNs

4.1 Dihydrorhodamine 123

The non-fluorescent compound dihydrorhodamine 123 is converted intracellularly into the fluorescent rhodamine 123 by several oxidants including superoxide anion radicals, hydrogen peroxide, hydroxyl radicals, hypochlorous acid, but not singlet oxygen (Bizyukin et al., 1995). This selection makes dihydrorhodamine 123 a suitable flow cytometric dye to follow the oxidative activity in neutrophils and other cells. The involvement of given enzymes in oxidant production can be determined by specific enzyme inhibitors.

Here we give an example for the application of dihydrorhodamine 123 to measure the oxidant generation in PMNs activated by the chemokine IL-8 (Fig. 7). In the presence of IL-8 significant higher fluorescence values were observed indicating a more promoted formation of oxidants in stimulated PMNs.

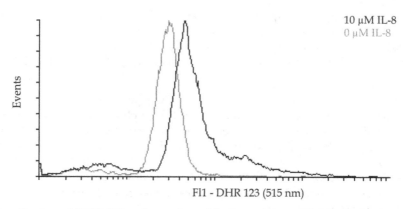

Fig. 7. Application of DHR 123 to detect the oxidative activity in PMNs. PMNs (2×10^6) were cultivated for 1 h at 37°C either in the presence (black) or in the absence (grey) of 10 µM IL-8. Afterwards they were stained with 10 µM DHR 123 and analyzed by flow cytometry. One representative example from four different experiments is shown

4.2 The APF/HPF system

In APF and HPF, the fluorescence is quenched by protection of the phenolic hydroxy group at the 6'-position of fluorescein with an electron-rich aromatic ring. The initial reaction with oxidants starts in both dyes by an attack on the aryloxyphenol group resulting in the cleavage of this group and the formation of highly emitting fluorescein. Hydroxyl radicals and peroxynitrite are able to oxidize both HPF and APF. HOCl activates only APF but not HPF. All other biologically relevant oxidants are insensitive against APF and HPF. This oxidant profile allows the application of APF and HPF to detect specifically the formation of HOCl in activated neutrophils (Setsukinai et al., 2003).

We successfully applied APF to visualize the production of HOCl in non-stimulated and PMA-activated neutrophils (Fig. 8). A significant stronger shift in the fluorescence intensity distribution of APF indicates a higher HOCl production in the stimulated cells. In both non-stimulated and PMA-stimulated PMNs the application of the MPO inhibitor 4-aminobenzoic acid hydrazide (4-ABAH) strongly inhibited the HOCl production as can be seen by much lower APF fluorescence values.

4.3 Further systems for oxidant detection

Among other unspecific dyes suitable for flow cytometric analysis of ROS generation in PMNs, we will mention only 2,7-dichlorofluorescein diacetate (Walrand et al., 2003). This non-fluorescent compound is converted by several oxidants into a fluorescent derivative. This system is often used to assess unspecific ROS formation in activated cells.

Besides the APF/HPF method for specific detection of HOCl generation in PMNs, a few further systems have recently been described to visualize the intracellular formation of HOCl. These novel approaches include the use of sulfonaphthoaminophenyl fluorescein (Shepherd et al., 2007), a rhodamin-hydroxamic acid-based system (Yang et al., 2009), and a thiol analogue of hydroxymethyltetramethylrhodamine (Kenmoku et al., 2007).

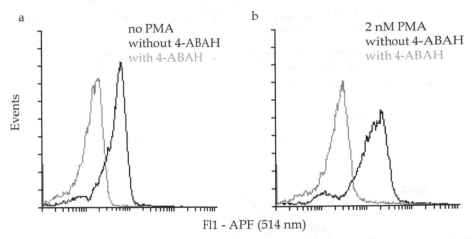

Fl1 - APF (514 nm)

Fig. 8. Detection of HOCl generation in non-stimulated and PMA-stimulated PMNs. PMNs (10^6) were cultivated for 3 h either in the absence (a) or presence (b) of 2 nM PMA. In order to analyze the formation of the myeloperoxidase metabolite HOCl, APF (5 μM) was present during the incubation. In some cases, the myeloperoxidase inhibitor 4-aminobenzoic acid hydrazide (4-ABAH) was also included in the incubation cocktail (grey). One representative example of five different experiments is shown

5. Concluding remarks

Polymorphonuclear leukocytes are key players in innate and acquired immune response. Upon invasion to inflammatory sites, they sense their local environment and contribute to regulation of the further inflammatory process. For successful resolution of inflammation, the initiation of apoptotic cell death in PMNs and the subsequent fast removal of apoptotic bodies by macrophages are crucial. Despite numerous investigations, fine mechanisms of apoptosis induction in PMNs and factors modulating apoptotic pathways are only poorly understood.

Flow cytometry approaches are an important tool in this challenging field. We summarized here advantages and limitations of a number of widely applied methods, but focused also our attention on newly developed approaches for oxidant detection. The myeloperoxidase-hydrogen peroxide-halide system is unique to PMNs and its products are apparently involved in regulation of immune responses. Thus, novel methods for detection of hypochlorous acid generation such as the APF/HPF system are highly necessary for further investigation of modulation of myeloperoxidase activity.

We also directed our focus on enforcement of PMN apoptosis by low-concentrated PMA. This approach allows the convenient study of early changes during cell activation. In granulocytes, mitochondria are mainly responsible for providing pro-apoptotic factors that initiate caspase activation. The application of the dye JC-1 is useful to follow the loss of mitochondria integrity. Mechanisms leading to the appearance of phosphatidylserine epitopes remain puzzling. In this regard, distortions of enzyme reactions maintaining the phospholipid asymmetry and fusion events with internal granule stores are under discussion. Last but not least, the importance of oxidant generation is acknowledged for apoptosis induction in PMNs however, their fine mechanisms are poorly understood.

Detailed knowledge about PMN functions is also necessary to better understand the role of these immune cells in the pathogenesis of diseases and for introducing new therapies helping to terminate inflammatory processes.

6. Acknowledgements

This work was supported by the German Research Foundation (Transregio 67, project A-06) and by the Development Bank of Saxony, Germany.

7. References

Arnhold, J. & Flemmig, J. (2010) Human myeloperoxidase in innate and acquired immunity. *Arch. Biochem. Biophys.* 500, 92-106.

Bergin, D.A.; Reeves, E.P.; Meleady, P.; Henry, M.; McElvaney, O.J.; Carroll, T.P.; Condron, C.; Chotirmall, S.H.; Clynes, M.; O'Neill, S.J. & McElvaney, N.G. (2010) α-1 Antitrypsin regulates human neutrophil chemotaxis induced by soluble immune complexes and IL-8. *J. Clin. Invest.* 120, 4236-4250.

Bizyukin, A.V.; Korkina, L.G. & Velichkovskii, B.T. (1995) Comparative use of 2,7-dichlorofluorescin diacetate, dihydrorhodamine 123, and hydroethidine to study oxidative metabolism in phagocytic cells. *Bull. Exp. Biol. Med.* 119, 347-351.

Blink, E.; Maianski, N.A.; Alnemri, E.S.; Zervos, A.S.; Roos, S. & Kuijpers, T.W. (2004) Intramitochondrial serine protease activity of Omi/Htr2A is required for caspase-independent cell death of human neutrophils. *Cell Death Diff.* 11, 937-939.

Chen, H,-C.; Wang, C.-J.; Chou, C.-L.; Lin, S.-M.; Huang, C.-D.; Lin, T.-Y.; Wang, C.-H.; Lin, H.-C.; Yu, C.-T.; Kuo, H.-P. & Liu, C.-Y. (2006) Tumor necrosis factor-α induces caspase-independent cell death in human neutrophils via reactive oxidants and associated with calpain activity. *J. Biomed. Sci.* 13, 261-273.

Cossarizza, A.; Baccarani-Contri, M.; Kalashnikova, G. & Franceschi, C. (1993) A new method for the cytofluorimetric analysis of mitochondrial membrane potential using the J-aggregate forming lipophilic cation 5,5′,6,6′-tetrachloro-1,1′,3,3′ tetraethylbenzimidazolylcarbocyanine iodide (JC-1). *Biochem. Biophys. Res. Commun.* 197, 40-45.

van Dalen, C.J.; Whitehouse, M.W.; Winterbourn C.C. & Kettle, A.J. (1997) Thiocyanate and chloride as competing substrates for myeloperoxidase. *Biochem. J.* 327, 487-492.

Das, S.T.; Rajagopalan, L.; Guerrero-Plata, A.; Sai, J.; Richmond, A.; Garafalo, R.P. & Rajarathman, K. (2010) Monomeric and dimeric CXCL8 are both essential for in vivo neutrophil recruitment. *PloS One* 5, e11754.

Degterev, A. & Yuan, J. (2008) Expansion and evolution of cell death programmes. *Nat. Rev. Mol. Cell. Biol.* 9, 378-390.

Emerson, D.K.; McCormick, M.L.; Schmidt, J.A. & Knudson, C.M. (2005) Taurine monochloramine activates a cell death pathway involving Bax and caspase-9. *J. Biol. Chem.* 280, 3233-3241.

Erwig, L.-P. & Henson, P.M. (2007) Immunological consequences of apoptotic cell phagocytosis. *Am. J. Pathol.* 171, 2-8.

Fanning, N.F.; Kell, M.R.; Shorten, G.D.; Kirwan, W.O.; Bouchier-Hayes, D.; Cotter, T.G. & Redmond, H.P. (1999) Circulating granulocyte macrophage colony-stimulating factor in plasma of patients with the systemic inflammatory response syndrome delays neutrophil apoptosis through inhibition of spontaneous reactive oxygen species generation. *Shock* 11, 167-174.

Fernandes, C.A.; Fievez, L.; Ucakar, B.; Neyrinck, A.M.; Fillee, C.; Huaux, F.; Delzenne, N.M.; Bureau, F. & Vanbever, R. (2011) Nicotinamide enhances apoptosis of G(M)-CSF-treated neutrophils and attenuates endotoxin-induced airway inflammation in mice. *Am. J. Physiol. Lung Cell. Mol. Physiol.* 300, L354-L361.

Fleddermann, J.; Pichert, A. & Arnhold, J. (2011) Interaction of serine proteases from polymorphonuclear leukocytes with the cell surface and heparin. *Inflammation*, in press.

Flemmig, J.; Leßig, J.; Reibetanz, U.; Dautel, P. & Arnhold, J. (2007) Non-vital polymorphonuclear leukocytes express myeloperoxidase on their surface. *Cell. Physiol. Biochem.* 21, 287-296.

Franz, S.; Herrmann, K.; Führnhorn, B.; Sheriff, A.; Frey, B.; Gaipl, U.S.; Voll, R.E.; Kalden, J.R.; Jäck, H.-M. & Herrmann, M. (2007) After shrinkage apoptotic cells expose internal membrane-derived epitopes on their plasma membranes. *Cell Death Differ.* 14, 733-742.

Hoogewerf, A.J.; Kuschert, G.S.V.; Proudfoot, A.E.I.; Borlat, F.; Clark-Lewis, I.; Power, C.A. & Wells, T.N.C. (1997) Glycosaminoglycans mediate cell surface oligomerization of chemokines. *Biochemistry* 36, 13570-13578.

Jiminez, M.F.; William, R.; Watson, G.; Parodo, J.; Evans, D.; Foster, D.; Steinberg, M.; Rotstein, O. & Marshall, J.C. (1997) Dysregulated expression of neutrophil apoptosis in the systemic inflammatory response syndrome. *Arch. Surg.* 132, 1263-1270.

Kasahara, Y.; Iwai, K.; Yachie, A.; Ohta, K.; Konno, A.; Seki, H.; Miyawaki, T. & Taniguchi, N. (1997) Involvement of reactive oxygen intermediates in spontaneous and CD95 (Fas/APO-1)-mediated apoptosis of neutrophils. *Blood* 89, 1748-1753.

Kenmoku, S.; Urano, Y.; Kojima, H. & Nagano, T. (2007) Development of a highly specific rhodamine-based fluorescence probe for hypochlorous acid and its application to real-time imaging of phagocytosis. *J. Am. Chem. Soc.* 129, 7313-7318.

Kern, P.M.; Herrmann, M.; Stockmeyer, B.; Kalden, J.R.; Valerius, T. & Repp, R. (2000) Flow cytometric discrimination between viable neutrophils, apoptotic neutrophils and eosinophils by double labelling of permeabilized blood granulocytes. *J. Immunol. Meth.* 241, 11-18.

Kessenbrock, K.; Dau, T. & Jenne, D.E. (2011) Tailor-made inflammation: how neutrophil serine proteases modulate the inflammatory response. *J. Mol. Med.* 89, 23-28.

Kinnula, V.L.; Soini, Y.S.; Kvist-Mäkelä, K.; Savolainen, E.-R. & Koistinen, P. (2002) Antioxidant defense mechanisms in human neutrophils. *Antioxid. Redox Signal.* 4, 27-34.

Klebanoff, S.J. (2005) Myeloperoxidase: friend and foe. *J. Leukoc. Biol.* 77, 598-625.

Kondo, T.; Takeuchi, K.; Dori, Y.; Yonemura, S.; Nagata, S.; Tsukita, S. & Tsukita, S. (1997) ERM (ezrin/radixin/moesin)-based molecular mechanism of microvilli breakdown at an early stage of apoptosis. *J. Cell Biol.* 139, 749-758.

Kontny, E.; Maslinki, W. & Marcinkiewicz, J. (2003) Anti-inflammatory activities of taurine chloramine. *Adv. Exp. Med. Biol.* 526, 845-852.

Korkmaz, B.; Moreau, T. & Gauthier, F. (2008) Neutrophil elastase, proteinase 3 and cathepsin G: physicochemical properties, activity and physiopathological functions. *Biochimie* 90, 227-242.

Kuschert, G.S.V.; Coulin, F.; Power, C.A.; Proudfoot, A.E.I.; Hubbard, R.E.; Hoogewerf, A.J. & Wells, T.N.C. (1999) Glycosaminoglycans interact selectively with chemokines and modulate receptor binding and cellular responses. *Biochemistry* 38, 12959-12968.

Lauber, K.; Blumenthal, S.G.; Waibel, M. & Wesselborg, S. (2004) Clearance of apoptotic cells: getting rid of the corpses. *Mol. Cell* 14, 277-287.

Leßig, J.; Spalteholz, H.; Reibetanz, U.; Salavei, P.; Fischlechner, M.; Glander, H.-J. & Arnhold, J. (2007) Myeloperoxidase binds to nonvital spermatozoa on phosphatidylserine epitopes. *Apoptosis* 12, 1803-1812.

Lin; F.; Nguyen, C.M.-C.; Wang, S.J.; Saadi, W.; Gross, S.P. & Jeon, N.L. (2004) Effective neutrophil chemotaxis is strongly influenced by mean IL-8 concentration. *Biochem. Biophys. Res. Commun.* 319, 576-581.

Lloyd, M.M.; van Reyk, D.M.; Davies, M.J. & Hawkins, C.L. (2008) Hypothiocyanous acid is a more potent induced of apoptosis and protein thiol depletion in murine macrophage cells than hypochlorous acid or hypobromous acid. *Biochem. J.* 414, 271-280.

Lortat-Jacob, H.M; Grosdidier, A. & Imberty, A. (2002) Structural diversity of heparan sulfate binding domains in chemokines. *Proc. Natl. Acad. Sci. USA* 99, 1229-1234.

Lundqvist-Gustafsson, H. & Bengtsson, T. (1999) Activation of the granule pool of the NADPH oxidase accelerates apoptosis in human neutrophils. *J. Leukoc. Biol.* 65, 196-204.

Maianski, N.A.; Geissler, J.; Srinivasula, S.M.; Alnemri, E.S.; Roos, D. & Kuijpers, T.W. (2004) Functional characterization of mitochondria in neutrophils: a role restricted to apoptosis. *Cell Death Diff.* 11, 143-153.

Meyer-Hoffert, U. (2009) Neutrophil-derived serine proteases modulate innate immune response. *Front. Biosci.* 14, 3409-3418.

Milla, C.; Yang, S.; Cornfield, D.N.; Brennan, M.-L.; Hazen, S.L.; Panoskaltsis-Mortari, A.; Blazar, B. & Haddad, I.Y. (2004) Myeloperoxidase deficiency enhances inflammation after allogenic marrow transplantation. *Am. J. Physiol. Lung Cell Mol. Physiol.* 287, L706-L714.

Mirnikjoo, B.; Balasubramanian, K. & Schroit, A.J. (2009) Suicidal membrane repair regulates phosphatidylserine externalization during apoptosis. *J. Biol. Chem.* 284, 22512-22516.

Mortier, A.; Gouwy, M.; van Damme, J. & Proost, P. (2011) Effect of posttransitional processing on the in vitro and in vivo activity of chemokines. *Exp. Cell Res.* 317, 642-654.

Muller, W.A. (2002) Leukocyte-endothelial cell interactions in the inflammatory response. *Lab. Invest.* 82, 521-533.

Ogino, T.; Hosako, M.; Hiramatsu, K.; Omori, M.; Ozaki, M. & Okada, S. (2005) Oxidative modification of IkappaB by monochloramine inhibits tumor necrosis factor alpha-induced NFkappaB activation. *Biochim. Biophys. Acta* 1746, 135-142.

Ogino, T.; Ozaki, M.; Hosako, M.; Omori, M.; Okada, S. & Matsukawa, A. (2009) Activation of c-Jun N-terminal kinase is essential for oxidative stress-induced Jurkat cell apoptosis by monochloramine. *Leuk. Res.* 33, 151-158.

Padrines, M.; Wolf, A.; Walz, A. & Baggiolini, M. (1994) Interleukin-8 processing by neutrophil elastase, cathepsin G and proteinase-3. *FEBS Lett.* 352, 231-235.

Payne, C.M.; Glasser, L.; Tischler, M.E.; Wyckoff, D.; Cromey, D.; Fiederlein, R. & Bohnert, O. (1994) Programmed cell death of the normal human neutrophil: an in vitro model of senescence. *Microsc. Res. Tech.* 28, 327-344.

Peng, S.L. (2006) Neutrophil apoptosis in autoimmunity. *J. Mol. Med.* 84, 122-125.

Pham, C.T.N. (2006) Neutrophil serine proteases: specific regulators of inflammation. *Nat. Rev. Immunol.* 6, 541-550.

Pichert, A.; Samsonov, S.; Theisgen, S.; Baumann, L.; Schiller, J.; Beck-Sickinger, A.G.; Huster, D. & Pisabarro M.T. (2012) Characterization of the interaction of interleukin-8 with hyaluronan, chondroitin sulfate, dermatan sulfate, and their sulfated derivatives by spectroscopy and molecular modelling. *Glycobiology*, 22, 134-145.

van Raam, B.J.; Verhoeven, A.J. & Kuijpers, T.W. (2006) Introduction: Mitochondria in neutrophil apoptosis. *Int. J. Hematol.* 84, 199-204.

Rajarathnam, K.; Sykes, B.D.; Kay, C.M.; Dewald, B.; Geiser, M.; Baggiolini, M. & Clark-Lewis, I. (1994) Neutrophil activation by monomeric interleukin-8. *Science* 264, 90-92.

Serhan, C.N.; Chiang, N. & van Dyke, T.E. (2008) Resolving inflammation: dual anti-inflammatory and pro-resolution lipid mediators. *Nat. Rev. Immunol.* 8, 349-361.

Setsukinai, K.; Urano, Y.; Kakinuma, K.; Majima, H.J. & Nagano, T. (2003) Development of novel fluorescence probes that can reliably detect reactive oxygen species and distinguish specific species. *J. Biol. Chem.* 278, 3170-3175.

Shatwell, K.P. Segal, A.W. (1996) NADPH oxidase. *Int. J. Cell Biol.* 28, 1191-1195.

Shepherd, J.; Hilderbrand, S.A.; Waterman, P.; Heinecke, J.W.; Weissleder, R. & Libby, P. (2007) A fluorescent probe for the detection of myeloperoxidase activity in atherosclerosis-associated macrophages. *Chem. Biol.* 14, 1221-1231.

Simon, H.-U. (2003) Neutrophil apoptosis pathways and their modifications in inflammation. *Immunol. Rev.* 193, 101-110.

Springer, T.A. (1994) Traffic signals for leukocyte recirculation and leukocyte emigration: the multistep paradigm. *Cell* 76, 301-314.

Stillie, R.M.; Farooq S.M.; Gordon, J.R. & Stadnyk, A.W. (2009) The functional significance behind expressing two IL-8 receptor types on PMN. *J. Leukoc. Biol.* 86, 529-543.

Taylor, K.R. & Gallo, R.L. (2006) Glycosaminoglycans and their proteoglycans: host-associated molecular pattern for initiation and modulation of inflammation. *FASEB J.* 20, 9-22.

Tsurubuchi, T.; Aratani, Y.; Maeda, N. & Koyama, H. (2001) Retardation of early-onset PMA-induced apoptosis in mouse neutrophils deficient in myeloperoxidase. *J. Leukoc. Biol.* 70, 52-58.

Vandivier, R.W.; Henson, P.M. & Douglas, I.S. (2006) Burying the dead. The impact of failed apoptotic cell removal (efferocytosis) on chronic inflammatory lung disease. *Chest* 129, 1673-1682.

Walker, A.; Ward, C.; Taylor, E.L.; Dransfield, I.; Hart, S.P.; Haslett, C. & Rossi, A.G. (2005) Regulation of neutrophil apoptosis and removal of apoptotic cells. *Curr. Drug Targets Inflamm. Allergy* 4, 447-454.

Walrand, S.; Valeix, S.; Rodriguez, C.; Ligot, P.; Chassagne, J. & Vasson, M.P. (2003) Flow cytometry study of polymorphonuclear neutrophil oxidative burst: a comparison of three fluorescent probes. *Clin. Chim. Acta* 33, 103-110.

Xu, Y.; Loison, F. & Luo, H.R. (2010) Neutrophil spontaneous death is mediated by down-regulation of autocrine signaling through GPCR, PI3Kγ, ROS, and actin. *Proc. Nat. Acad. Sci.* 107, 2950-2955.

Yamamoto, A.; Taniuchi, S.; Tsuji, S.; Hasui, M. & Kobayashi, Y. (2002) Role of reactive oxygen species in neutrophil apoptosis following ingestion of heat-killed Staphylococcus aureus. *Clin. Exp. Immunol.* 129, 479-484.

Yang, Y.-K.; Cho, H.J.; Lee, J.; Shin, I. & Tae, J. (2009) A rhodamine-hydroxamic acid-based fluorescent probe for hypochlorous acid and its application to biological imagings. *Org. Lett.* 11, 859-861.

Yoshida, H.; Kawane, K.; Koike, M.; Uchiyama, Y. & Nagata, S. (2005) Phosphatidylserine-dependent engulfment by macrophages of nuclei from erythroid precursor cells. *Nature* 437, 754-758.

Zernecke, A. & Weber, C. (2010) Chemokines in the vascular inflammatory response of atherosclerosis. *Cardiovasc. Res.* 86, 192-201.

7

Median Effect Dose and Combination Index Analysis of Cytotoxic Drugs Using Flow Cytometry

Tomás Lombardo, Laura Anaya, Laura Kornblihtt and Guillermo Blanco
Laboratory of Immunotoxicology (LaITo), Hospital de Clínicas San Martín,
University of Buenos Aires
Argentina

1. Introduction

Targeted therapy is a strategy of anticancer treatment that aims to interfere with processes of tumorigenesis, cancer progression and metastasis by selectively affecting key molecules of tumor cells (Armand et al., 2007; Favoni & Florio, 2011; Gross-Goupil & Escudier, 2010). Targeted therapies are directed to small molecules participating in different mechanisms that control cell survival through cellular proteins or signalling pathways (Mueller et al., 2009; Zahorowska et al., 2009). Targeted therapies may offer enhanced efficacy, enhanced selectivity, and less toxicity. However, targeting selective molecules and pathways often induces the activation of redundant mechanisms and enhances the emergence of resistant cells due to selective pressure (Woodcock et al., 2011). This is one of the reasons why the effects of targeted agents are not durable when used alone, and often result in drug resistance and clinical relapse.

Except for specific cases the use of these targeted drugs as monotherapy is often discouraged due to lack of efficacy. However, combined therapy with drugs targeting several mechanisms of tumor cell death can greatly improve efficacy and may overcome resistance. Several genomic and epigenetic alterations have been identified in tumor cells that lead to unrestrained proliferation, evasion of proapoptotic signals, metastasis, and resistance to drug-induced cell death. These alterations are critical for cancer progression and therefore combination strategies employing multiple targeted agents can be a successful therapeutic strategy. In vertical combination strategies two or more drugs target a same pathway at two different points, while in horizontal combinations drugs are directed towards different intracellular signalling pathways and have the potential advantage of combining agents with non-overlapping toxicities (Gross-Goupil & Escudier, 2010).

Novel treatments require the investigation of mechanisms of action and synergy of combination treatments to enhance the role of the targeted pharmacological agents (Carew et al., 2008; Mitsiades et al., 2011). Evaluating combinations of targeted drugs, including investigational agents, are an essential part of this effort (Dancey & Chen, 2006). An interesting example is represented by Bortezomib, a drug currently effective as single agent in multiple myeloma and mantle cell lymphoma (Bross et al., 2004; Kane et al., 2007; Wright,

2010). Bortezomib is a proteasome inhibitor that selectively triggers apoptosis in various types of neoplastic cells. It has been tested in a wide variety of solid tumors but has generally been ineffective as monotherapy (Boccadoro et al., 2005). However Bortezomib has shown increased activity when combined with several novel targeted agents including protein deacetylase inhibitors, kinase inhibitors, farnesyltransferase inhibitors, HSP-90 inhibitors, pan-Bcl-2 family inhibitors, and other classes of targeted inhibitors (Dai et al., 2003; Karp & Lancet, 2005; Pei et al., 2004; Perez-Galan et al., 2008; Trudel et al., 2007; Workman et al., 2007; Yanamandra et al., 2006). Thus, Bortezomib in combination with novel targeted therapies increase antitumor activity and overcome specific cellular antiapoptotic mechanisms (Wright, 2010).

Two-drug combination therapies are being assessed in a variety of tumors, usually testing agents that have different targets, nonoverlapping toxicity, and some rationale for evaluation (Belinsky et al., 2011; Castaneda & Gomez, 2009; Eriksen et al., 2009; Klosowska-Wardega et al., 2010). An increase in the number of these studies is expected in coming years, on the basis of emerging data with new agents, which is expanding our understanding of the molecular pathways important in cancer progression. (Woodcock et al., 2011). Tumor intracellular signalling pathway dependencies are being increasingly analyzed, and patients treated on the basis of resistance profile detected for specific drug combinations (Busch et al.; Derenzini et al., 2009; Michiels et al., 2011). This approach may facilitate the development of combination regimens optimized for specific tumor subtypes, thus providing the potential for tailored therapy in individual patients on the basis of certain molecular and genetic characteristics of their disease.

1.1 Targeted drugs often induce programmed cell death as their main mechanism of anti-tumor activity

Many of the classic chemotherapeutic agents (alkylating agents, antimetabolites, antibiotics, topoisomerase inhibitors) are known to block cell division by compromising DNA replication and halting cell cycle progression, or inhibit mitosis, eventually leading to cell death (Foye, 1995; Goodman et al., 2010; Shuck & Turchi, 2010). Indicators of cell proliferation are suitable effect biomarkers to assess whether a combination of these agents is synergic, additive or antagonistic. Biomarkers frequently used for this purpose include incorporation of nucleotide analogues such as bromodeoxyuridine, or metabolic indicators of cell number such as tetrazolium salt-based assays (Karaca et al., 2009; Olszewska-Slonina et al., 2004; Sims & Plattner, 2009). However many of the new targeted agents interfere with constitutively active survival pathways or initiate apoptosis by directly influencing pro-apoptotic signals (Citri et al., 2004; Kim et al., 2005; Larsen et al., 2011; Vega et al., 2009; Zhang et al., 2009). In addition autophagic cell death and programmed necrosis are being actively investigated as alternative and pharmacologically relevant forms of programmed cell death (Berghe et al., 2010; Bijnsdorp et al., 2011; Duan et al., 2010; Gozuacik & Kimchi, 2007; McCall, 2010; Notte et al., 2011; Paglin et al., 2005; Platini et al., 2010). Combination studies should be conducted using effect biomarkers that are as close as possible to the known mechanisms targeted by single agents, and biomarkers specifically related to drug-induce tumor cell death appear more adequate for the assessment of new targeted agents (Cameron et al., 2001; Facoetti et al., 2008; Wesierska-Gadek et al., 2005). A straightforward approach is to determine the proportion of live and dead cells in viability studies scoring thousands of cells through flow cytometry which ensures exceptional precision for dose-effect cytotoxicity studies.

2. Assessment of viability through flow cytometry

The strength of flow cytometry when compared to other methods available to assess the proportion of live and dead cells is the accuracy and precision brought by single cell multiparametric assessment. A variety of fluorescent probes may be chosen to use in viability assessment through flow cytometry. These probes are based on cell functions and biological conditions that are differentially preserved in live cells and lost in dead cells. It is usual to select at least two probes measuring independent functional conditions. For example, a probe evaluating membrane integrity of cells and another probe evaluating enzymatic activity. Probes should be selected to match specific experimental requirements such as biological variability, duration of the experiments, whether cells exposed to drugs are adherent or non-adherent, and illumination lines available in the flow cytometer. Some probes may be released after being retained within the cells for a short time and require immediate assessment through the flow cytometer after labelling, while others may be retained for several hours or may be even retained indefinitely by being covalently linked to cellular components. In addition some specimens may require fixation due to biohazard issues, so another kind of probes should be chosen and combined in these cases (De Clerck et al., 1994).

2.1 Fluorescent staining of live and dead cells

Viability is not easily defined in terms of a single physiological or morphological parameter. No single parameter fully defines cell death in all systems; therefore, it is often advantageous to use more than one cell death indicators based on different parameters such as membrane damage, and enzymatic or metabolic activity. A considerably large number of fluorescent probes have been introduced in the recent years that are dedicated to the assessment of viability on a single cell basis. Many of these new probes have features that are useful under specific experimental circumstances. The two conditions most often detected are increased cell membrane permeability in dead cells and the presence of enzymatic activity in live cells. The former is assessed with probes that become fluorescent when bound to DNA but are not able to pass through cell membrane if selective permeability is preserved, while the later is determined by fluorogenic substrates. However other conditions occurring only in live cells may be used to demonstrate viability such as enzymatic oxidation, reduction and mitochondrial membrane potential (Callewaert et al., 1991). It is important to underscore this concept because these probes are often used for assessment of specific cellular functions and it may be erroneously assumed that they have no contribution to the assessment of viability.

2.2 Enzymatic activity in live cells, use of tracker dyes

One of the first probes introduced and most frequently used to stain live cells has been fluorescein-diacetate (FDA) (Jones & Senft, 1985; Ross et al., 1989). This non-fluorescent cell-permeant esterase substrate penetrates the cell and is converted by nonspecific intracellular esterases into fluorescein.

Thus it becomes a more polar compound and is retained within those cells that have intact plasma membranes. In contrast, nonfluorescent FDA and fluorescein rapidly leak from those cells that have a damaged cell membrane because it is no longer retained due to increased permeability (Prosperi et al., 1986). This property ensures that dead cells will never retain FDA or fluorescein, even if cell death occurs after the labelling procedure. This is one of the reasons why it is recommended to analyze cells rapidly through the flow cytometer after staining with FDA and why they should be kept in low incubation temperatures to minimize potential fluorescein leakage.

Calcein-acetoxymethyl-ester (Calcein-AM) is a derivative of calcein that has several improvements over FDA (Duan et al., 2010; Papadopoulos et al., 1994). Calcein-AM is also a substrate of nonspecific intracellular esterases. The fluorescent product is calcein and is better retained in cells because it is a polyanionic compound that has six negative charges and two positive charges at pH 7. Calcein-Blue-AM and Calcein-Violet-AM are similar to Calcein-AM but have different excitation and emission wavelengths (Fuchs et al., 2007). Calcein-Blue-AM is excited with UV lasers while Calcein-Violet-AM is excited with 405 nm violet diode lasers, although both dyes emit blue fluorescence (Prowse et al., 2009). They can be used when the green fluorescence channel from the 488 nm excitation line is needed for other purpose and a UV or violet illumination line is available. Chloromethyl-fluorescein-diacteate (CM-FDA), is a FDA derivative that is retained within the cell even after damage to the plasma membrane due to its ability to bind thiol groups (Lantz et al., 2001; Sarkar et al., 2009). The weakly thiol-reactive chloromethyl moieties of this compound react with intracellular thiols and the acetate groups are cleaved by cytoplasmic esterases (West et al., 2001). This compound will not stain dead cells but the label will be preserved in those cells that die after the labelling procedure because the fluorescent product will be bound to SH groups within the cells (Sebastia et al., 2003). Chloromethyl SNARF-1 acetate is similar to CM-FDA but exhibits red fluorescence when excited with 488 nm blue laser. Thus it can be used when the green fluorescence channel is needed for other purpose and a UV or violet illumination line is not available (Hamilton et al., 2007). Carboxi-fluorescein-succinimidyl-ester (CFSE) is converted to fluorescent compound by intracellular esterases but covalently

Probe	Excitation line	Fluorescence emission	Intracellular retention
FDA	Blue	Green	Poor
Calcein-AM	Blue	Green	Good
Calcein Blue-AM	UV	Blue	Good
Calcein Violet-AM	Violet	Blue	Good
5-Cl-M-FDA	Blue	Green	Excellent
5-Cl-M-SNARF	Blue	Orange	Excellent
CFSE	Blue	Green	Excellent

Table 1. Fluorogenic substrates of intracellular esterases that are commonly used as viability probes

binds amino groups of proteins and is completely retained within cells, even after damage of cell membrane (Fujioka et al., 1994; Li et al., 2003). This dye is also used as cell tracker because it is retained in daughter cells after several rounds of cell division (Parish & Warren, 2002). It is worth to note that probes like FDA may give poor results with trypsinized cells owing to potential leakage of fluorescein during the staining and washing procedures (Zamai et al., 2001). Thus probes like CM-FDA, CM-SNARF-1, and CFSE may be a better choice for staining adherent cells.

2.3 DNA labelling in live and dead cells

Many polar nucleic acid stains are able to enter eukaryotic cells only when the plasma membrane is damaged. These stains are known as cell-impermeant dyes and include propidium iodide (PI) which is the most frequently used probe for assessing viability in flow cytometry (Yeh et al., 1981). This dye is excluded from live cells because it is negatively charged but readily penetrates the membrane of damaged cells and binds DNA. When excited at 488 nm DNA-bound PI increases orange-red fluorescence emission more than 1000 fold. Another commonly used cell-impermeant dye excited with 488 nm laser is 7-aminoactinomycin-D (7AAD). This dye binds DNA only in dead cells but emits fluorescence beyond 610 nm and allows the usage of the yellow-orange fluorescence channel for other purpose (Pallis et al., 1999).

Both PI and 7AAD have large Stokes shifts and can be used in 488 nm laser flow cytometers with green fluorescent tracker dyes such as FDA, CM-FDA, Calcein-AM and CFSE. Cells with damaged membranes may be identified with other cell-impermeant DNA fluorescent dyes that emit fluorescence in different wavelengths than that of PI.

The SYTOX series includes SYTOX-green (excited with 488 nm laser), SYTOX-red (excited with 633 and 635 nm lasers) and SYTOX-blue (excited with UV or 405 nm violet diode laser) (Haase, 2004; Lebaron et al., 1998; Yan et al., 2005). In contrast to SYTOX dyes, the SYTO series of nucleic acid stains can enter live cells and are thus cell-permeant DNA dyes (Ullal et al., 2010). The SYTO probes bind DNA with low affinity in live or dead cells (Eray et al., 2001; Poot et al., 1997). They are combined with high affinity cell-impermeant dyes to discriminate live from dead cells (Wlodkowic & Skommer, 2007). For example cell-permeant SYTO-red will stain live and dead cells with red fluorescence binding with low affinity to DNA, but if used together with SYTOX-green dead cells will be green fluorescent, because SYTOX-green has much higher affinity for DNA and will displace the low affinity SYTO-red. In addition, SYTOX-green will never stain live cells because it is cell-impermeant.

2.4 Biohazardous specimens

Viability staining of biohazardous specimens often requires fixation procedures that inactivate pathogens but produce minimal distortion of cellular characteristics. Some combinations of cell permeant and cell impermeant DNA dyes can be treated with fixatives such as 4% glutaraldehyde or formaldehyde to allow safer handling during analysis, without disrupting the distinctive staining pattern. An example is provided by cell-permeant, green-fluorescent DNA probe SYTO-10 and the cell-impermeant, red-fluorescent DNA probe ethidium homodimer-2 (Barnett et al., 2004; Poole et al., 1996). Using these two probes cells can be stained and fixed at various times during an experiment, and the results can be analyzed several hours later. This method may be applied to viability assessment of any non-adherent cells, as well as trypsinized adherent cells. Tables 1, 2, and 4 summarize the main features of viability probes based on enzymatic activity and DNA labelling discussed above that may be considered to meet specific experimental requirements.

Probe	Excitation line	Fluorescence emission	Membrane Permeant	DNA Affinity	Fixable
SYTOX-Blue	UV-Violet	Blue	NO	High	NO
SYTOX-Green	Blue	Green	NO	High	NO
SYTOX-Red	Red	Red	NO	High	NO
Propidium iodide	Blue	Orange-Red	NO	High	NO
7AAD	Blue	Red	NO	High	NO
SYTO-10	Blue	Green	YES	Low	YES
SYTO-Red	Blue	Red	YES	Low	NO
Ethidium homodimer-2	Blue	Green	NO	High	YES

Table 2. DNA probes used for viability assessment

2.5 Two parameter assessment of viability through flow cytometry

Identification of live and dead cells is often conducted with simultaneous use of two probes. The combination may include a cell-impermeant DNA probe and either a fluorogenic substrate or a cell-permeant DNA probe. It should be highlighted that viability may be also indicated by probes that have been designed to assess other cellular functions. For example generation of hydrogen peroxide and superoxide anion occurs in live cells due to normal function of mitochondrial electron transport chain and does not occur in dead cells. The superoxide anion probe dihydroethidine (HE) and the hydrogen-peroxide probe dihydro-dichloro-fluoresceindiacetate (DH-DCFDA) will stain live cells red fluorescent and green fluorescent respectively (Eruslanov & Kusmartsev, 2010; Zanetti et al., 2005). Both probes may be appropriately combined with cell-impermeant DNA dyes to discriminate between live and dead cells.

Similarly potentiometric dyes stain live cells with preserved mitochondrial membrane potential, but not dead or compromised live cells where the mitochondrial membrane potential has collapsed (Marchetti et al., 2004). Thus they may also be combined with cell-impermeant DNA dyes to discriminate live and dead cells. For example, rhodamine 123 has been used in combination with propidium iodide for viability assessment with two-color flow cytomety (Darzynkiewicz et al., 1982). Metabolically active cells undergo normal oxidation-reduction reactions and thus can also reduce a variety of probes, providing a measure of cell viability and overall cell health (Callewaert et al., 1991; Radcliff et al., 1991). Resazurin and dodecylresazurin (C12-resazurin) have been extensively used as oxidation–reduction indicators to detect viable cells (Czekanska, 2011). Reduction of resazurin yields the red fluorescent product resorufin while C12-resazurin yields C12-resorufin which is better retained by single cells (Talbot et al., 2008).

Again these probes may be combined with cell-impermeant DNA probes like SYTOX-green to discriminate live and dead cells with two color flow cytometry.

2.6 Viability assessment with single-color fixable dyes

In some experimental circumstances only one fluorescence channel may be dedicated to assessment of cell viability. In this case amine-reactive fluorescent dyes can be used to evaluate mammalian cell viability.

In cells with compromised membranes, these dyes react with free amines both in the cell interior and on the cell surface, yielding intense fluorescent staining. In viable cells, the dyes only stain cell-surface amines, resulting in less intense fluorescence (Elrefaei et al., 2008). The difference in intensity between the live and dead cell populations is preserved following formaldehyde fixation, using conditions that inactivate pathogens (Burmeister et al., 2008). There are several options of fluorescence excitation (UV, violet, blue, and red lasers) and emission wavelength (blue, green, yellow, red).

Probe	Excitation line	Fluorescence emission
LIVE/DEAD® fixable Blue	UV	Blue-450 nm
LIVE/DEAD ® Fixable Aqua	UV	Green-526 nm
LIVE/DEAD ® Fixable Yellow	Violet	Yellow-575 nm
LIVE/DEAD ® Fixable Violet	Violet	Blue-450 nm
eFluor® 450	Violet	Blue-450 nm
Fixable Viability Stain 450 ®	Violet	Blue-450 nm
LIVE/DEAD ® Fixable Green	Blue	Green-520 nm
eFluor® 506	Blue	Green-506 nm
LIVE/DEAD ® Fixable Red	Red	Red-615 nm
LIVE/DEAD ® Fixable Far Red	Red	Far Red-665 nm
eFluor® 660	Red	Far Red-660 nm

Table 3. Fixable amine-reactive fluorescent probes used for single-color assessment of cell viability. The wavelengths indicated correspond to the emission peaks as specified by the probe manufacturer

2.7 Viability vs. apoptosis

Most targeted cytotoxic drugs have been shown to induce apoptosis or other modes of programmed cell death, including autophagic cell death or programmed necrosis. These mechanisms of cell death are often contrasted to necrosis where a passive, sudden and uncontrolled disintegration of the cell occurs. Physiological consequences of apoptosis and passive necrosis are different, and thus it is important to determine the cell death phenotype. When assessed through flow cytometry, cells undergoing apoptosis or other forms of programmed cell death show a decrease in cell volume and forward light scatter (FSC), and an increase in side light scatter (SSC) mainly due to cytoplasmic and nuclear changes such as blebbing, and nuclear fragmentations (Dive et al., 1992; Pheng et al., 2000). In contrast necrosis often shows increased cell volume and FSC without changes in SSC (Healy et al., 1998). Viability assessment after cytotoxic drug exposure does not address the cell death phenotype, thus any kind of cell death phenotype may be induce by drug treatment including passive necrosis (Healy et al., 1998). However studies determining the median cytotoxic dose will require exposure to increasing doses from sub-lethal levels to the minimal doses approaching 100% cell death. In this scenario, programmed cell death phenotypes are more frequently observed than passive necrosis.

Probe	Membrane Permeant	Excitation line	Fluorescence emission
YOYO-1	NO	Blue	Green
TOTO	NO	Blue	Green
TO-PRO	NO	Blue	Green
POPO-1	NO	UV-Violet	Blue
BOBO-1	NO	UV-Violet	Blue
YOYO-3	NO	Red	Red
TOTO-3	NO	Red	Far Red
BOBO-3	NO	Red	Red
JOJO-1	NO	Green	Orange
JO-PRO-1	NO	Green	Orange

Table 4. Membrane–impermeant dimeric and monomeric cyanine dyes are nonfluorescent unless bound to nucleic acids and have extinction coefficients 10–20 times greater than that of DNA-bound propidium iodide

2.8 FDA-PI staining and the "cell death pathway": Frequency distributions of graded and abrupt transitions

When two fluorescent probes are used to determine the proportion of live and dead cells after exposure to cytotoxic drugs over an extended dose range data analysis would be better analyzed on a bivariate plot.

An example is the pair represented by FDA as an indicator of esterase activity in live cells and PI as an indicator of cell membrane damage (Fig. 1). In this case a bivariate plot of green fluorescence against red fluorescence will aid in determining the percentage of live and dead cells (Fig. 1A,C). In addition, the bivariate plot will provide useful data about the biological processes evaluated with FDA and PI.

A concept frequently present in flow cytometry, particularly when analyzing bivariate plots, is the presence of graded transitions or abrupt changes (Shapiro, 2003). These patterns in bivariate distributions are determined by the underlying biological process that is being studied. For example damage of cell membrane allows staining by PI probe so that cells may be classified as dead with a permeable membrane or live having preserved selective permeability, depending on whether they are red fluorescent or not. Cells are observed to be bright stained or having no stain at all, but very rarely they are observed to have dim red fluorescence. Thus membrane damage and PI staining is an example of an abrupt transition or discrete process represented by membrane damage that produces a sudden change in the frequency distribution. This frequency distribution is symmetric, bell-shaped, and has low variability around the peak (Fig. 1C,E.).

By contrast when analyzing drug-induced effects on esterase activity through green fluorescence we will observe a graded transition from bright fluorescence to dim fluorescence (Fig. 1C). Thus a graded biological process represented by progressively decreased esterase activity determines a skewed frequency distribution with higher variability around the peak (Fig. 1F). In this case there will be a higher probability of observing cells within any level of metabolic activity represented by the amount of green fluorescence: bright, intermediate and dim. Note also that cells with damaged membrane no longer retain fluorescein (very few events are seen in upper right quadrant, Fig. 1C).

When analysis is restricted to live cells without damaged membrane (PI negative) it is more evident that the probability of finding live cells with low metabolic activity in the drug-treated population decreases gradually (Fig. 1H). By contrast, when the analysis is restricted to cells without metabolic activity a narrow bell-shape distribution is observed meaning that the probability of finding cells with damaged membrane in cells without metabolic activity increases abruptly (Fig. 1G). When combined in a bivariate plot the gradual decrease in metabolic activity in live cells is observed as a continuous distribution or pathway, while the abrupt transition from membrane-impermeable to membrane-permeable is observed as a discrete transition to a main single cluster of PI-positive cells with very low or no metabolic activity (Fig. 1C). The probability of finding cells with low or no metabolic activity is very low as shown by the few cells in an intermediate position along this "death-pathway". Changes in FSC and SSC are also graded transitions and define a "death-pathway" in bivariate plots (Fig. 1D). Most cells having membrane damage have low FSC and high SSC, those cells without membrane damage and with metabolic activity have high FSC and low SSC, while intermediate positions may be occupied by either of these populations (Fig. 1D). Thus the death pathway defines a whole range of changes occurring in all four parameter FSC, SSC, FDA, and PI. However the main result is characterizing cells as either dead or alive and this difference is brought by PI staining and membrane damage. Thus applying quadrant analysis we would add the fraction of cells in both upper quadrants and the fraction of cells in the lower quadrants as live cells (Fig 1C). The remaining parameters will work as internal quality controls.

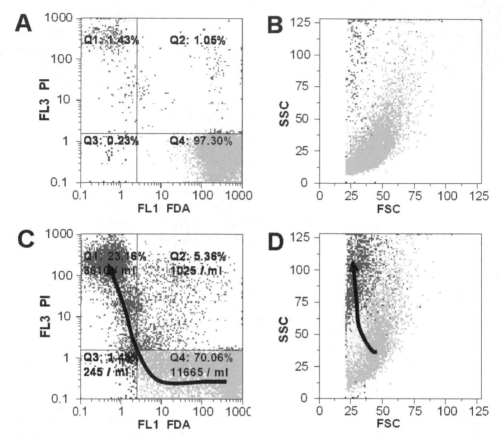

Fig. 1. (continues on next page) The FDA vs. PI bivariate plot and the cell death pathway. A. Sample of human U937 leukemic cells labelled with FDA and PI. Lower right quadrant shows that most cells (97.30%) have esterase enzymatic activity and preserved membrane permeability because they exclude PI staining. Only 1.43% of cells have PI staining without FDA fluorescence, while 1.05% are double positive indicating both enzymatic activity and damaged membrane. B. FSC-SSC profile of live cells is show in green and corresponds to the 97.30% of cells shown in the lower right quadrant of panel A. The small amount of single PI positive (red) and double positive cells (blue) are also observed. C. Sample of human U937 cells exposed to 5 µM sodium arsenite (AsNaO2) for 72h stained with FDA and PI showing a "slow" transition from high FDA fluorescence to low FDA fluorescence (green) and a further "abrupt" transition to a PI positive FDA negative cluster of dead cells (red). A minority of cells are double positive (blue). The whole transition is indicated with the black arrow. D. FSC vs. SSC plot of the sample shown in C. Green color represents FSC-SSC paired values only occupied by live cells (lower right quadrant shown in C), red color represents FSC-SSC paired values only occupied by dead cells (upper left quadrant in C), while yellow color represents FSC-SSC paired values occupied by both live and dead cells. The black arrow shows the graphical death pathway transition in the FSC SSC plot. The FSC-SSC values of the minority of double positive cells are shown in blue

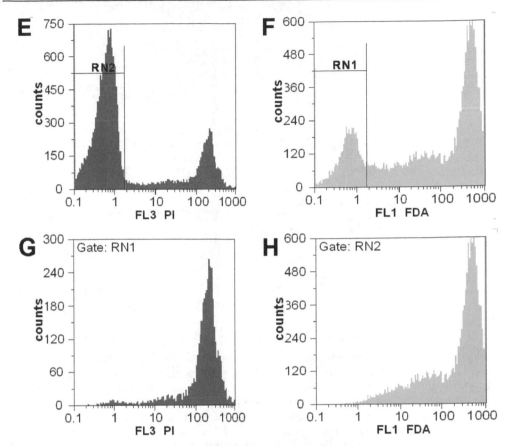

Fig. 1. (continued) E. Frequency distribution of PI fluorescence corresponding to the sample shown in C and D. Note that positive and negative cell populations are bell-shaped, symmetrical, with low variability around the peaks, and well separated from each other F. Frequency distribution of FDA fluorescence of the same sample shown in C and D. The population of positive cells shows asymmetrical left-skewed distribution with great variability to the left of the peak. G. The sample shown in E with live cells excluded. The probability of finding positive cells with intermediate and dim PI fluorescence decreases abruptly to the left. H. The sample shown in F with dead cells excluded. The probability of finding positive cells with intermediate and dim FDA fluorescence decreases slowly to the left

3. Building a cytotoxic dose response curve

Theoretically, if a population of cells were homogenously sensitive to cell death induced by a certain drug there would be a single dose D at which 100% cell death would be observed (Casarett et al., 2008; Goodman et al., 2010). However in any given sample of drug-treated cells a random proportion of cells will die at doses lower or higher than D due to experimental and biological variability. This random divergence from D follows a Gaussian distribution (Fig. 2A,C).

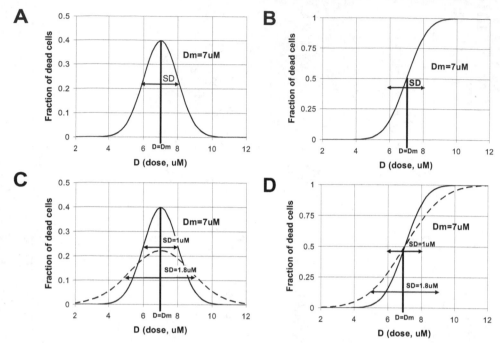

Fig. 2. Quantal dose response model. A. Single cells will show differences on the minimal drug dose required to induce cell death. Dm is the most frequent minimal dose required to elicit cell death. Variations around this value have a normal distribution. In this particular example Dm is 7uM and the standard deviation (SD) is 1uM. B. Dm is the drug dose that is estimated to kill half of the cells in a sample. When running experiments exposing cells to incremental doses of a cytotoxic drug the fraction of dead cells observed will follow a normal cumulative function. C. Differences between two cytotoxic drugs in the variability observed around Dm. The drug represented by the dotted line has larger variability (SD=1.8 uM) than the drug represented in full line (SD=1 uM). D. Increased variability around Dm is observed as a dose response curve with a smaller slope as shown by the dotted line of the cumulative normal distribution

The median dose Dm represents a dose D where half of the cells are killed and half of the cells remain live (Fig. 2 B). However, as indicated by the bell shape of the Gaussian distribution many of the 50% of cells killed at a dose Dm may have required less than Dm to be killed. In fact as shown in figure 2A only a fraction of cells will require strictly a dose Dm (those in the bell peak), while a minority of cells will require a dose much lower than Dm to be killed (those in the left tail of the bell-shaped curve). If we conduct experiments evaluating the cytotoxic effect of increasing doses we will observe a sigmoid curve that follows a cumulative Gaussian frequency distribution (Fig. 2B). Doses lower than Dm will show decreased probability of cell death approaching 0% while doses higher than Dm will show increased probability of cell death approaching 100% (Fig. 2B). This model is known as quantal dose-response because it is based on the scoring of all members of a sample population for having or not having a certain condition at a given applied dose (Casarett et al., 2008). This is precisely what is done through flow cytometry assessing a sample on a cell by cell basis for being dead or alive. Using flow cytometry we can measure the fraction of cells killed (fa) at a dose D with high accuracy and precision due to the large number of cells analyzed which ensures an extremely low standard error (SE). However, high accuracy and precision apply to a single sample and not to replicates. The source of variability between replicates will be both biological and experimental. For example cells overgrown in culture may respond with higher variability than cells in exponential growth when estimated from replicates. Similarly any problems around drug exposure or the staining procedure will add to the variability of replicates although the precision and accuracy of each sample determination will be very high due to the large amount of cells scored in each sample tube. Regarding calculation of the median cytotoxic dose Dm this replicates will have a critical impact on the statistical precision of the Dm estimate.

3.1 Calculating the median dose: The median effect equation

Cells cultured in vitro can be exposed to increasing doses of a cytotoxic drug during a certain time interval (e.g., 48h or 72h) to determine a median cytotoxic dose Dm. Several doses should be tested to extend over a dose range. The lower doses should induce a fraction of dead cells close to that of unexposed cells, while the higher doses should induce death values approaching 100% or achieve a plateau of maximal effect. In between these boundary doses the more intervals assessed the more precision we will get in the estimates. For example seven doses assessed in triplicate that yield cell death fractions between 5% and 95% would be enough to obtain regression estimates with an adequate precision. The dose response sigmoidal curve can be fit to a two parameter logistic function of the type:

$$fa = 1 / [1+1/(D/Dm)^m] \tag{1}$$

where D is the dose, Dm is the dose required to achieve the median cytotoxic effect, fa is the fraction of dead cells, and m is a measurement of the sigmoidicity of the curve. When m=1 the dose-effect curve is hyperbolic, when m>1 the curve is sigmoidal, while m<1 indicates a negative sigmoidal shape (Fig. 3A).

To estimate Dm and m the median effect equation is written as:

$$fa/(1-fa) = (D/Dm)^m \tag{2}$$

The factor (1-fa) is the fraction of live cells. Applying a log transformation the following linear function is obtained:

$$\log (fa/(1-fa) = m . \log (D) - m . \log (Dm) \tag{3}$$

Thus plotting log values of experimental doses against log values of the ratio of dead/live cells will show a linear trend that is often referred to as median effect plot (Fig. 3B).

This is a linear function of the type y= m.x + b, where y=fa/(1-fa), x=log(D), and b=-m.log (Dm). A linear regression can be applied to these data to obtain estimates for m and Dm. The m coefficient can be readily determined by the slope of the regression, and Dm is derived from the estimate of the intercept -m.log(Dm).

The squared correlation coefficient or R2 value is an estimate of the precision of the overall regression (Fig. 3B). In this particular application to data representing cell death vs. dose of cytotoxic drug, R2 is affected mainly by the scattering of replicate values, which in turn depend on the experimental and biological variability, and also on the number of sample doses in between the lowest and highest effect values. The standard error (SE) of m and the intercept can also be obtained from regression analysis to get a 95% confidence interval around log(Dm). The formula for manual calculation is rather complicated and involves computing SE(log(D)) when D=Dm (shown in eq. 14, Table 5) but may be obtained using any software that implements this calculation such as Calcusyn or Compusyn (Bijnsdorp et al., 2011; Ikeda et al., 2011; Ramachandran et al., 2010).

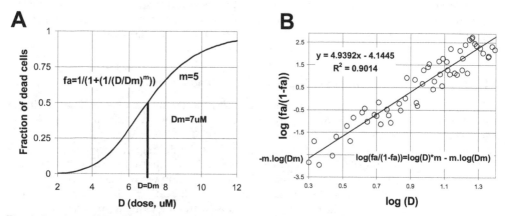

Fig. 3. Median effect plot. A. The two parameter dose-response sigmoidal curve for a particular example with Dm=7uM and m=5. B. Algebraic and log transformation to obtain a linear function. A linear regression can be applied to experimental data in order to obtain estimates of the two parameters Dm and m. The squared correlation coefficient R2 is a measure of the overall precision of the linear regression and thus the Dm and m estimates

3.2 Threshold, median dose and maximal efficacy

Analysis of the dose-effect plot can be informative about threshold values and maximal efficacy values.

In practice the threshold value will be the minimal dose where the fraction of dead cells is higher than that of untreated cells. The maximal efficacy would be the fraction of dead cells where the sigmoidal curve approaches a plateau. Quite often the maximal efficacy approaches 100%. However a cell population may exhibit a differential response and a fraction of cells may require quite larger doses. In these cases the maximal efficacy will be much less than 100%.

3.3 Comparing two drugs

To compare two drugs and evaluate whether combination results in synergy or not a first step is to calculate Dm for the two drugs. Thus, the same approach described above should be applied to the second drug. The procedure will include evaluating several doses in replicates with the same exposure time interval as the first drug, obtaining data to create a median effect plot and estimating m and Dm by linear regression (Chang et al., 1987; Chou & Talalay, 1984; Sugiyama et al., 1998).

Results from assessment of the two drugs can be analyzed together in a combined median effect plot where log doses (in molar units) are depicted against log (fa/1-fa). In this plot the relative potency of the two drugs can be easily appreciated (Fig. 4A). A drug a is said to be more potent than a second drug b regarding the cytotoxic effect when less dose of dug a is needed to achieve the same cytotoxic effect with drug b on a molar basis. In addition when the slopes of the two drugs are different it suggests that the drugs have different mechanisms to induce cytotoxicity.

The dose effect equation can be re-written to calculate the dose required to induce a given cytotoxic effect:

$$D=Dm \, (fa/(1-fa))^{1/m} \tag{4}$$

For example the effective cytotoxic dose 50% (EC50) is the dose D that is estimated to kill 50% of the cells. In this case fa=0.5 and EC50 is coincident with the median dose Dm. Similarly EC25 is the dose D that is estimated to kill 25% of the cells.

3.3.1 Assessing the combined effect of two drugs under fixed molar ratio

Once obtained the dose-response curve for two drugs a and b, a third experiment with combination of a + b will be needed to determine if the interaction of these drugs is additive, synergic or antagonistic.

Assuming that the drug b is less potent than the drug a, fixed molar ratio could be used in the combination based on the relative potency EC50(a)/EC50(b). For example if EC50(a) is 10µM and EC50(b) is 30µM, the molar ratio of the combination would be 1/3. An empirical approach is to start the combination experiment with a combination of a+b calculated as

$$EC50(a+b)=10^{\{[\log(EC50(a))+\log(EC50(b))]/2\}} \tag{5}$$

In this example this estimated value would be 17.3 µM. Assuming the fixed molar ratio 1/3 this combination would have 4.3µM of drug a and 13.0 µM of drug b.

Next we should treat this combination as a new drug a+b and evaluate several doses above and below 17.3µM in replicates to span a dose range of the combination. Thus, we will experimentally obtain a new data set of doses and cytotoxic effects that we should evaluate by the same procedure with the median effect plot, and conduct a linear regression to obtain estimates for m and Dm with the combination of a+b.

In particular we will obtain an effect-dose equation as shown above (eq. 4) to determine the dose of the combination a+b to achieve a desired cytotoxic effect level (EC)

$$D= Dm_{a+b} \, (fa/(1-fa))^{1/m}a+b \tag{6}$$

For example applying (eq.5) EC50 (a+b) will be equal to the median dose estimated from the regression in the combined experiment (Dm$_{a+b}$)

3.4 Graphical analysis

A first approach is to plot this result together with results of single drug effects to depict some relevant information (Pegram et al., 1999). When the combined-drug curve lies in a midpoint between the two single drug curves it suggests an additive effect (Fig. 4A). It indicates that the potency of the combined drugs is at an intermediate point between the potency of each drug. When the combined drug curve is shifted to the left it would be closer to the more potent single drug and thus suggests synergism (Fig. 1B). On the other side when the combined drug curve is shifted to the right and closer to the low potency drug it is suggestive of antagonism (Fig. 1C). Another hint that could aid in generating hypothesis is the curve shape and particularly the slope. The variability around Dm is represented by the standard deviation (SD) of the Gaussian distribution underlying the quantal dose-response model discussed above. A large SD is in accordance with a flat curve while a small SD is in accordance with a steep curve (Fig. 2C,D). This variability has a biological significance and two drugs having different mechanisms of inducing cell death in a certain cell line may have different slopes.

3.5 Calculating the combination index (CI)

The graphical analysis gives some clues about what kind of interaction results from the combination of drugs a and b and depicts useful information but is less conclusive in quantitative terms. A more thorough conclusion can be derived from computing the combination index for each cytotoxic effect level (Chou, 2008; 2010). Computing the combination index (CI) for each effect level provides an answer to what kind of interaction occurs between drug a and drug b throughout the whole dose range.

The combination index method takes data provided by single and combined dose-effect equations to provide an estimate at the whole range of cytotoxic effects. The combination index is defined for a given effect level i by the following equation:

$$CI\ (i) = Dac(i)\ /Das(i) + Dbc(i)\ /Dbs(i) + \alpha\ Dac(i)\ .\ Dbc(i)\ /\ Das(i)\ .\ Dbs(i) \qquad (7)$$

Where Dac(i) and Dbc(i) are the doses of drugs a and b respectively required in the combination a+b to produce an effect level i.

Das(i) and Dbs(i) are the doses of drug a and b respectively, required to produce an effect level i when used as single drugs. For any level i, these values are obtained from the three dose response curves defined by (eq. 4) (two single and one combined) obtained with parameters Dm and m that in turn were obtained from regression analysis with (eq.3) applied to experimental data. It is often assumed the conservative criteria that cytotoxic drugs are mutually non exclusive and $\alpha=1$. If the three lines are strictly parallel and both drugs have a similar molecular target it could be assumed that they are mutually exclusive and in that case $\alpha=0$. If the fixed molar ratio of drug a and b in the combined treatment is p/q, then for an effect level i:

$$Das(i) = Dm_a\ (fa(i)/(1-fa(i)))^{1/m}a \qquad (8)$$

$$Dbs(i) = Dm_b\ (fa(i)/(1-fa(i)))^{1/m}b \qquad (9)$$

$$Dac(i) = p/(p+q)\ Dm_{a+b}\ (fa(i)/(1-fa(i)))^{1/m}a+b \qquad (10)$$

$$Dbc(i) = q/(p+q)\ Dm_{a+b}\ (fa(i)/(1-fa(i)))^{1/m}a+b \qquad (11)$$

Where fa(i) is the fraction of dead cells at effect level i, Dm_a and m_a are the median dose and the slope estimated for drug a, Dm_b and m_b are the median dose and the slope estimated for drug b, and Dm_{a+b} and m_{a+b} are the median dose and the slope estimated for the combined treatment with drugs a and b. Thus the combination index is calculated for any effect level above 0 and below 1.

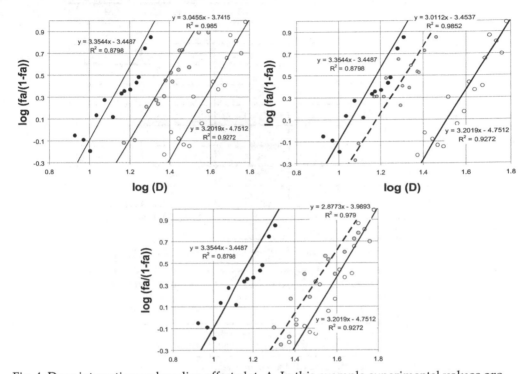

Fig. 4. Drug interaction and median effect plot. A. In this example experimental values are represented by circles and linear regression is applied to obtain estimates for Dm and m. Experimental values obtained for a drug a are shown in black circles. The values of Dm and m for drug a were 10 μM and and 3.0 respectively. Experimental values for a less potent drug b are shown in open circles. The values of Dm and m for drug a were 30 uM and and 3.0 respectively. A combined experiment was run with a+b with a constant mass ratio of 1/3 based on EC50(a)/EC50(b) and assuming that under additive effect EC50(a+b)=10{[log(EC50(a))+log(EC50(b))]/2}. Experimental values for the combination are shown in grey circles. The values of Dm and m for the combination a+b were 17.0 uM and and 3.0 respectively. In this example drugs a and b have an additive interaction and the median effect plot of the combination lies in a mid position between the lines corresponding to the single drugs. B. The same experiment shown in A, but in this case the drugs a and b have synergic effect. The values of Dm and m for the combination a+b were 14.0 uM and and 3.0 respectively. The median effect plot of the combination is shifted towards the drug a, which has the highest potency. C. The same experiment shown in A, but in this example the drugs a and b have antagonistic effect. The values of Dm and m for the combination a+b were 24.4 uM and and 2.9 respectively. The median effect plot of the combination in this case is shifted towards the drug b which has the lowest potency

			Three alternative results of the experimental assay with combination *a+b* (considering α=1, mutually non-exclusive condition)		
			Additive	Synergic	Antagonistic
	Single drug a	Single drug b	Combined drugs a+b	Combined drugs a+b	Combined drugs a+b
Dm (uM) [#(1)]	10	30	17	14	24.4
m [#(1)]	3	3	3	3	3
fa(i)	0.5				
Das(i) [#(2)]	10				
Dbs(i) [#(3)]		30			
Dac(i) [#(4)]			4.25	3.5	6.1
Dbc(i) [#(5)]			12.75	10.5	18.3
p/(p+q) [#(6)]			0.25	0.25	0.25
q/(p+q) [#(6)]			0.75	0.75	0.75
CI (i) [#(7)]			1.03	0.82	1.59

$SE(CI(i))=\{\{ Dac(i)/ Das(i) . [SE(Dac(i)/ Dac(i)+ SE(Das(i)/ Das(i))]\}^2 +\{ Dbc(i)/ Dbs(i) . [SE(Dbc(i)/ Dbc(i)+ SE(Dbs(i)/ Dbs(i))]\}^2\}^{1/2}$ **(eq. 12)**

$SE(D) = 1/2 . \{10^{[log(D)+SE(log(D))]} -10^{[log(D)-SE(log(D))]}\}$ **(eq. 13)**

$SE(log (D)) =\{log(D) . [SE(b)/log(fa/(1-fa)-b]^2+[SE(m)/m]^2+2[-(logD)^{1/2} . SE(m)/SE/(b)] . SE(b)/b . SE(m)/m\}^{1/2}$ **(eq. 14)**

where $b=-m.log(Dm)$

A 95% confidence interval around D in general and around Dm in particular, can be computed using the formulas for standard error (SE, eq. 13 and eq. 14).

A 95% confidence interval around CI at any effect level i can be computed from the standard error formulas presented in eq. 12, 13, and 14.

[#(1)] Obtained from linear regression of experimental data

[#(2)] $Das(i)=Dm_a (fa(i)/(1-fa(i)))^{1/m}a$

[#(3)] $Dbs(i)=Dm_b (fa(i)/(1-fa(i)))^{1/m}b$

[#(4)] $Dac(i)=p/(p+q) Dm_{a+b} (fa(i)/(1-fa(i)))^{1/m}a+b$

[#(5)] $Dbc(i)=q/(p+q) Dm_{a+b} (fa(i)/(1-fa(i)))^{1/m}a+b$

[#(6)] Molar ratio p/q = Dm(a) /Dm(b) = 10/30 = 1/3

[#(7)] $CI(i) = Dac(i)/Das(i) +Dbc(i)/Dbs(i) + α Dac(i) . Dbc(i) / Das(i) . Dbs(i)$

Table 5. CI calculation between two drugs a and b at the 50% effect level under three alternative conditions: additive, synergic, antagonistic. To obtain CI as a function of the effect level i, the calculation has to be repeated for each arbitrary level i between 0 and 1. A 95% confidence interval around D in general and around Dm in particular, can be computed using the formulas for standard error (SE, eq. 13 and eq. 14). A 95% confidence interval around CI at any effect level i can be computed from the standard error formulas presented in eq. 12, 13 and 14

Table 5 summarizes a manual calculation of CI of two drugs a and b using these formulas for the 50% effect level under three hypothetical results: additive, synergic or antagonistic effect.

The same calculation can be applied to any effect level to plot CI as a function of effect level. When the interaction is additive CI =1. In this case it can be interpreted that one of the drugs (the less potent one, i.e. drug b in the example) is acting as though it is merely a diluted form of the other (drug a in the example). When CI<1 the combination of a+b is synergic while CI>1 indicates antagonism. Synergy, implies that the combination of the two drugs achieves a cytotoxic effect greater than that expected by the simple addition of the effects of the drugs a and b, while antagonism achieves a cytotoxic effect lower than that expected by additive effects of drugs a and b.

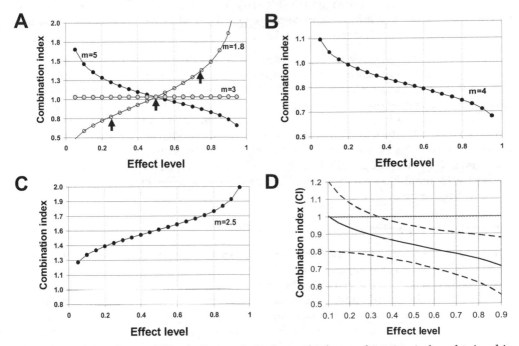

Fig. 5. Drug interaction and CI calculation. A. Only rarely the combination index obtained is constant for all effect levels. Here it is shown how different values of the slope m obtained through linear regression in the combination experiment (a+b) would affect the shape of the curve representing the CI as a function of the effect level. Similarly, differences between the slopes obtained for drugs a and b through the single drug experiments will contribute to the uneven shape of the CI function. Note that depending on the effect level the interaction a+b with m=1.8 would be synergism, additive or antagonism at EC25. EC50, and EC75 respectively (arrows). B. Results of CI calculation for the example where a+b results in synergism considering m=4 in the regression of the combined-drug experiment. C. Results of CI calculation for the example where a+b results in antagonism considering m=2.5 in the regression of the combined-drug experiment. D. 95% confidence level intervals around CI, using an algebraic approximation (eq. 12, Table 5) in an example where combination of a+b is synergic

The horizontal line corresponding to CI(i)=1, where i is any effect level in the interval (0,1), is often call the additive effect line. A combination of drugs a and b may result in CI values above or below the additive line at different effect levels. Thus, the CI as a function of effect level is not constant or linear and it may be decreasing or increasing (Fig 5A). If data from the combination experiment in the example of figure 4A resulted in m_{a+b}=5 or m_{a+b}=1.8, even still with Dm_{a+b} =17 the CI line would be inclined downwards or upwards respectively (Fig. 5A). Only at effect levels close to EC50 the result would be strictly additive. An important conclusion is that for some drug combinations, experiments conducted at different single dose-effect levels may yield opposing results. For example if the combination a+b with m=1.8 shown in figure 5A were experimentally evaluated only at effect level 0.25 the single dose analysis would conclude on synergism. However if it were evaluated at effect level 0.75 it would conclude on antagonism (Fig. 5A). This exemplifies why the assessment of combination index over the whole dose range will show all kinds of interactions that may result from combination at different effect levels.

Computing a standard error of CI allows plotting confidence intervals at all effect levels providing a further assurance over the computation. A 95% confidence interval will indicate that if we repeat the experiment 100 times, 95 out 100 times the CI would be within this interval. For example, observing whether or not confidence limits are above or below the additive line will allow concluding with further statistical support on antagonism or synergism respectively. Computation of the standard error of CI and confidence intervals at all levels should be better obtained through specialized software such as Calcusyn or Compusyn (Bijnsdorp et al., 2011; Chou, 2010). It may also require approaches such as Monte Carlo simulation based on the estimated parameter for m and Dm in single and combined equations.

4. Concluding remarks

A thorough assessment of drug interaction is an essential step in targeted combined therapy. The new targeted agents are seldom useful as single agents but may be effective when used in specific combinations. The median effect and combination index calculation are well founded methods traditionally used in pharmacological and toxicological studies. Since new cytotoxic drugs target mechanisms eliciting cell death, biomarkers related to viability assessment are preferred to biomarkers of cell proliferation. Flow cytometry is an ideal technology to provide massive data from cell death biomarkers to build dose response curves of cytotoxic effect. When these data is further used to determine the combination index a full characterization of drug interaction over the cytotoxic effect is obtained at all effect levels. This approach can be applied to tumor cell lines in preclinical studies and also in patient-derived tumor cells, thus providing useful information as prospective indicators of the potential therapeutic response to combined-drug antitumor treatment.

5. References

Armand, J. P., Burnett, A. K., Drach, J., Harousseau, J. L., Lowenberg, B. & San Miguel, J. (2007). The emerging role of targeted therapy for hematologic malignancies: update on bortezomib and tipifarnib. Oncologist Vol. 12, No. 3, (Mar, 2007), pp. 281-290

Barnett, M. J., McGhee-Wilson, D., Shapiro, A. M. & Lakey, J. R. (2004). Variation in human islet viability based on different membrane integrity stains. Cell Transplant Vol. 13, No. 5, 2004), pp. 481-488

Belinsky, S. A., Grimes, M. J., Picchi, M. A., Mitchell, H. D., Stidley, C. A., Tesfaigzi, Y., Channell, M. M., Liu, Y., Casero, R. A., Jr., Baylin, S. B. et al. (2011). Combination therapy with vidaza and entinostat suppresses tumor growth and reprograms the epigenome in an orthotopic lung cancer model. Cancer Res Vol. 71, No. 2, (Jan 15, 2011), pp. 454-462

Berghe, T. V., Vanlangenakker, N., Parthoens, E., Deckers, W., Devos, M., Festjens, N., Guerin, C. J., Brunk, U. T., Declercq, W. & Vandenabeele, P. (2010). Necroptosis, necrosis and secondary necrosis converge on similar cellular disintegration features. Cell Death Differ Vol. 17, No. 6, (Jun, 2010), pp. 922-930

Bijnsdorp, I. V., Giovannetti, E. & Peters, G. J. (2011). Analysis of drug interactions. Methods Mol Biol Vol. 731, No., 2011), pp. 421-434

Boccadoro, M., Morgan, G. & Cavenagh, J. (2005). Preclinical evaluation of the proteasome inhibitor bortezomib in cancer therapy. Cancer Cell Int Vol. 5, No. 1, (Jun 1, 2005), pp. 18

Bross, P. F., Kane, R., Farrell, A. T., Abraham, S., Benson, K., Brower, M. E., Bradley, S., Gobburu, J. V., Goheer, A., Lee, S. L. et al. (2004). Approval summary for bortezomib for injection in the treatment of multiple myeloma. Clin Cancer Res Vol. 10, No. 12 Pt 1, (Jun 15, 2004), pp. 3954-3964

Burmeister, Y., Lischke, T., Dahler, A. C., Mages, H. W., Lam, K. P., Coyle, A. J., Kroczek, R. A. & Hutloff, A. (2008). ICOS controls the pool size of effector-memory and regulatory T cells. J Immunol Vol. 180, No. 2, (Jan 15, 2008), pp. 774-782

Busch, C., Geisler, J., Knappskog, S., Lillehaug, J. R. & Lonning, P. E. Alterations in the p53 pathway and p16INK4a expression predict overall survival in metastatic melanoma patients treated with dacarbazine. J Invest Dermatol Vol. 130, No. 10, (Oct, pp. 2514-2516

Callewaert, D. M., Radcliff, G., Waite, R., LeFevre, J. & Poulik, M. D. (1991). Characterization of effector-target conjugates for cloned human natural killer and human lymphokine activated killer cells by flow cytometry. Cytometry Vol. 12, No. 7, 1991), pp. 666-676

Cameron, D. A., Ritchie, A. A. & Miller, W. R. (2001). The relative importance of proliferation and cell death in breast cancer growth and response to tamoxifen. Eur J Cancer Vol. 37, No. 12, (Aug, 2001), pp. 1545-1553

Carew, J. S., Giles, F. J. & Nawrocki, S. T. (2008). Histone deacetylase inhibitors: mechanisms of cell death and promise in combination cancer therapy. Cancer Lett Vol. 269, No. 1, (Sep 28, 2008), pp. 7-17

Casarett, L. J., Doull, J. & Klaassen, C. D. (2008). Casarett and Doull's toxicology: the basic science of poisons. 7th Edition. McGraw-Hill. ISBN (9780071470513-0071470514) New York

Castaneda, C. A. &Gomez, H. L. (2009). Targeted therapies: Combined lapatinib and paclitaxel in HER2-positive breast cancer. Nat Rev Clin Oncol Vol. 6, No. 6, (Jun, 2009), pp. 308-309

Citri, A., Kochupurakkal, B. S. & Yarden, Y. (2004). The achilles heel of ErbB-2/HER2: regulation by the Hsp90 chaperone machine and potential for pharmacological intervention. Cell Cycle Vol. 3, No. 1, (Jan, 2004), pp. 51-60

Czekanska, E. M. (2011). Assessment of cell proliferation with resazurin-based fluorescent dye. Methods Mol Biol Vol. 740, No., 2011), pp. 27-32

Chang, T. T., Gulati, S., Chou, T. C., Colvin, M. & Clarkson, B. (1987). Comparative cytotoxicity of various drug combinations for human leukemic cells and normal hematopoietic precursors. Cancer Res Vol. 47, No. 1, (Jan 1, 1987), pp. 119-122

Chou, T. C. (2008). Preclinical versus clinical drug combination studies. Leuk Lymphoma Vol. 49, No. 11, (Nov, 2008), pp. 2059-2080

Chou, T. C. (2010). Drug combination studies and their synergy quantification using the Chou-Talalay method. Cancer Res Vol. 70, No. 2, (Jan 15, 2010), pp. 440-446

Chou, T. C. &Talalay, P. (1984). Quantitative analysis of dose-effect relationships: the combined effects of multiple drugs or enzyme inhibitors. Adv Enzyme Regul Vol. 22, No., 1984), pp. 27-55

Dai, Y., Rahmani, M. & Grant, S. (2003). Proteasome inhibitors potentiate leukemic cell apoptosis induced by the cyclin-dependent kinase inhibitor flavopiridol through a SAPK/JNK- and NF-kappaB-dependent process. Oncogene Vol. 22, No. 46, (Oct 16, 2003), pp. 7108-7122

Dancey, J. E. &Chen, H. X. (2006). Strategies for optimizing combinations of molecularly targeted anticancer agents. Nat Rev Drug Discov Vol. 5, No. 8, (Aug, 2006), pp. 649-659

Darzynkiewicz, Z., Traganos, F., Staiano-Coico, L., Kapuscinski, J. & Melamed, M. R. (1982). Interaction of rhodamine 123 with living cells studied by flow cytometry. Cancer Res Vol. 42, No. 3, (Mar, 1982), pp. 799-806

De Clerck, L. S., Bridts, C. H., Mertens, A. M., Moens, M. M. & Stevens, W. J. (1994). Use of fluorescent dyes in the determination of adherence of human leucocytes to endothelial cells and the effect of fluorochromes on cellular function. J Immunol Methods Vol. 172, No. 1, (Jun 3, 1994), pp. 115-124

Derenzini, M., Brighenti, E., Donati, G., Vici, M., Ceccarelli, C., Santini, D., Taffurelli, M., Montanaro, L. & Trere, D. (2009). The p53-mediated sensitivity of cancer cells to chemotherapeutic agents is conditioned by the status of the retinoblastoma protein. J Pathol Vol. 219, No. 3, (Nov, 2009), pp. 373-382

Dive, C., Gregory, C. D., Phipps, D. J., Evans, D. L., Milner, A. E. & Wyllie, A. H. (1992). Analysis and discrimination of necrosis and apoptosis (programmed cell death) by multiparameter flow cytometry. Biochim Biophys Acta Vol. 1133, No. 3, (Feb 3, 1992), pp. 275-285

Duan, X. F., Wu, Y. L., Xu, H. Z., Zhao, M., Zhuang, H. Y., Wang, X. D., Yan, H. & Chen, G. Q. (2010). Synergistic mitosis-arresting effects of arsenic trioxide and paclitaxel on human malignant lymphocytes. Chem Biol Interact Vol. 183, No. 1, (Jan 5, 2010), pp. 222-230

Elrefaei, M., Baker, C. A., Jones, N. G., Bangsberg, D. R. & Cao, H. (2008). Presence of suppressor HIV-specific CD8+ T cells is associated with increased PD-1 expression on effector CD8+ T cells. J Immunol Vol. 180, No. 11, (Jun 1, 2008), pp. 7757-7763

Eray, M., Matto, M., Kaartinen, M., Andersson, L. & Pelkonen, J. (2001). Flow cytometric analysis of apoptotic subpopulations with a combination of annexin V-FITC, propidium iodide, and SYTO 17. Cytometry Vol. 43, No. 2, (Feb 1, 2001), pp. 134-142

Eriksen, K. W., Sondergaard, H., Woetmann, A., Krejsgaard, T., Skak, K., Geisler, C., Wasik, M. A. & Odum, N. (2009). The combination of IL-21 and IFN-alpha boosts STAT3 activation, cytotoxicity and experimental tumor therapy. Mol Immunol Vol. 46, No. 5, (Feb, 2009), pp. 812-820

Eruslanov, E. &Kusmartsev, S. (2010). Identification of ROS using oxidized DCFDA and flow-cytometry. Methods Mol Biol Vol. 594, No., 2010), pp. 57-72

Facoetti, A., Ranza, E. & Nano, R. (2008). Proliferation and programmed cell death: role of p53 protein in high and low grade astrocytoma. Anticancer Res Vol. 28, No. 1A, (Jan-Feb, 2008), pp. 15-19

Favoni, R. E. &Florio, T. (2011). Combined chemotherapy with cytotoxic and targeted compounds for the management of human malignant pleural mesothelioma. Trends Pharmacol Sci Vol. 32, No. 8, (Aug, 2011), pp. 463-479

Foye, W. O. (1995). Cancer chemotherapeutic agents.American Chemical Society. ISBN (9780841229204-0841229201) Washington, DC

Fuchs, T. A., Abed, U., Goosmann, C., Hurwitz, R., Schulze, I., Wahn, V., Weinrauch, Y., Brinkmann, V. & Zychlinsky, A. (2007). Novel cell death program leads to neutrophil extracellular traps. J Cell Biol Vol. 176, No. 2, (Jan 15, 2007), pp. 231-241

Fujioka, H., Hunt, P. J., Rozga, J., Wu, G. D., Cramer, D. V., Demetriou, A. A. & Moscioni, A. D. (1994). Carboxyfluorescein (CFSE) labelling of hepatocytes for short-term localization following intraportal transplantation. Cell Transplant Vol. 3, No. 5, (Sep-Oct, 1994), pp. 397-408

Goodman, L. S., Brunton, L. L., Chabner, B. & Knollmann, B. C. (2010). Goodman & Gilman's pharmacological basis of therapeutics.12th Edition.McGraw-Hill. ISBN (9780071624428-0071624422) New York

Gozuacik, D. &Kimchi, A. (2007). Autophagy and cell death. Curr Top Dev Biol Vol. 78, No., 2007), pp. 217-245

Gross-Goupil, M. &Escudier, B. (2010). [Targeted therapies: sequential and combined treatments]. Bull Cancer Vol. 97, No., 2010), pp. 65-71

Haase, S. B. (2004). Cell cycle analysis of budding yeast using SYTOX Green. Curr Protoc Cytom Vol. Chapter 7, No., (Nov, 2004), pp. Unit 7 23

Hamilton, D., Loignon, M., Alaoui-Jamali, M. A. & Batist, G. (2007). Novel use of the fluorescent dye 5-(and-6)-chloromethyl SNARF-1 acetate for the measurement of intracellular glutathione in leukemic cells and primary lymphocytes. Cytometry A Vol. 71, No. 9, (Sep, 2007), pp. 709-715

Healy, E., Dempsey, M., Lally, C. & Ryan, M. P. (1998). Apoptosis and necrosis: mechanisms of cell death induced by cyclosporine A in a renal proximal tubular cell line. Kidney Int Vol. 54, No. 6, (Dec, 1998), pp. 1955-1966

Ikeda, H., Taira, N., Nogami, T., Shien, K., Okada, M., Shien, T., Doihara, H. & Miyoshi, S. (2011). Combination treatment with fulvestrant and various cytotoxic agents (doxorubicin, paclitaxel, docetaxel, vinorelbine, and 5-fluorouracil) has a synergistic effect in estrogen receptor-positive breast cancer. Cancer Sci Vol., No., (Jul 30, 2011),

Jones, K. H. &Senft, J. A. (1985). An improved method to determine cell viability by simultaneous staining with fluorescein diacetate-propidium iodide. J Histochem Cytochem Vol. 33, No. 1, (Jan, 1985), pp. 77-79

Kane, R. C., Dagher, R., Farrell, A., Ko, C. W., Sridhara, R., Justice, R. & Pazdur, R. (2007). Bortezomib for the treatment of mantle cell lymphoma. Clin Cancer Res Vol. 13, No. 18 Pt 1, (Sep 15, 2007), pp. 5291-5294

Karaca, B., Atmaca, H., Uzunoglu, S., Karabulut, B., Sanli, U. A. & Uslu, R. (2009). Enhancement of taxane-induced cytotoxicity and apoptosis by gossypol in human breast cancer cell line MCF-7. J Buon Vol. 14, No. 3, (Jul-Sep, 2009), pp. 479-485

Karp, J. E. &Lancet, J. E. (2005). Development of the farnesyltransferase inhibitor tipifarnib for therapy of hematologic malignancies. Future Oncol Vol. 1, No. 6, (Dec, 2005), pp. 719-731

Kim, D., Cheng, G. Z., Lindsley, C. W., Yang, H. & Cheng, J. Q. (2005). Targeting the phosphatidylinositol-3 kinase/Akt pathway for the treatment of cancer. Curr Opin Investig Drugs Vol. 6, No. 12, (Dec, 2005), pp. 1250-1258

Klosowska-Wardega, A., Hasumi, Y., Ahgren, A., Heldin, C. H. & Hellberg, C. (2010). Combination therapy using imatinib and vatalanib improves the therapeutic efficiency of paclitaxel towards a mouse melanoma tumor. Melanoma Res Vol., No., (Oct 21, 2010),

Lantz, R. C., Lemus, R., Lange, R. W. & Karol, M. H. (2001). Rapid reduction of intracellular glutathione in human bronchial epithelial cells exposed to occupational levels of toluene diisocyanate. Toxicol Sci Vol. 60, No. 2, (Apr, 2001), pp. 348-355

Larsen, A. K., Ouaret, D., El Ouadrani, K. & Petitprez, A. (2011). Targeting EGFR and VEGF(R) pathway cross-talk in tumor survival and angiogenesis. Pharmacol Ther Vol. 131, No. 1, (Jul, 2011), pp. 80-90

Lebaron, P., Catala, P. & Parthuisot, N. (1998). Effectiveness of SYTOX Green stain for bacterial viability assessment. Appl Environ Microbiol Vol. 64, No. 7, (Jul, 1998), pp. 2697-2700

Li, X., Dancausse, H., Grijalva, I., Oliveira, M. & Levi, A. D. (2003). Labeling Schwann cells with CFSE-an in vitro and in vivo study. J Neurosci Methods Vol. 125, No. 1-2, (May 30, 2003), pp. 83-91

Marchetti, C., Jouy, N., Leroy-Martin, B., Defossez, A., Formstecher, P. & Marchetti, P. (2004). Comparison of four fluorochromes for the detection of the inner mitochondrial membrane potential in human spermatozoa and their correlation with sperm motility. Hum Reprod Vol. 19, No. 10, (Oct, 2004), pp. 2267-2276

McCall, K. (2010). Genetic control of necrosis - another type of programmed cell death. Curr Opin Cell Biol Vol. 22, No. 6, (Dec, 2010), pp. 882-888

Michiels, S., Potthoff, R. F. & George, S. L. (2011). Multiple testing of treatment-effect-modifying biomarkers in a randomized clinical trial with a survival endpoint. Stat Med Vol. 30, No. 13, (Jun 15, 2011), pp. 1502-1518

Mitsiades, C. S., Davies, F. E., Laubach, J. P., Joshua, D., San Miguel, J., Anderson, K. C. & Richardson, P. G. (2011). Future directions of next-generation novel therapies, combination approaches, and the development of personalized medicine in myeloma. J Clin Oncol Vol. 29, No. 14, (May 10, 2011), pp. 1916-1923

Mueller, M. T., Hermann, P. C., Witthauer, J., Rubio-Viqueira, B., Leicht, S. F., Huber, S., Ellwart, J. W., Mustafa, M., Bartenstein, P., D'Haese, J. G. et al. (2009). Combined targeted treatment to eliminate tumorigenic cancer stem cells in human pancreatic cancer. Gastroenterology Vol. 137, No. 3, (Sep, 2009), pp. 1102-1113

Notte, A., Leclere, L. & Michiels, C. (2011). Autophagy as a mediator of chemotherapy-induced cell death in cancer. Biochem Pharmacol Vol. 82, No. 5, (Sep 1, 2011), pp. 427-434

Olszewska-Slonina, D., Drewa, T., Musialkiewicz, D. & Olszewski, K. (2004). Comparison of viability of B16 and Cl S91 cells in three cytotoxicity tests: cells counting, MTT and flow cytometry after cytostatic drug treatment. Acta Pol Pharm Vol. 61, No. 1, (Jan-Feb, 2004), pp. 31-37

Paglin, S., Lee, N. Y., Nakar, C., Fitzgerald, M., Plotkin, J., Deuel, B., Hackett, N., McMahill, M., Sphicas, E., Lampen, N. et al. (2005). Rapamycin-sensitive pathway regulates mitochondrial membrane potential, autophagy, and survival in irradiated MCF-7 cells. Cancer Res Vol. 65, No. 23, (Dec 1, 2005), pp. 11061-11070

Pallis, M., Syan, J. & Russell, N. H. (1999). Flow cytometric chemosensitivity analysis of blasts from patients with acute myeloblastic leukemia and myelodysplastic syndromes: the use of 7AAD with antibodies to CD45 or CD34. Cytometry Vol. 37, No. 4, (Dec 1, 1999), pp. 308-313

Papadopoulos, N. G., Dedoussis, G. V., Spanakos, G., Gritzapis, A. D., Baxevanis, C. N. & Papamichail, M. (1994). An improved fluorescence assay for the determination of lymphocyte-mediated cytotoxicity using flow cytometry. J Immunol Methods Vol. 177, No. 1-2, (Dec 28, 1994), pp. 101-111

Parish, C. R. &Warren, H. S. (2002). Use of the intracellular fluorescent dye CFSE to monitor lymphocyte migration and proliferation. Curr Protoc Immunol Vol. Chapter 4, No., (Aug, 2002), pp. Unit 4 9

Pegram, M., Hsu, S., Lewis, G., Pietras, R., Beryt, M., Sliwkowski, M., Coombs, D., Baly, D., Kabbinavar, F. & Slamon, D. (1999). Inhibitory effects of combinations of HER-2/neu antibody and chemotherapeutic agents used for treatment of human breast cancers. Oncogene Vol. 18, No. 13, (Apr 1, 1999), pp. 2241-2251

Pei, X. Y., Dai, Y. & Grant, S. (2004). Synergistic induction of oxidative injury and apoptosis in human multiple myeloma cells by the proteasome inhibitor bortezomib and histone deacetylase inhibitors. Clin Cancer Res Vol. 10, No. 11, (Jun 1, 2004), pp. 3839-3852

Perez-Galan, P., Roue, G., Lopez-Guerra, M., Nguyen, M., Villamor, N., Montserrat, E., Shore, G. C., Campo, E. & Colomer, D. (2008). BCL-2 phosphorylation modulates sensitivity to the BH3 mimetic GX15-070 (Obatoclax) and reduces its synergistic interaction with bortezomib in chronic lymphocytic leukemia cells. Leukemia Vol. 22, No. 9, (Sep, 2008), pp. 1712-1720

Pheng, S., Chakrabarti, S. & Lamontagne, L. (2000). Dose-dependent apoptosis induced by low concentrations of methylmercury in murine splenic Fas+ T cell subsets. Toxicology Vol. 149, No. 2-3, (Aug 21, 2000), pp. 115-128

Platini, F., Perez-Tomas, R., Ambrosio, S. & Tessitore, L. (2010). Understanding autophagy in cell death control. Curr Pharm Des Vol. 16, No. 1, (Jan, 2010), pp. 101-113

Poole, C. A., Brookes, N. H., Gilbert, R. T., Beaumont, B. W., Crowther, A., Scott, L. & Merrilees, M. J. (1996). Detection of viable and non-viable cells in connective tissue explants using the fixable fluoroprobes 5-chloromethylfluorescein diacetate and ethidium homodimer-1. Connect Tissue Res Vol. 33, No. 4, 1996), pp. 233-241

Poot, M., Gibson, L. L. & Singer, V. L. (1997). Detection of apoptosis in live cells by MitoTracker red CMXRos and SYTO dye flow cytometry. Cytometry Vol. 27, No. 4, (Apr 1, 1997), pp. 358-364

Prosperi, E., Croce, A. C., Bottiroli, G. & Supino, R. (1986). Flow cytometric analysis of membrane permeability properties influencing intracellular accumulation and efflux of fluorescein. Cytometry Vol. 7, No. 1, (Jan, 1986), pp. 70-75

Prowse, A. B., Wilson, J., Osborne, G. W., Gray, P. P. & Wolvetang, E. J. (2009). Multiplexed staining of live human embryonic stem cells for flow cytometric analysis of pluripotency markers. Stem Cells Dev Vol. 18, No. 8, (Oct, 2009), pp. 1135-1140

Radcliff, G., Waite, R., LeFevre, J., Poulik, M. D. & Callewaert, D. M. (1991). Quantification of effector/target conjugation involving natural killer (NK) or lymphokine activated killer (LAK) cells by two-color flow cytometry. J Immunol Methods Vol. 139, No. 2, (Jun 3, 1991), pp. 281-292

Ramachandran, C., Resek, A. P., Escalon, E., Aviram, A. & Melnick, S. J. (2010). Potentiation of gemcitabine by Turmeric Force in pancreatic cancer cell lines. Oncol Rep Vol. 23, No. 6, (Jun, 2010), pp. 1529-1535

Ross, D. D., Joneckis, C. C., Ordonez, J. V., Sisk, A. M., Wu, R. K., Hamburger, A. W. N. R. E. & Nora, R. E. (1989). Estimation of cell survival by flow cytometric quantification of fluorescein diacetate/propidium iodide viable cell number. Cancer Res Vol. 49, No. 14, (Jul 15, 1989), pp. 3776-3782

Sarkar, A., Mandal, G., Singh, N., Sundar, S. & Chatterjee, M. (2009). Flow cytometric determination of intracellular non-protein thiols in Leishmania promastigotes using 5-chloromethyl fluorescein diacetate. Exp Parasitol Vol. 122, No. 4, (Aug, 2009), pp. 299-305

Sebastia, J., Cristofol, R., Martin, M., Rodriguez-Farre, E. & Sanfeliu, C. (2003). Evaluation of fluorescent dyes for measuring intracellular glutathione content in primary cultures of human neurons and neuroblastoma SH-SY5Y. Cytometry A Vol. 51, No. 1, (Jan, 2003), pp. 16-25

Shapiro, H. M. (2003). Practical flow cytometry. 4th Edition. Wiley-Liss. ISBN (9780471411253-0471411256) Hoboken, N.J.

Shuck, S. C. & Turchi, J. J. (2010). Targeted inhibition of Replication Protein A reveals cytotoxic activity, synergy with chemotherapeutic DNA-damaging agents, and insight into cellular function. Cancer Res Vol. 70, No. 8, (Apr 15, 2010), pp. 3189-3198

Sims, J. T. & Plattner, R. (2009). MTT assays cannot be utilized to study the effects of STI571/Gleevec on the viability of solid tumor cell lines. Cancer Chemother Pharmacol Vol. 64, No. 3, (Aug, 2009), pp. 629-633

Sugiyama, K., Shimizu, M., Akiyama, T., Ishida, H., Okabe, M., Tamaoki, T. & Akinaga, S. (1998). Combined effect of navelbine with medroxyprogesterone acetate against human breast carcinoma MCF-7 cells in vitro. Br J Cancer Vol. 77, No. 11, (Jun, 1998), pp. 1737-1743

Talbot, J. D., Barrett, J. N., Barrett, E. F. & David, G. (2008). Rapid, stimulation-induced reduction of C12-resorufin in motor nerve terminals: linkage to mitochondrial metabolism. J Neurochem Vol. 105, No. 3, (May, 2008), pp. 807-819

Trudel, S., Li, Z. H., Rauw, J., Tiedemann, R. E., Wen, X. Y. & Stewart, A. K. (2007). Preclinical studies of the pan-Bcl inhibitor obatoclax (GX015-070) in multiple myeloma. Blood Vol. 109, No. 12, (Jun 15, 2007), pp. 5430-5438

Ullal, A. J., Pisetsky, D. S. & Reich, C. F., 3rd. (2010). Use of SYTO 13, a fluorescent dye binding nucleic acids, for the detection of microparticles in in vitro systems. Cytometry A Vol. 77, No. 3, (Mar, 2010), pp. 294-301

Vega, M. I., Martinez-Paniagua, M., Huerta-Yepez, S., Gonzalez-Bonilla, C., Uematsu, N. & Bonavida, B. (2009). Dysregulation of the cell survival/anti-apoptotic NF-kappaB pathway by the novel humanized BM-ca anti-CD20 mAb: implication in chemosensitization. Int J Oncol Vol. 35, No. 6, (Dec, 2009), pp. 1289-1296

Wesierska-Gadek, J., Gueorguieva, M., Ranftler, C. & Zerza-Schnitzhofer, G. (2005). A new multiplex assay allowing simultaneous detection of the inhibition of cell proliferation and induction of cell death. J Cell Biochem Vol. 96, No. 1, (Sep 1, 2005), pp. 1-7

West, C. A., He, C., Su, M., Swanson, S. J. & Mentzer, S. J. (2001). Aldehyde fixation of thiol-reactive fluorescent cytoplasmic probes for tracking cell migration. J Histochem Cytochem Vol. 49, No. 4, (Apr, 2001), pp. 511-518

Wlodkowic, D. & Skommer, J. (2007). SYTO probes: markers of apoptotic cell demise. Curr Protoc Cytom Vol. Chapter 7, No., (Oct, 2007), pp. Unit7 33

Woodcock, J., Griffin, J. P. & Behrman, R. E. (2011). Development of novel combination therapies. N Engl J Med Vol. 364, No. 11, (Mar 17, 2011), pp. 985-987

Workman, P., Burrows, F., Neckers, L. & Rosen, N. (2007). Drugging the cancer chaperone HSP90: combinatorial therapeutic exploitation of oncogene addiction and tumor stress. Ann N Y Acad Sci Vol. 1113, No., (Oct, 2007), pp. 202-216

Wright, J. J. (2010). Combination therapy of bortezomib with novel targeted agents: an emerging treatment strategy. Clin Cancer Res Vol. 16, No. 16, (Aug 15, 2010), pp. 4094-4104

Yan, X., Habbersett, R. C., Yoshida, T. M., Nolan, J. P., Jett, J. H. & Marrone, B. L. (2005). Probing the kinetics of SYTOX Orange stain binding to double-stranded DNA with implications for DNA analysis. Anal Chem Vol. 77, No. 11, (Jun 1, 2005), pp. 3554-3562

Yanamandra, N., Colaco, N. M., Parquet, N. A., Buzzeo, R. W., Boulware, D., Wright, G., Perez, L. E., Dalton, W. S. & Beaupre, D. M. (2006). Tipifarnib and bortezomib are synergistic and overcome cell adhesion-mediated drug resistance in multiple myeloma and acute myeloid leukemia. Clin Cancer Res Vol. 12, No. 2, (Jan 15, 2006), pp. 591-599

Yeh, C. J., Hsi, B. L. & Faulk, W. P. (1981). Propidium iodide as a nuclear marker in immunofluorescence. II. Use with cellular identification and viability studies. J Immunol Methods Vol. 43, No. 3, 1981), pp. 269-275

Zahorowska, B., Crowe, P. J. & Yang, J. L. (2009). Combined therapies for cancer: a review of EGFR-targeted monotherapy and combination treatment with other drugs. J Cancer Res Clin Oncol Vol. 135, No. 9, (Sep, 2009), pp. 1137-1148

Zamai, L., Canonico, B., Luchetti, F., Ferri, P., Melloni, E., Guidotti, L., Cappellini, A., Cutroneo, G., Vitale, M. & Papa, S. (2001). Supravital exposure to propidium iodide identifies apoptosis on adherent cells. Cytometry Vol. 44, No. 1, (May 1, 2001), pp. 57-64

Zanetti, M., d'Uscio, L. V., Peterson, T. E., Katusic, Z. S. & O'Brien, T. (2005). Analysis of superoxide anion production in tissue. Methods Mol Med Vol. 108, No., 2005), pp. 65-72

Zhang, N., Wu, Z. M., McGowan, E., Shi, J., Hong, Z. B., Ding, C. W., Xia, P. & Di, W. (2009). Arsenic trioxide and cisplatin synergism increase cytotoxicity in human ovarian cancer cells: therapeutic potential for ovarian cancer. Cancer Sci Vol. 100, No. 12, (Dec, 2009), pp. 2459-2464

8

New Insights into Cell Encapsulation and the Role of Proteins During Flow Cytometry

Sinéad B. Doherty and A. Brodkorb

Teagasc Food Research Centre Moorepark, Fermoy, Co. Cork,
Ireland

1. Introduction

Modern approaches to science tend to follow divergent paths. On one hand, instruments and technologies are developed to capture as much information as possible with the need for complex data analysis to identify problematic issues. On the other hand, formulation focused, minimalistic approaches that gather only the most pertinent data for specific questions also represent a powerful methodology. This chapter will provide many examples of the latter by integrating Flow Cytometry (FACS - Fluorescence-Activated Cell Sorting) technology with high throughput screening (HTS) of encapsulation systems with extensive utility of one-dimensional (1-D) imaging for protein localisation. In this regard, less information is acquired from each cell, data files will be more manageable, easier to analyse and throughput screening will be significantly enhanced beyond traditional HTS analysis, irrespective of the protein concentration present in the background or delivery media.

1.1 Real-time on-line monitoring of bioprocesses

The production of heterologous therapeutic proteins is well established in today's biotechnology industry; however their presence during cytometric screening poses complex analytical obstacles for food and biotechnologists alike. The Process Analytical Technology (PAT) initiative, launched by the Food and Drug Administration (FDA) encourages extensive process understanding to achieve the desired quality of pharmaceutical and bioactive protein products rather than a 'quality-by-testing' approach. Thus, elucidating and monitoring variations within protein production, purification and encapsulation systems is fundamental for quality control, protein detection and localisation (Glassey et al., 2011). One key issue for PAT is the use of on-line process analysers; in this context, 'on-line' is used in the sense of 'fully automatic' or 'without any manual interaction'.

The foundation of early microscopy and its transition to the modern FACS nowadays, proved that cell growth within a population differs significantly from one cell to another. These heterogeneities are caused by either fluctuations in the micro-environment of individual cells (Dunlop and Ye, 1990); phenotypic changes during the cell cycle (Münch et al., 1992); or by mutations resulting in genotypic variations in the population (Hall, 1995) leading to different protein expression levels. Thus, the most important variable in a bioprocess - cell productivity of a heterologous protein product - is distributed over a wide range within the cell population. Protein analytical methods (i.e. SDS-PAGE, Western blot,

MALD-TOF-MS) are off-line, time and cost intensive, require human interaction and are not capable of exhibiting protein expression levels in single cell level. Earlier studies demonstrated that tagging the target protein with a fluorescent reporter molecule (green fluorescent protein GFP) permits detection of low protein concentrations by measuring the reporter molecule's specific fluorescence (Broger et al., 2011). Specific protein fluorescence signals can be measured through on-line, *in situ* fluorescence sensors (Jones et al., 2004, Reischer et al., 2004); but the main drawback of these measuring techniques is that they all result in population averaged data; hence analysis of single cell productivity was not possible. As mentioned FACS is one of the available methods for measurement of population distribution and has been exploited in biotechnology (Rieseberg et al., 2001). To overcome this disadvantage of FACS as an off-line analyser, a number of flow injection (FI) flow cytometer systems have been designed for the determination of proteins (Kelley, 1989). Ruzicka and Lindberg (1992) were the first to utilise FI as an interface between a bioreactor and FACS. Further progress in FI-FACS has been made by Srienc and co-workers (Zhao et al., 1999), examining population dynamics of *Saccharomyces cerevisiae* (Kacmar and Srienc, 2005) and growth dynamics of Chinese Hamster Ovary (CHO) cells and their associated proteins (Sitton and Srienc, 2008a, b). None of them, however, measure single-cell productivity using fluorescent reporter tag co-expressed with target protein, which restricts the purification of secreted protein products.

1.2 Efficacious phenotypic analyses of cell encapsulation systems

Cellular detection and characterization usually involves cell harvesting for more than 20 generations in order to achieve the formation of macro-colonies on solid media. What is more, encapsulation provides an innovative solution for the handling and use of live cells through the use of three-dimensional scaffolds. These scaffolds - commonly known as beads or capsules - can be used in a number of ways, such as for the isolation of individual cells to form clonal populations, for establishing partial barrier for cells from their environment, and for creating a matrix allowing the formation of 3D-cell cultures or clusters. These methods have important implications for the biomedical engineering field and the areas of biomanufacturing and bioprocessing. Nonetheless, a myriad of information remains to be discovered. For example, the chemical composition and engineered functionalities of these scaffolds affect the biology of the encapsulated cells. One particular system that has recently achieved significant recognition are the dairy proteins. A variety of cell types can be enclosed in casein or whey beads scaffolds in a range of sizes. While maintaining their morphology and function, these encapsulated cells can proliferate, form cell clusters, and even lay down extracellular matrix components. Furthermore, cells can be encapsulated in small particles that can then be handled, characterized and analysed with ease. These features make milk proteins an attractive encapsulation material for three dimensional scaffolds of live cells. Meanwhile, HTS of these encapsulated cells is restricted by several practical aspects including low sample throughput, protein matrix limitations and the absence of sorting capability for mixed cell cultures. Hence, a unique reconsideration of conventional cytometric approaches for accelerated food and/ or bioprocess intensification represents a timely research strategy for encapsulation technologists and cytometrists alike. It was recently discovered (Delgado et al., 2010) that microbial growth can be monitored in encapsulated cell systems by following the development of micro-colonies. Therefore, development and dissemination of a FACS technology capable of analysing encapsulated cells within intact micro-capsules encapsulation systems would represent a novel cytometric strategy for academic and industrial applications. This approach would require knowledge management of cytometric control parameters, which would generate reduced capital cost, human error and environmental deterrents associated with conventionally high-throughput screening strategies. However, there is a need for addressing physico-chemical properties of micro-capsules for efficient cytometric screening without i) occlusion of capsular flow in the fluidics system during hydrodynamic focusing or ii) adverse interactions between protein (originating from the micro-capsule or cell metabolic activity) and cytometric buffer media during the screening process.

Delgado et al., (2010) demonstrated that cell proliferation within encapsulation systems can be detected via FACS, provided that the population of capsules exhibit appropriate optical and mechanical properties and are mono-dispersed in size and shape. Encapsulated cells can be further utilized for a variety of applications: from characterizing secreted enzymes to detection of thermo-sensitive mutants. Delgado et al., (2010) successfully revealed the application of Flow Focusing® technology for microencapsulated cells of different types in mono-disperse hydrogel microspheres. Using a CellENA® Flow Focusing® microencapsulator, monodisperse capsules were fabricated containing one single cell with sizes ranging from

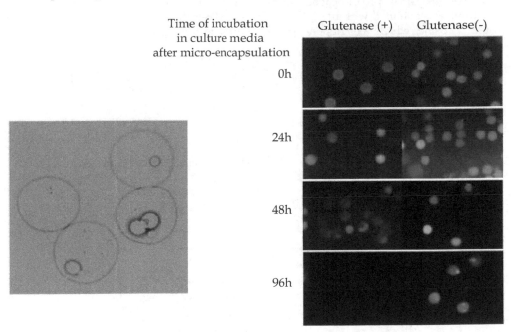

Source: (Delgado et al., 2011). http://www.cellena.net/en/documents/CellenaPoster.pdf

Fig. 1. Data detecting glutenases in micro-encapsulated bacteria. Colonies of bacteria were grown in gliadin-containing capsules. Images on the right illustrate the change in fluorescent intensity as a function of incubation time in culture media after micro-encapsulation. Gliadin content was detected by incubating the particles with the monoclonal antibody G12, conjugated to FITC. The micrograph on the left illustrated colonies of bacteria growing gliadin-containing micorparticles (Delgado et al., 2011)

100 μm to over 600 μm diameter. This offers a plethora of applications including the characterization of secreted therapeutic proteins to detect encapsulated cells or thermo-sensitive mutants. More importantly, cell proliferation inside the micro-capsules was detected by FACS without the need for fluorescent labelling, which represents a significant development (Delgado et al., 2011). Furthermore, bacteria expressing glutenase activity, isolated from agrochemical samples were detected by their ability to degrade gliadin when growing inside capsules as shown in Figure 1. Gliadin content was detected by incubating the particles with the monoclonal antibody G12, conjugated to FITC (Delgado et al., 2010). Further information relating to COPAS analysis can be obtained from Union Biometrica (http://www.unionbio.com/).

Figure 2 illustrates capsules containing yeast colonies incubated for different times. Data revealed that capsules had a similar size (Time of Flight; TOF) but differed in optical complexity (Extinction; EXT). Hence, different yeast concentrations were predicted as a function of time. This Flow Focusing® Technology from Union Biometrica, Belgium, will inevitably bring additional utility and commercial recognition to FACS as an economically viable biochemical application for improved analytical monitoring of downstream processing during fermentation and drug/ protein generation. Figure 3 illustrates the successful analysis of protein capsules based on size (TOF) and extinction (EXT). The best resolution was obtained with the 561nm solid state laser due to higher penetration into the protein capsule, which will depend upon membrane thickness, coating or presence of a

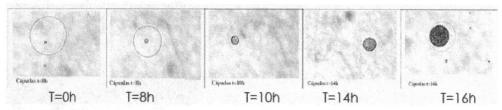

Source: (Delgado et al., 2011). http://www.cellena.net/en/documents/CellenaPoster.pdf

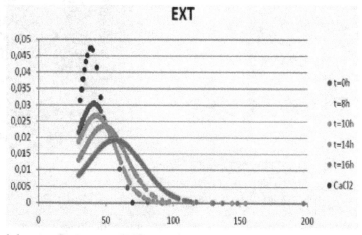

Fig. 2. Optical density (Extinction; EXT; x-axis) measurements of encapsulated yeast following the time points indicated (above, the images of the encapsulated yeast)

surface cross-linker. It was also demonstrated that maximum peak height can be used as a sorting parameter for cytometric sorting of encapsulated cells in protein capsules. Further research is currently being conducted in this promising area of cytometric analysis.

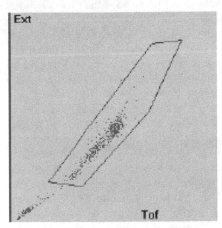

Fig. 3. Displays the dot plots TOF vs. EXT for empty protein capsules. TOF is an indicator of size, while EXT represents the density of the detected object

These developments may aid bioprocessing technology, improve process development and analytical control by sorting cells based on i) micro-capsule size; ii) cell type; iii) cell density; iv) monoclonality and v) protein expression. In this way, fluorescent tagging is not an essential requirement for the characterisation of encapsulated cell cultures and their respective recombinant proteins, which demonstrates a significant development for cost- and time-efficient cytometric screening. Moreover, encapsulation evaluation using Flow Focusing® Technology, can be used for phenotypic analysis of encapsulated cells, such as the expression of specific characteristics of microbial colonies (glutenase expression by bacteria) or growth-related phenotypes (antibiotic resistance and proliferation in thermo-sensitive yeast). Mammalian cells are delicate, difficult to handle and manipulate individually. However, these cells can be encapsulated within a polymer material such as protein or alginate, and these encapsulated scaffolds remain intact throughout all steps of cytometric analysis and dispensing. Furthermore, fragile cells like adipocyte stem cells remain viable for extended periods of time (2 weeks) (Delgado et al., 2010) and can be released and recovered to permit further analysis and use, which endorses an exclusive cytometric development.

1.3 High throughput processing of encapsulated bacterial libraries

Industrial interest in encapsulation has grown exponentially in the fields of bioprocessing, fermentation and elucidation of bacterial libraries, especially in high-throughput environments exceeding 10^6 samples per day. Fundamental pre-requisites reveal the necessity for a one-to-one relationship between individual cells and analytical algorithms. Essentially, each micro-carrier (i.e. capsule) would therefore contain exactly one cell or colony. However, synthesis of larger numbers of capsules containing exactly one cell is not feasible as cells are randomly distributed during capsule production. The problem is clear -

high dilution conditions will yield an adequate degree of monoclonality; however, this will be coupled with the generation of a significant fraction of empty micro-capsules. Conversely, distribution under low dilution conditions will generate unacceptable numbers of polyclonal capsules for whose removal no satisfactory technologies exists to date. Recent finding demonstrated (Walser et al., 2008, 2009) that hydrogel micro-carriers can be applied as growth milieu for individual cell colonies. *Escherichia coli* cells expressing green fluorescent protein (GFP) were encapsulated at low dilution thereby intentionally producing a considerable amount of polyclonal micro-carriers. Empty and polyclonal micro-carriers were then removed from the desired monoclonal fraction using a particle analyzer. Data was compared to model predictions in order to investigate possible limitations in the analysis and sorting of monoclonal micro-carriers. Fluorescent *E. coli* cells (GFP) were randomly distributed throughout the micro-carrier population and cells successfully propagated to colonies in the micro-carrier with enrichment to 95% monoclonality. Interestingly, colony diameter represents a limiting factor for enrichment-efficiency in encapsulation systems. With increasing colony size, two antagonistic effects are associated with the cytometric approach: First, improved sorting efficiency due to increased fluorescence intensity and thus higher detection efficiency, and second, deterioration of sorting efficiency due to occlusion occurring in polyclonal micro-capsules. Hence, encapsulation under low dilution conditions with high-throughput sorting via FACS represents a practical economically viable initiative for isolating large quantities of monoclonal micro-capsules from bacterial libraries and at the same time, keeping the amounts of empty micro-capsules at a moderate level.

1.4 Utilization of micro-capsules for next generation cytometric assays

It is evident that cytometric screening is a pragmatic approach for integration of cell encapsulation within continuous bioprocesses. Subsequent segregation and sorting of microcapsules containing high cell densities of animal cells and associated proteins has recently catalysed interest in cytometric screening. Research performed by Union Biometrica Inc. (www.unionbiometrica.com) applied this technology and found COPAS™ (Complex Object Parametric Analyzer and Sorter) suitable for automated analysis, sorting, and dispensing of 'large' objects such as capsules. The COPAS™ PLUS is capable of analysing particles with diameters of 30 - 800 µm in a continuously flowing stream at a rate of 25-50 objects/second. Physical properties such as size, optical density and intensity of fluorescent markers are taken into consideration. To avoid damaging or changing the fragile biological samples, a gentle pneumatic device located after the flow cell is used for sorting encapsulated cells and therefore makes the instrument suitable for handling live biological materials or sensitive proteins (Figure 4). It is interesting to note that the fluid pressures of the instrument are also significantly lower than those of traditional flow cytometers. The COPAS™ XL instrument has a 2,000 µm flow cell, which allows the analysis of larger beads (30-1,500 µm) compared to the COPAS™ PLUS. If an encapsulated sample contains certain fluorophores that can be excited by light of 488 nm, the emission levels can be detected for each of the objects in the encapsulated system.

Figure 4 details the analysis of intact encapsulation capsules inside the flow cells. Objects are carried through the flow cell by a liquid stream while their physical properties are being measured. Convergence of the sheath and sample fluid allows "hydrodynamic focusing" of the micro-capsules, forcing them to go through the centre of the flow cell along their

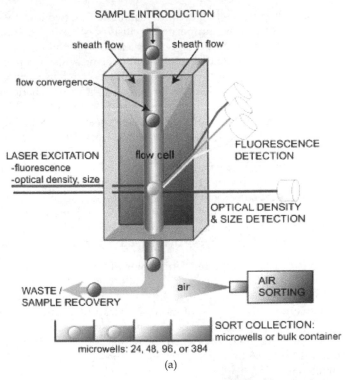

(a)

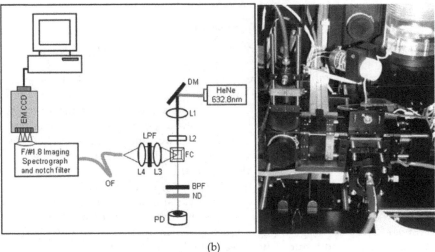

(b)

Fig. 4. Schematic of Flow Focus Technology™ shown in (a). Figure (b) illustrates a schematic (left) and an image (right) of the spectral detection leg constructed on the open face of the COPAS™ flow cell (Watson et al., 2009). DM: dichroic mirror; L1: 80 mm spherical lens; L2: 6 mm cylindrical lens; FC: flow cell; L3, L4: 21 mm aspherical lenses; OF: optical fiber; ND: 0.8 OD neutral density filter

longitudinal axes. Inside the flow cell objects are illuminated by two low energy lasers that measure the microcapsule optical properties inclusive of size, optical density and fluorescence. Micro-capsules that meet the sort criteria are permitted to drop into sort collection, while those that do not are diverted to waste recovery using a pneumatic sorting device. Cells may be classified according to their fluorescent intensity or optical density, which will accommodate a greater efficiency for detection of recombinant proteins during encapsulation procedures. Preliminary studies performed using COPAS™ instrumentation has demonstrated the discrimination of hydrogel microcapsules containing mixed cell cultures. During initial sampling, micro-capsules were selected based on their size (TOF), monoclonality and optical density (EXT). Preparation of a standard curve using micro-capsules of known size and TOF measurements provided a practical correlation between TOF measurements generated by the flow cytometer and the actual size. Hence, it is plausible to select micro-capsules that meet specific size and optical density criteria. Furthermore, encapsulated cells can also be chosen to meet criteria of size, optical density and levels of green and red fluorescence. Hence, it is also possible to sort micro-capsules with various magnitudes of cell growth, cell density and recombinant protein content.

Watson et al., (2009) successfully adapted Union Biometrica COPAS™ Plus instrument to allow red excitation and optical fiber-based light collection and spectral analysis using a spectrograph and CCD array detector (Figure 4B). These modifications did not compromise the ability of the instrument to resolve different sized capsules based on their extinction (EXT) and time of flight (TOF) signals. The modified instrument has the sensitivity and spectral resolution to measure the fluorescence and Raman signals from individual particles with signal integration times of 10 usecond. The high speed spectral analysis of individual particles in flow will enable new applications in biological encapsulation systems

2. High content screening – getting more for less

High content screening (HCS) can identify proteins via automated image analysis and, in general, is designed to capture image information regarding cells and associated protein products. The availability of commercial imaging systems has made HCS increasingly practical for protein determination and has placed HCS into a standard tool for protein and drug discovery (Zanella et al., 2010). The value of this protein determination approach is apparent since a vast number of cellular characteristics can be captured in large data files. Over the past decade advances in information technology and biological probes have led to practically automated image analysis with throughput far above that possible by manual cell analysis in protein-rich environments. Several commercial instruments are now available for HCS of proteins, for example the *BD Pathway* (BD Biosciences), *In cell Analyzer* (GE healthcare), *ImageXpress Ultra* (Molecular Devices) and *Scan^R* (Olympus), to name a few. The *Imagestream* (Amnis) is a relatively new instrument that combines aspects of FACS with image analysis, resulting in a device capable of imaging cells at multiple fluorescent wavelengths with a throughput of hundreds of cells per second (Reardon et al., 2009). In HCS devices, the presence of environmental protein presents trade-offs between the quantity (and quality) of data acquired and throughput. Hence, environmental protein is a hindrance to the efficacy of cytometric screening. The *Imagestream*, for example, trades ultrahigh image resolution for speed, and many cellular images are captured in one data file, the size of which can easily exceed 3 gigabytes.

3. Cytometric limitations associated with environmental protein

Contemporary approaches to food fermentations and drug discoveries frequently use HTS to measure the fluorescence of cellular features tagged with fluorochrome-conjugated antibodies or other fluorescent labels in a very rapid manner. This is quite often performed by FACS, a technique well-suited to this purpose. However, tagged cells are often entrapped within a dense protein matrix or lattice, which represents an obstacle against true microbial counts. Hence, applications can be significantly hindered by the presence of proteins since environmental protein particles, matching the size range of cells, are often recognised as cellular bodies during hydrodynamic focusing. An appropriate methodology is therefore required to eliminate interfering environmental protein - non-cellular in origin - from cytometric analysis for successful application of cytometric screening. Traditional FACS can easily capture emissions from ten or more different fluorochromes on a cell at rates exceeding 25,000 cells per second. Although this is appropriate for many assays, FACS cannot provide accurate details on cell morphology or subcellular localisation of fluorescence in the presence of obtrusive environment protein. In the interim, manual microscope examination of fluorescent staining can provide information for these characteristics. However, this approach can be too subjective and lowers throughput substantially. Hence, environmental protein originating from e.g. encapsulation matrices or carrier media may exhibit adverse effects for accurate cytometric analysis.

4. Flow Cytometry for cell viability assessment in complex protein matrices

Proof of principle that product quality can be assessed within complex systems using FACS may raise awareness and further develop cytometric technologies within the industrial domains of quality management and product /process optimization. Rapid and efficient viability assessment is essential for regulation and legislation on therapeutic bioactives or drug product quality. Hence, there is a clear need for real-time cell enumeration techniques. Data procurement in minutes rather than days may identify a problem faster than usual. Within the food industry, FACS represents a major development for high throughput screening (HTS) of high cell density cultures in addition to providing real-time quality control. Numerous technologies, inclusive of FACS have been developed to accelerate data acquisition compared to traditional culture methods. In the food industry, techniques utilizing antibodies for cell labelling prior to FACS analysis have been developed for *Salmonella enterica* serovar Typhimurium and *Listeria monocytogenes* in milk and other dairy products (Patchett et al., 1991). Moreover, Doherty et al., (2010) demonstrated that FACS, coupled with fluorescent techniques, can be successfully applied for the assessment of cell viability in seven different protein matrices ranging in structural complexity. Food containing complex protein matrices can frequently generate unpredictable results regarding cell viability. Cell entrapment within protein networks can severely affect accurate cell enumeration, an issue which requires special attention as it has an impact on both quality and safety of the product. Cell viability can be accurately determined by FACS, specialised for cell encapsulation and protein systems. The distinctive features of this strategy can be summarised as follows: while Delgado et al., (2010) performed cytometric screening on encapsulated cells within intact micro-carriers, Doherty et al., (2010) enumerated cell viability following complete protein matrix digestion. Cell extraction and digestive pre-treatments were designed to liberate cells from the scaffold in order to minimise the protein background, the predominant compositional obstacles for efficient FACS analysis. It is interesting to compare the success of these reverse strategies for use in encapsulation systems. Cell extraction by Doherty et al., (2010) required 40-minute sample preparation and distinct functional cell populations were discriminated based on fluorescent labelling by of Thiazole Orange (TO) and Propidium Iodide (PI). This assay yielded 45–50 samples/hour, a detection range of 10^2–10^{10} cfu mL^{-1} of homogenate and generated correlation coefficients (r) of 0.95, 0.92 and 0.93 in relation to standard plate counts during heat, acid and storage trials, respectively. This cytometric approach could also alleviate problems relating to environmental compatibility during the production of nutraceutical products; a formulation problem generating a strong current of industrial activity. However, uptake of this technology is dependent on cost-efficiency and the scope for extension of product applications. Both of these pre-requisites are satisfactory for food and pharmaceutical manufacture environments due to the i) multi-disciplinary function of the assay for cell viability assessment; ii) minimal personnel training required for instrument commission and iii) rapid, reproducible cytometric signature responses in a variety of encapsulation matrices in the presence of protein. The timely availability of cytometric results also provides manufacturers with the necessary skills to promote problem-solving investigations in bioactive and/ drug development for enhanced performance of therapeutic cultures with subsequent detection and utilisation of recombinant proteins.

4.1 Significance of a protein clearing procedure

Flow Cytometry is commonly associated with inaccuracies due to its basic operating principle involving 'hydrodynamic focusing' whereby each particle, either cell aggregate or single bacterium, is counted as one cell (Maukonen et al., 2006). Hence, fluorescent techniques like FACS are not universal and successful application necessitates detailed tailoring of pre-treatments and buffer compositions for cell lines and product types. Previous research (Bunthof and Abee, 2002, Gunasekera et al., 2000) failed to generate a procedure capable of consistent cellular discrimination within a diverse range of protein environments inclusive of dairy scaffolds, clinical protein supplements and encapsulation polymers. This challenge is associated with the fluctuating proteolysis response of various protein environments due to differing structural orientation and protein complexity. Doherty et al., (2010) optimised sample pre-treatments (Figure 5) buffer composition and probe concentrations in order accurately detect live, injured and dead cells within encapsulation systems. Moreover, this strategy liberated cells from encapsulation and protein networks without any adverse cellular injury as visualised in Figure 6.

The assay advocates rapid, reproducible cell liberation compared to lengthy extraction procedures previously reported for immobilized systems (Sun and Griffiths, 2000). Since cells are in the micron range, the signal-to-noise ratio in FACS analysis is of paramount importance especially in clinical protein supplements, which are normally viscous in nature. More importantly, encapsulation polymers and protein matrices will inevitably generate increased particle scatter due to high concentrations of colloidal particles in the cell size range. It is evident that protein clearing strategy is fundamental for the achievement of reproducible reliable cytometric viability screening in dense environmental milieu commonly encountered in drug and cell delivery models.

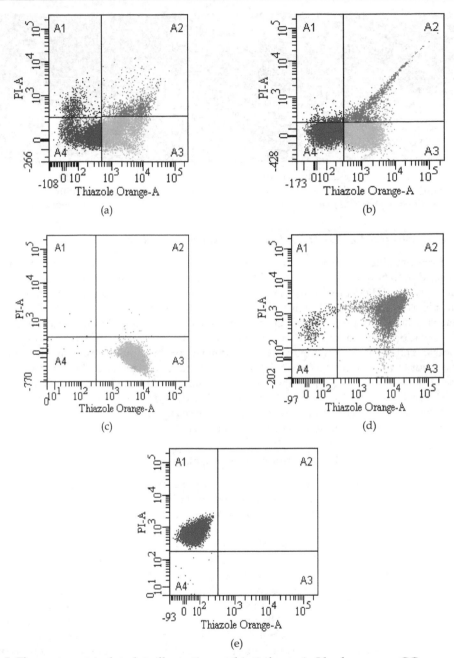

Fig. 5. Flow cytometric dot plots illustrating probiotic bacteria Lb. rhamnosus GG encapsulated in protein gel matrix and pre-sample extraction (a), post-homogenization (b) and after proteolysis digestion of the protein capsule (c). Results also demonstrated the clear distinction between live (c), injured (d) and dead probiotic cells (e)

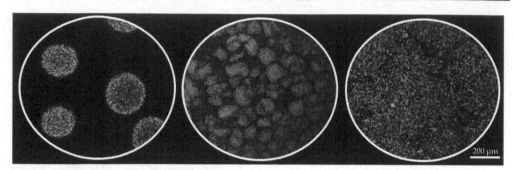

Fig. 6. Confocal image visualising the release of live probiotic bacteria from intact encapsulation scaffolds (left) and their progressive digestion during the protein clearing procedure (left- right)

4.2 Cell release mechanism for enhanced cytometric screening

Current research illustrates a poor correlation between standard plate counts and cytometric screening due to non-specific binding of fluorescent dyes to protein particles (Gunasekera et al., 2000), which failed to be adequately removed by enzymatic treatments or commercial protein-clearing agents (McClelland and Pinder, 1994a, b). Meanwhile, Doherty et al., (2010) further resolved the negative effect of environmental proteins by introducing a mild homogenisation step in order to break-down cellular chains for the provision of true cytometric cell counts. Essentially, this two-stage pre-treatment substantially reduced particle counts or cytometric 'events' that were similar in size to, or larger than typical cellular dimensions i.e. 1 to 5 µm (Figure 7). Interestingly, further physico-chemical analysis e.g. zeta potential and hydrophobicity, provided details of charge interactions within protein-cell systems. This procedure allows to release of clean cell populations from complex or encapsulated protein matrices (Doherty et al., 2010).

4.3 Optimum compensation and fluorescent staining

Spectral overlap of the different fluorochromes used during FACS, if uncorrected, will lead to misinterpretation of data from false positives or artifactual populations. Compensation values in excess of 90% have been reported (Gunasekera et al., 2000, Gunasekera et al., 2003), which can be substantially reduced by pre-treatment to values as low as 28%, which may alleviate distortion of true viable counts. Pre-treatment of the preparations can influence fluorescence compensation values via optimization of instrument settings and equilibration of overlapping spectral channels. Despite dense protein environments in some products, the regression between FACS and plate counts can closely match the guidelines of Feldsine et al., (2002) for validation of qualitative and quantitative microbiological methods. Since all matrix material in the size range of cells can be adequately digested, cell overestimation as a result of matrix staining during FACS analysis was reduced. However, these findings may be interpreted as evidence of a dormant or an active but non-culturable cell condition, since Lahtinen et al., (2005) discovered a subpopulation of non-culturable cells with a functional cell membrane typical of viable cells. Interestingly, non-viable probiotic cells can also illustrate adherence to intestinal mucus for subsequent conveyance of immunomodulatory effects (Ouwehand et al., 2000, Vinderola et al., 2005). Therefore, the detection of compromised

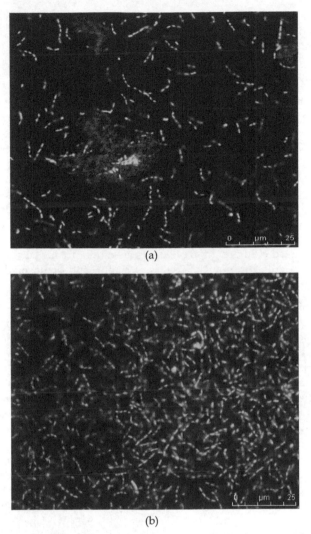

Fig. 7. Fluorescent Confocal Scanning Laser Microscope (CSLM) images showing the incomplete protein digestion (a) generated after single protein digestion and the homogenous cell suspension generated (b) for FACS analysis following cell extraction and double proteolysis

cells or possibly dead cells may be screened by FACS for the provision of therapeutic cell cultures. Furthermore, Maukonen et al., (2006) highlighted the choice of fluorescent stain, sample pre-treatment and product matrix as additional cytometric factors influencing cell stainability, contrary to general credence based solely on strain characteristics. Retention of fluorescent dyes is synonymous with hydrophilic cell surface properties, both of which were illustrated for probiotic bacteria *Lactobacillus rhamnosus* GG by Doherty et al., (2010). Interestingly, surfaces of lactic acid bacteria studied in literature are also hydrophilic

(Boonaert and Rouxhet, 2000, Pelletier et al., 1997). This acquired knowledge relating to cytometric screening of protein-cell systems will diversify the choice of cell lines and the product range applicable for efficacious cell delivery via encapsulation.

4.4 Next generation flow cytometry for encapsulation applications

As conventional flow cytometers with advanced capabilities mature as large powerful laboratory systems for obtaining highly complex information, a new generation of 'personal flow cytometers' have evolved. These bench-top user-friendly cytometers focus on specific functions (Vignali, 2000) and include less flexible laser selection systems (Shapiro, 1995), less sensitive detectors (www.accuricytometers.com) and/or elimination of sheath fluid (www.millipore.com). These smaller systems can accomplish most customary tasks, including cell counting, measurement of cell viability, antibody quantification and detection of cell death. The Luminex® family of cytometers, in particular, have advanced the use of micro-particle-based assays to provide multiplexed analytical capability while keeping the optics relatively simple. With recent developments in micro-fabrication and microfluidics, such as small, optimised chip design, developers are miniaturising current cytometers even further to create systems for point-of-use applications and some of these systems will continue to employ particle-based assays (Kim et al., 2010).

Micro-particle assays often rely on antibodies as capture molecules on the surface of the capsule. Although, other types of capture molecules can be used, the high specificity and affinity of antibodies in the presence of complex sample matrices has made them a reagent of choice. In 1977, Horan and Wheeless published a manuscript in *Science* detailing the first microsphere-based immunoassay (Horan and Wheeless, 1977). Since then, others have followed with technology of increasing sophistication (Vignali, 2000). Capture antibodies are generally immobilized onto the micro particles using available reactive groups, for example amines, hydroxyls or thiols but carboxyls are the most frequently used. Functionalisation of carboxyl is followed by exposure to ethydimethylaminopropylcarbodiimide and N-hydroxysuc-cinimide, which provides a mild procedure for antibody attachment to the surface of micro-particles. The attached antibody can captures antigen from a sample on to the surface for the microsphere. The signal can then be generated as an aggregation event measured using light scattering or electrical or magnetic properties. More frequently, a fluorescent tracer is included in the particle-capture antibody-antigen complex, and fluorescence is measured.

Light scattering in flow cytometers is a staple phenomenon for detecting and characterising particles in FACS and more recently micro-capsules. A light beam directed at a particle can interact through reflective, refractive and diffractive effects. Then, information about the particle, aggregate or micro-capsule can be derived from the change in direction and intensity of a scattered light beam. Collecting scattered light at various angles from the incident beam has been reported to provide different types of information about the particle, e.g. size and density (Zharinov et al., 2001). The diameter of the particle/ capsule should be within the range of 1-50 wavelengths of the incident light beam.

Typically, forward scatter can provide approximate information about size of a protein gel micro-bead or capsule (Shapiro, 2003). It should however, be noted that an increase in the intensity of forward-scattered light does not always correlate with increasing particle size.

Side scattered light is often collected at 90° and provides information about smaller particles and structures within protein particles, scaffold, capsule or gel aggregates. Proportionally more light is scattered by micro-capsules or encapsulated structures at a wide angle than at a small angle, and thus side-scattered light can provide information about the relative roughness or shape of micro-particles and capsules in addition to the granularity of their internal structures. Measuring side-scattered light and forward scattered light has become a standard FACS for biomedical research because this behaviour enables cells to be distinguished by size and granularity, providing insight into mixed populations, viability, or change in internal structure of encapsulation entities.

Using information derived from scatter light at different angles, particles can be classified and studied. Zharinov, et al., (2001) used light scattering data from a scanning FACS to distinguish lymphocytes, erthyrocytes, encapsulation capsules and milk-fat particles of various size and refractive index (Zharinov et al., 2001). These cells and particles generated different scattered light profiles dependent on scattering angle. Steen et al., (1990) custom-built a flow cytometer to characterise viruses of different size using light scattering (Steen et al., 1990).This device could easily distinguish particles with diameters in the range of 30-700 (Steen et al., 1990). In the 1980 Masson and co-workers presented a strong body of work describing particle-counting agglutination immunoassays (PACIA) (Sindic et al., 1981). Initially, PACIA was publicised as a replacement for expensive assay utilizing radioactive labels for characterisation of antigen and antibody interactions. In these assays, polystyrene particles coated with antigens were incubated with antibodies to cause agglutination or aggregation of the particles. Key aspects of this assay were:

-   Use of an antibody with multivalent binding sites to enable particle-particle interaction
-   Determination of particle concentrations that would allow aggregation; and
-   Prevention of non-specific interactions between particles

The samples before and after agglutination were measured in an optical particle counter based on light scattering. The aggregated particles were larger in size than unaggregated particles and resulted in more side-scattered light. In 2003, Pamme, Koyama and Manz described a microfluidic device that used light scattering to analyse agglutination immuno assays (Pamme et al., 2003). The device used a design that focused particles in two dimensions into an optical interrogation region. Particles ranging from 2-9 μm in diameter were distinguished, which is significantly larger than the 70-300 nm range reported previously (Steen et al., 1990). Light scattering is extensively studied with many advantages to developers of microflow cytometry assays for encapsulation purposes. Using just a beam of light of suitable wavelength (relative to the capsule) and detectors at various angles, information regarding size, shape and granularity of a capsule are easily derived. Additionally, scattered light signals tend to be strong and do not need the most advanced or expensive detectors. Yet, distinguishing differing in diameter of a few microns can be challenging and represents the fundamental pre-requisite for encapsulation screening assays.

4.5 High content encapsulation screening

Literature also shows that high resolution two-dimensional (2-D) images consume limited detector bandwidth, introduce a data-acquisition delay that is a barrier for real-time decisions needed for sorting capsules and introduce noise via inaccuracies in image segmentation. This impediment is further addressed to provide the foundation knowledge for a parallel microfluidic cytometer (PMC) using a high-speed scanning photomultiplier-based detector. Development of parallel flow channels within this model PMC would inevitably decouple count rates from signal-to-noise ratio for cytometric analysis of encapsulation systems. Essentially, this approach would demonstrate the feasibility for high throughput visual screening of encapsulated animal cells and stem cell clusters with concomitant determination of recombinant protein generation and localisation.

Imaging flow cytometers based on wide-field-charge-coupled device (CCD) imagers have demonstrated throughput limitations similar to those of microscopes (George et al., 2006). To address these restrictions, industry has developed a multi-channel parallel microfluidic cytometer (PMC) based on analog detection with parallel microfluidics. This design may potentially reduce i) data-buffering and storage requirements and ii) simplify the classification algorithm required to differentiate encapsulated animal or stem cells. Hence, parallel microfluidics will bypass sample-changeover restrictions commonly encountered with single channel flow cytometers. This novel strategy will enhance process efficiency of the PMC detector by increasing the diameter of a laser spot from 1 μm and 4 μm. Furthermore, sample flowing into the focal volume will potentially eliminate the focusing and stage volumes that limit high content screening on a microscope to approx. 2-6 wells min[-1]. This new cytometric design circumvents many throughput limitations for both high cell density encapsulation and flow cytometer analysis by combining the best features of each technology to achieve efficient cell encapsulation screening.

Although much can be borrowed from the methods of 2-D high content screening (de Vos et al., 2010), 1-D algorithms are fundamentally different. Notably, microscope or flow cytometers draw boundaries around 'primary and secondary objects' (known as segmentation). However, this user-defined aspect of segmentation is a source of assay variability and is often considered the most challenging and time-consuming step. However, the recent application of 1-D cytometric imaging strategy for encapsulation systems integrated resolution issues experience within the standard hardware systems in order to eliminate potential segmentation problems. This was achieved by accepting any resolution element as an 'object'. Hence, this demonstrates the feasibility to develop efficient high content screening algorithms for micro-capsules using relatively low input.

5. Conclusion

Incorporation of bioactive compounds such as probiotics, animal cells and stem cell clusters into encapsulation matrices for food and pharmaceutical purposes can provide a simple way to develop novel nutraceuticals and drug treatment delivery systems with physiological benefits. Exact elucidation of cell proliferation and stability during encapsulation is of utmost importance for industrial or academic application. Hence, this chapter outlines the importance of cytometric analysis of encapsulation systems, in addition to the role of protein during this screening process. Proteins present various challenges for cytometric analysis; however, several techniques have been developed to overcome these limitations for commercial development of this novel technology.

6. Acknowledgements

The support provided by the staff and experts in Union Biometrica, Geel, Belgium is gratefully acknowledged. Some of the cytometric work presented in this chapter was supported by the Irish Dairy Research Trust project NU518 "Probiotic Protection", the Irish National Development Plan 2007 to 2013 and Science Foundation Ireland (SFI). The support provided by C. Stanton and R.P. Ross as well as the National Food Imaging Centre in Moorepark, Teagasc, A.E. Auty, L. Wang and V.L. Gee in particular, is gratefully acknowledged.

7. References

Boonaert, C. J. and P. G. Rouxhet. 2000. Surface of lactic acid bacteria: relationships between chemical composition and physicochemical properties. Applied and environmental microbiology 66(6):2548-2554.

Broger, T., R. P. Odermatt, P. Huber and B. Sonnleitner. 2011. Real-time on-lineflow cytometry for bioprocessmonitoring. Journal of Biotechnology 154(4):240-247.

Bunthof, C. J. and T. Abee. 2002. Development of a flow cytometric method to analyze subpopulations of bacteria in probiotic products and dairy starters. Applied and environmental microbiology 68(6):2934-2942.

de Vos, P., M. M. Faas, M. Spasojevic and J. Sikkema. 2010. Encapsulation for preservation of functionality and targeted delivery of bioactive food components. International Dairy Journal 20(4):292-302.

Delgado, L., G. Jurado, G. Galayo, E. Ogalla, L. Moreno, J. C. Rodríguez-Aguilera, A. Cebolla, M. Flores and S. Chávez. 2010. Encapsulation in monodispersed hydrogel microspheres enables fast and sensitive phenotypic analyses using COPAS large particle flow cytometry. Vol. QTN-019. COPAS™, COPAS™ QUICK TECH NOTES.

Delgado, L., G. Jurado, G. Galayo, E. Ogalla, L. Moreno, J. C. Rodríguez-Aguilera, A. Cebolla, C. Sousa, C. Flores and S. Chávez. 2011. Microbial encapsulation in monodisperse hydrogel microspheres enables fast and sensitive phenotypic analyses using flow cytometers. in Application of Flow Cytometry in Microbiology. Geel, Belgium.

Doherty, S. B., L. Wang, R. P. Ross, C. Stanton, G. F. Fitzgerald and A. Brodkorb. 2010. Use of viability staining in combination with flow cytometry for rapid viability assessment of Lactobacillus rhamnosus GG in complex protein matrices. Journal of Microbiological Methods 82(3):301-310.

Dunlop, E. H. and S. J. Ye. 1990. Micromixing in fermentors: Metabolic changes in Saccharomyces cerevisiae and their relationship to fluid turbulence. Biotechnology and Bioengineering 36(8):854-864.

Feldsine, P., C. Abeyta and W. H. Andrews. 2002. AOAC International methods committee guidelines for validation of qualitative and quantitative food microbiological official methods of analysis. Journal of AOAC International 85(5):1187-1200.

George, T. C., S. L. Fanning, P. Fitzgeral-Bocarsly, R. B. Medeiros, S. Highfill, Y. Shimizu, B. E. Hall, K. Frost, D. Basiji, W. E. Ortyn, P. J. Morrissey and D. H. Lynch. 2006. Quantitative measurement of nuclear translocation events using similarity analysis of multispectral cellular images obtained in flow. Journal of Immunological Methods 311(1-2):117-129.

Glassey, J., K. V. Gernaey, C. Clemens, T. W. Schulz, R. Oliveira, G. Striedner and C.-F. Mandenius. 2011. Process analytical technology (PAT) for biopharmaceuticals. Biotechnology Journal 6(4):369-377.

Gunasekera, T. S., P. V. Attfield and D. A. Veal. 2000. A flow cytometry method for rapid detection and enumeration of total bacteria in milk. Applied and Environmental Microbiology 66(3):1228-1232.

Gunasekera, T. S., D. A. Veal and P. V. Attfield. 2003. Potential for broad applications of flow cytometry and fluorescence techniques in microbiological and somatic cell analyses of milk. International Journal of Food Microbiology 85(3):269-279.

Hall, B. G. 1995. Adaptive mutations in Escherichia coli as a model for the multiple mutational origins of tumors. Proceeedings of the National Academy of Sciences of the United States of America 92(12):5669-5673.

Horan, P. K. and L. L. Wheeless. 1977. Quantitative single cell analysis and sorting. Science 14(198(4313):):149-157.

Jones, J. J., A. M. Bridges, A. P. Fosberry, S. Gardner, R. R. Lowers, R. R. Newby, P. J. James, R. M. Hall and O. Jenkins. 2004. Potential of real-time measurement of GFP-fusion proteins. Journal of Biotechnology 109(1-2):201-211.

Kacmar, J. and F. Srienc. 2005. Dynamics of single cell property distributions in Chinese hamster ovary cell cultures monitored and controlled with automated flowcytometry. Journal of Biotechnology 120(4):410-420.

Kelley, K. A. 1989. Sample station modification providing on-line reagent addition and reduced sample transit time for flow cytometers. Cytometry 10(6):796-800.

Kim, J. Y., W. Choi, Y. H. Kim, G. T. Tae, S. Y. Lee, K. Kim and I. C. Kwon. 2010. In-vivo tumor targeting of pluronic-based nano-carriers Journal of Controlled Release 147(1):109-117.

Lahtinen, S. J., M. Gueimonde, A. C. Ouwehand, J. P. Reinikainen and S. J. Salminen. 2005. Probiotic bacteria may become dormant during storage. Applied and Environmental Microbiology 71(3):1662-1663.

Maukonen, J., H. L. Alakomi, L. Nohynek, K. Hallamaa, S. Leppämäki, J. Mättö and M. Saarela. 2006. Suitability of the fluorescent techniques for the enumeration of probiotic bacteria in commercial non-dairy drinks and in pharmaceutical products. Food Research International 39(1):22-32.

McClelland, R. G. and A. C. Pinder. 1994a. Detection of low levels of specific Salmonella species by fluorescent antibodies and flow cytometry. The Journal of applied bacteriology 77(4):440-447.

McClelland, R. G. and A. C. Pinder. 1994b. Detection of Salmonella typhimurium in dairy products with flow cytometry and monoclonal antibodies. Applied and environmental microbiology 60(12):4255-4262.

Münch, T., B. Sonnleitner and A. Fiechter. 1992. The decisive role of the Saccharomyces cerevisiae cell cycle behaviour for dynamic growth characterization. Journal of Biotechnology 22:329-352.

Ouwehand, A. C., S. Tolkko, J. Kulmala, S. Salminen and E. Salminen. 2000. Adhesion of inactivated probiotic strains to intestinal mucus. Letters in Applied Microbiology 31(1):82-86.

Pamme, N., R. Koyama and A. Manz. 2003. Counting and sizing of particles and particle agglomerates in a microfluidic device using laser light scattering: application to a particle-enhanced immunoassay Lab on chip 3(3):187-192.

Patchett, R. A., J. P. Back, A. C. Pinder and R. G. Kroll. 1991. Enumeration of bacteria in pure cultures and in foods using a commercial flow cytometer. Food Microbiology 8(2):119-125.

Pelletier, C., C. Bouley, C. Cayuela, S. Bouttier, P. Bourlioux and M. N. Bellon-Fontaine. 1997. Cell surface characteristics of Lactobacillus casei subsp. casei, Lactobacillus paracasei subsp. paracasei and Lactobacillus rhamnosus strains. Applied and environmental microbiology 63(5):1725-1731.

Reardon, A. J., J. A. W. Elliott and L. E. McGann. 2009. Fluorescence as a better approach to gate cells for cryobiological studies with flow cytometry. Crybiology 63(3):317-329.

Reischer, H., I. Schotola, G. Striedner, F. Pötschacher and K. Bayer. 2004. Evaluation of the GFP signal and its aptitude for novel on-linemonitoring strategies of recombinant fermentation processes. Journal of Biotechnology 108(2):115-125.

Rieseberg, M., C. Kasper, K. F. Reardon and T. Scheper. 2001. Flowcytometry in biotechnology. Applied Microbiology and Biotechnology 56(3-4):350-360.

Ruzicka, J. and W. Lindberg. 1992. Flow injection cytoanalysis. Analytical Chemistry 69(9):A537-A545.

Shapiro, H. M. 1995. Practical Flow Cytometry. Vol. 3. No. 3. A.R. Liss, New York.

Sindic, C. J. M., M. P. Chalon, C. L. Cambiaso, D. Collet-Cassart and M. P.L. 1981. Particle counting immunoassay (PACIA) – VI. The determination of rabbit IgG antibodies against myelin basic protein using IgM rheumatoid factor. Molecular Immunology 18(293-299).

Sitton, G. and F. Srienc. 2008a. Growth dynamics of mammalian cells monitored with automated cell cycle staining and flowcytometry. Cytometry A 6:538-545.

Sitton, G. and F. Srienc. 2008b. Mammalian cell culture scale-up and fed-batch control using automated flowcytometry. Journal of Biotechnology 32(2):174-180.

Steen, P. D., E. R. Ashwood, K. Huang, R. A. Daynes, H. T. Chung and W. E. Samlowski. 1990. Mechanisms of pertussis toxin inhibition of lymphocyte-HEV interactions: I. Analysis of lymphocyte homing receptor-mediated Cellular Immunology 131(1):67-85.

Sun, W. and M. W. Griffiths. 2000. Survival of bifidobacteria in yogurt and simulated gastric juice following immobilization in gellan-xanthan beads. International Journal of Food Microbiology 61(1):17-25.

Vignali, D. A. A. 2000. Multiplexed particle-based flow cytometric assays Journal of Immunological Methods 243(1-2):243-255.

Vinderola, G., C. Matar and G. Perdigon. 2005. Role of intestinal epithelial cells in immune effects mediated by gram-positive probiotic bacteria: involvement of toll-like receptors. Clinical and Diagnostic Laboratory Immunology 12(9):1075-1084.

Walser, M., R. M. Leibundgut, R. Pellaux, S. Panke and M. Held. 2008. Isolation of monoclonal microcarriers colonized by fluorescent E. coli. Cytometry 73(9):788-798.

Walser, M., R. Pellaux, A. Meyer, M. Bechtold, H. Vanderschuren, Reinhardt R., J. Magyar, S. Panke and M. Held. 2009. Novel method for high-throughput colony PCR screening in nanoliter-reactors. Nucleic Acid Research 37(8):e57/51-e57/58.

Watson, D. A., D. F. Gaskill, L. O. Brown, S. K. Doorn and J. P. Nolan. 2009. Spectral Measurements of Large Particles by Flow Cytometry. Cytometry A 75(5):460-464.

Zanella, F., J. B. Lorens and W. Link. 2010. High content screening: seeing is believing Trends in Biotechnology 28(5):237-245.

Zhao, R., A. Natarajan and F. Srienc. 1999. A flow injection flowcytometry system for on-linemonitoring of bioreactors. Biotechnology and Bioengineering 65(5):609-617.

Zharinov, A., P. Tarasov, A. Shvalov, K. Semyanov, D. R. van Bockstaele and V. Maltsev. 2001. A study of light scattering of mononuclear blood cells with scanning flow cytometry Journal of Quantitative Spectroscopy and Radiative Transfer 102(1):121-128.

9

Flow Cytometry Analysis of Intracellular Protein

Irena Koutná[1], Pavel Šimara[1], Petra Ondráčková[2] and Lenka Tesařová[1]

Masaryk University/ Centre for Biomedical Image Analysis, FI
Veterinary Reasearch Institute/ Department of Immunology
Czech Republic

1. Introduction

The technique of measuring intracellular protein levels using flow cytometry is a very rapid method to detect protein on a single-cell level.

The demand for determining protein levels in a continuously decreasing amount of input cells is currently being developed by scientists and medical doctors. Another demanding area is the splitting of active and inactive (phosphorylated and unphosphorylated, respectively) forms of the protein being monitored, e.g., during different phases of the cell cycle, differentiation or carcinogenesis. The Western blot technique has been typically used for this purpose. The utilization of multi color flow cytometry allows for measurements of multiple proteins in parallel, regardless of the protein length. As the technique evolves, it is possible to obtain information on over 13 parameters per cell (Krutzik et al., 2004).

Measuring the phosphorylation status of specific proteins using this technique is very efficient. It eliminates a number of problems related either to Western blotting or ELISA.

Compared to the Western blot, this progressive method has much lower requirements on the number of input cells (only 1×10^4 cells are needed for determining the concentration of a given protein), it requires less time for processing and, last but not least, is also much more cost effective. Furthermore, it allows researchers to distinguish the various cell populations in the sample without complicated cell separation, if the appropriate antibodies are used. This is beneficial for searching for cell populations and their reactions on external stimuli, such as changed cultivation conditions during in vitro experiments or influencing the organism during in vivo experiments.

In this chapter, the application and pitfalls of flow cytometric intracellular protein measurements are illustrated by the data of our original research of porcine monocytes and CD34+ hematopoietic stem/progenitor cells. In short, a successful utilization of this progressive technique depends on correctly performing the 4 following steps:

1. Precise experimental planning to obtain a defined cell population using this technique.
2. Effective fixation and permeabilization of the cell population.
3. Choosing the optimal isotype control.
4. Selecting the optimal system of primary and secondary antibodies.

2. Fixation and permeabilization

Intracellular flow cytometry, in comparison with conventional cell surface labeling methods, requires fixation and permeabilization of the cells before staining of intracellular antigens (Robinson et al., 1993). Moreover, a variety of commercial kits for fixation and permeabilization are available.

The first step in the population analysis is high quality fixation. A crosslinking reagent (typically formaldehyde) could be used for this purpose. For every given cell population, it is necessary to test the adequate concentration of the solution. The most common formaldehyde concentration is about 4%. Other fixation limiting factors are time and temperature. The incubation time, according to the cell population type, varies between 8 and 15 minutes. The incubation temperature is 37 °C. A variety of commercial kits for fixation and permeabilization are available.

The following step is for permeabilization with detergents (Triton X 100 or saponin) or alcohol (ethanol or methanol). Although the protocols that have been used to stain phosphoepitopes for flow cytometry differ from one to another, they rely on two primary permeabilization reagents – saponin or methanol.

Saponin permeabilization

Saponin is a mixture of terpenoid molecules and glycosides that permeabilize cells by interacting with cholesterol present in the cell membrane (Melan et al. 1999). This creates pores in the plasma membrane that are large enough for entry of fluorophore-conjugated antibodies. Because the intracellular proteins can leak out of saponin-treated cells, they must be first exposed to a crosslinking reagent, such as formaldehyde, to cross-link proteins and nucleic acids into a cohesive unit within the cell. Saponin has become the detergent of choice for cytokine staining, and several groups have utilized it for permeabilization in phospho-epitope staining protocols (Pala et al., 2000). It is typically used at concentrations ranging from 0.1% to 0.5%, similar to cytokine-staining procedures.

Three commercially available kits (Leukoperm, Serotec, UK; Fix & Perm, An Der Grub, Austria; IntraStain, DAKO Cytomation, Denmark) along with combinations of 2 or 4% paraformaldehyde with 0.1 or 0.05% saponin were tested for fixation and permeabilization of isolated pig's peripheral blood mononuclear cells or whole blood leukocytes (Zelnickova et al., 2007).

The fixation and permeabilization process could lead to non-specific binding of primary or secondary antibodies. In comparison to all three tested commercial kits, a combination of paraformaldehyde and saponin caused an increase in non-specific binding of antibodies. The intensity of fluorescence of the negative peak of paraformaldehyde/saponin fixed cells was evidently higher in comparison with the negative peak of cells fixed with the use of commercial kits.

The fixation and permeabilization process could lead to elevation of autofluorescence of cells. The autofluorescence of cells was at the lowest level in all tested kits. In contrast, the combination of paraformaldehyde and saponin in all concentrations caused an increase of autofluorescence. The autofluorescence in samples fixed with 4% paraformaldehyde was higher than with 2% paraformaldehyde.

The fixation and permeabilization process could damage the light scatter properties of the cells. The light scatter characteristics of all tested cells were comparable after any type of fixation and permeabilization. The light scatter characteristics were always changed in comparison with fresh preparations. However, well distinguishable and bounded subpopulations of mononuclear cells and neutrophils were obtained by adjusting the settings in the side and forward scatters.

Some protocols use labeling of intracellular molecules in the whole blood instead of isolated blood leukocytes. In our laboratory, the lysis of red blood cells (RBC) is always performed before starting the staining procedure. However, in order to make the processing of large amounts of samples as quick as possible, the hemolysis is commonly performed in the 96-well plate in which the cultivation was previously performed. The volume of the hemolytic reagent is therefore relatively small and the hemolysis is not complete. Therefore, it is advantageous if the fixation and permeabilization procedure leads to the lysis of these contaminating erythrocytes. It was found that only the IntraStain kit and paraformaldehyde/saponin fixation and permeabilization completely lysed RBC in the samples. The other two kits did not induce a lysis of porcine RBC (although complete lysis of human RBC was obtained by these kits in preliminary experiments, which is in accordance with the manufacturer's instructions).

The problem can occur when the lysis of RBC is performed before fixation and permeabilization. The lysis of RBC induced by ammonium chloride solution (the lysing reagent commonly used in our laboratory to lyse porcine RBC) before fixation and permeabilization with commercial kits strongly changed the light scatter characteristics of white blood cells (WBC). This was apparent, especially in neutrophils, which completely fused with the lymphocyte population. Moreover, the population of lymphocytes was much more dispersed as shown by the side scatter measurements. If the lysis of RBC was performed before their fixation and permeabilization with the IntraStain kit, and the cells were washed twice in CWS solution immediately after lysis, the scatters were not altered. If these washing steps were tested with the other two kits, the light scatters were changed. Thus, the lysis of RBC cannot be achieved with the Fix&Perm or Leukoperm kits without alteration of light scatter characteristics of WBC.

Finally, the effect of different fixation/permeabilization reagents on IFN-γ and TNF-α staining was tested. Generally, we can say that commercial kits mostly gave better results compared with paraformaldehyde/saponin. IntraStain and Leukoperm gave better results than Fix&Perm in some cases. If the use of a combination of paraformaldehyde/saponin is considered, 0.05% saponin should be avoided, especially in combination with 2% paraformaldehyde. Saponin in a concentration of 0.1% in combination with 4% paraformaldehyde slightly increased the percentages of positive cells in comparison with 2% paraformaldehyde. However, these differences were nonsignificant.

Consistent with the above mentioned parameters, the IntraStain kit gave the best results compared to the other two tested kits, as well as when compared to paraformaldehyde/saponin. Therefore, this kit was chosen for fixation and permeabilization of porcine leukocytes for experimentation.

Since a wide range of different fixation and permeabilization reagents that were not tested in our study are currently available, the above mentioned parameters can serve as a particular protocol of intracellular cytokine detection and also as a suggestion for optimization of the fixation, permeabilization and cell surface labeling procedures for any laboratory.

2.1 Methanol permeabilization

Alcohol permeabilization has typically been used for the analysis of DNA by flow cytometry (Ormerod et al., 2002), but can be successfully applied to phospho-epitope staining as well (Krutzik et al., 2003). It is thought that alcohols fix and permeabilize cells by dehydrating them and solubilizing molecules out of the plasma membrane. Proteins may be made more accessible to antibodies during the process and cells are permeabilized to a greater extent than with saponin, allowing efficient access to the nuclear antigens.

Another option is to use commercially available kits. Currently, there are a large number of them available on the market. A fact that needs to be taken into account when using the commercial kits is that even if the kit works well, the method cannot be excessively modified. This could be reflected in the scale of results.

By using one of the best available kits for permeabilization, BD Fix&Perm, with methanol permeabilization during "The decrease in p-CrkL levels upon imatinib treatment" experiment, we came to the following conclusion:

The method (permeabilization using BD Fix&Perm) is less sensitive. It does not recognize the difference between 0 µM and 5 µM imatinib (IM). The speed (2 hours) is an advantage and there is a smaller amount of necessary input material (cells) as well. Although the peaks are sharp, it seems to be more difficult for the antigens to get into the cell, because of less aggressive permeabilization. The methanol permeabilization is much more sensitive. It is able to recognize differences between 0 µM and 5 µM IM. It is more time consuming (5 hours) and requires a higher amount of input material (cells). The significant advantage is the customizability of the method according to user needs (Figure 1).

3. Isotype control

The selection of the appropriate isotype control is an important element in flow cytometry experiments. Isotype controls are antibodies of the same isotype as the target primary antibody. They are of unknown specificity or are raised against antigens known to be absent in target cells. Isotype controls are used to estimate non-specific staining of primary antibodies. Several factors can contribute to the levels of this "background" staining, including Fc receptor binding, non-specific protein interactions, and cell autofluorescence. These factors may vary depending on the target cell type and the isotype of the primary antibody. Therefore, isotype controls need to be properly chosen. Isotype control antibodies ideally match the primary antibody's host species, isotype, and conjugation format. For example, if the primary antibody is an APC-conjugated mouse IgG2a, then it will be necessary to choose an APC-conjugated mouse IgG2a isotype control. Thus, isotype control is supposed to have all the non-specific characteristics of the target primary antibody and it is able to accurately determine the level of specific staining. Various monoclonal antibody idiotypes are used in flow cytometry applications: most frequently, IgG1, IgG2a, IgG2b, IgG3, IgM, and IgA; less frequently IgD, IgE, IgG2c, Ig kappa, and Ig lambda are applied. Designing the experiment, this isotype and origin species of the primary antibody must be known to find a suitable isotype control. The appropriate isotype control is subsequently looked up in a company product list according to the desired isotype (IgG1, IgG2a, IgM, etc.), reactivity (mouse, human, rat, etc.), and conjugate (FITC, PE, APC, etc.). In addition, recommended isotype controls can often be found on the Data Sheets for primary

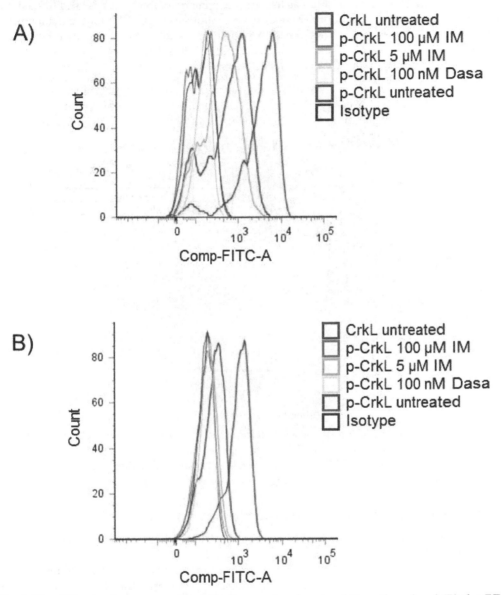

Fig. 1. The difference between permeabilization protocols using (A) methanol and (B) the BD Fix&Perm kit. Permebilization with methanol is more sensitive and allows researchers to distinguish between individual MFI peaks. Fix&Perm kit is less time consuming with relatively low sensitivity

antibodies. Isotype control antibodies are commercially available for both direct and indirect immunofluorescence in the form of fluorochrome-conjugated antibodies and unconjugated antibodies, respectively. During the flow cytometry analysis, the idiotype control antibody is diluted to the same concentration as the specific primary antibody, and is used to stain the sample of negative control cells. This negative control serves to determine the amount of non-specific "background" fluorescence. It allows for setting a threshold of negativity of stained cells. Any event generating a signal above this baseline is considered to be specifically labeled with the target primary antibody.

The isotype control plays an important role during the processing of final measurements. The level of monitored protein is determined as the geometric mean of fluorescent intensity (MFI) of labeled sample, minus the isotype control (O'Gorman et al., 1999, Holden et al., 2006,Hulspas et al., 2009).

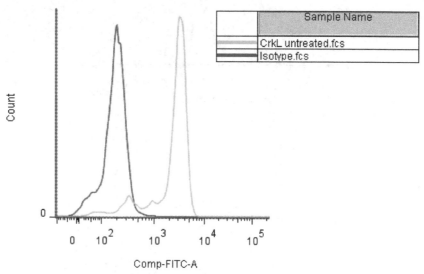

Fig. 2. Representative FACS plot showing isotype and CrkL MFI peaks visualized by a FITC-conjugated secondary antibody in CD34+ cells isolated from peripheral blood of a newly diagnosed CML patient. The MFI peak of isotype indicates the FITC unspecific background fluorescence. These were visualized using FlowJo software

4. Intracellular antigens for flow cytometry

The most common way to visualize of the complex between a monoclonal antibody and an antigen in flow cytometry is to covalently bind the antibody to different fluorescent molecules (fluorophores). After exposure to radiation from an excitation source, these fluorophores emit photons with longer wavelengths. Currently, there is a wide range of commercially available fluorophores starting from the small polycyclic molecules, such as fluorescein isothiocyanate (FITC), cyanines, and dyes of the Alexa series, through fluorescent phycobiliproteins whose best-known representatives are phycoerythrin (PE), allophycocyanin (APC). New way in flow cytometry is also Qdot® nanocrystals –

nanometer-scale semiconductor particles with unique fluorescence properties (eBioscience , Molecular Probes®)..

The world leader in bringing innovative diagnostic and research tools to different specialists are Beckman Coulter and eBioscience as the major companies producing unconjugated and conjugated antibodies for flow cytometry. BD Biosciences offers a number of Alexa Flouro (AF®) 488/647/700-, APC-, FITC-, PE-, and Pacific BlueTM- conjugated antibodies of BD PharmingenTM and BD PhosflowTM brands used for multicolor flow cytometry (BD Biosciences, 2011).

A particularly strong area of leukemia immunophenotyping that contains a broad panel of research products has been built up by (DAKO, 2010). DAKO offers various types of reagents for use in flow cytometry, including primary single-color antibodies conjugated with a single fluorochrome; MultiMix™ Dual-Color Reagents based on the combination of two or more antibodies labeled with FITC and RPE; and MultiMix™ Triple-Color Reagents based on the combination of three antibodies labeled with fluorescein isothiocyanate (FITC), R-phycoerythrin (RPE) and allophycocyanin (APC) or FITC, RPE and RPE-Cy5. Other products for use in flow cytometry are isotype reagents, secondary antibody conjugates, streptavidine conjugates and other accessories. There are also several kits available, e.g. an apoptosis kit used for flow cytometric distinction between viable cells in single cell suspensions or an Enumeration of Stem Cells Kit used for optimal enumeration of CD34+ hematopoietic stem/progenitor cells (DAKO, 2010).

Another strong company in the production of high-quality activation-state antibodies is Cell Signaling Technology. They offer a number of primary and conjugated antibodies as well as antibody-related kits. Conjugated antibodies are conjugated with AF®, PE or biotin (Cell Signaling Technology, 2011)

There are many other companies not mentioned in the table, including R&D Systems, Miltenyi Biotec and BioLegend, that offer a wide range of unconjugated and conjugated monoclonal antibodies and kits. For example, R&D Systems offers a wide range of biotin-, fluorescein-, PE-, PerCP-, AF® 488-, or APC- conjugated monoclonal antibodies specifically designed to monitor protein expression by flow cytometry (R&D Systems, 2011).

Generally, the problem with labeling various molecules for flow cytometry in animal species other than mice is the poor availability of directly conjugated primary antibodies. Therefore, indirect labeling is commonly used. This includes labeling cells with a non-conjugated antibody and subsequent visualization with a secondary fluorochrome-conjugated antibody. This indirect labeling is limited by the subclasses of primary antibodies, which should not share the same subclass. This limits the use of the antibodies, especially in the case of multicolor labeling.

When intracellular labeling is combined with cell surface marker labeling in pigs, the problem with sharing the same subclasses is more noticeable since all the anti-porcine cytokine antibodies used share the most common mouse IgG1 subclass. The cell surface molecules are always labeled prior to the labeling of the intracellular cytokine in our protocol. Accordingly, the anti-cytokine antibody must be directly conjugated with fluorochrome (such as anti-TNF-α or anti-IFN-γ) or it must be biotinylated (such as anti-IL-2 or anti-IL-10) or pre-labeled in some other way (Zelnickova et al., 2008).

Intracellular marker	Type	Clone name	Isotype	Reactivity	Labelling	Source	Ref.
Cytokines							
IL-2, IL-4, TNF- α, etc.	mAb	5344.111, 3010.211, 6401.1111, ...	Ms IgG$_1$,κ	Hu	FITC/ PE/APC/...	BD Biosciences	Karanikas et al., 2000
IFN- γ	mAb	P2G10	Ms IgG$_1$,κ	Pig	PE-conjugate	BD Biosciences	Gerner et al., 2008
Cell cycle or Apoptosis							
p53	mAb	DO7	Ms IgG$_2$b	Hu	nonconjugated	Dako	Millard et al., 1998
bcl-2	mAb	124	Ms IgG$_1$	Hu	FITC-conjugated	Dako	Millard et al., 1998
rb	mAb	G3-245	Ms IgG$_1$	Ms, Qua, Mn, Mk, Rat, Hu	nonconjugated	BD Biosciences	Millard et al., 1998
MDR1	mAb	JSB-1	Ms IgG$_1$	Hu, Rat, Ms	nonconjugated	Millipore	Millard et al., 1998
PCNA	mAb	PC10	Ms IgG$_2$a	Hu	nonconjugated	Dako	Millard et al., 1998
caspases	pAb	-	Rab IgG	Hu, Rat, Ms	PE-conjugated	BD Biosciences	Belloc et al., 2000
perforin	mAb	dG9	Ms IgG$_2$b	Hu	PE-conjugated	BD Biosciences	Gerner et al., 2008
Ki67	mAb	B56	Ms IgG$_1$,κ	Ms	FITC-conjugated	BD Biosciences	Gerner et al., 2008
Bax, Bcl-x$_L$, Mcl-1	pAb	-	Rab IgG	Ms/ Rab/ Hu	unconjugated	Santa Cruz Biotechnology	van Stijn et al., 2003
Viral particles							
HIV	mAb	unknown	MsIgG$_{1b12}$	Hu	unknown	Denis Burton	Mascola et al., 2002
Receptors or their proteins, enzymes							
Steroid hormone receptor proteins	mAb pAb	(various)	(various)	(various)	(various)	Afinity bioreagents	Butts et al., 2007
Cox1/2	mAb	AS70, AS67	Ms IgG$_1$,κ	Hu	FITC, PE	BD Biosciences	Ruitenberg et al., 2003
Estrogen Receptor α	mAb	1D5	Ms IgG$_1$	Hu	unconjugated	Dako	Cao et al., 2000
Phospho-proteins							
Akt	mAb pAb	5G3 -	Ms IgG$_1$ Rab IgG	Hu, Ms, Rat, Hm Hu, Ms, Rat, Hm, Mk, Chick, Dm, Bov, Pig, Dog	unconjugated	Cell Signaling Technology	Tazzari et al., 2002
MEK1/2	pAb	-	Rab IgG	Hu, Ms, Rat, Mk, Sc	unconjugated	Cell Signaling Technology	Chow et al., 2001
ERK1/2	mAb	20A	Ms IgG$_1$	Hu, Ms, Rat	AF-conjugated	BD Biosciences	Krutzik et al., 2003
JNK	mAb	41	Ms IgG$_1$	Hu	unconjugated	BD Biosciences	Krutzik et al., 2003
p38 MAPK	mAb	30	Ms IgG$_1$	Hu, Rat, Ms	unconjugated	BD Biosciences	Krutzik et al., 2003
Stat1	mAb	14	Ms IgG$_1$,κ	Hu, Ms	unconjugated	BD Biosciences	Krutzik et al., 2003
Stat3	mAb	D3A7	Rab IgG	Hu, Ms, Rat, Mk	AF-conjugated	Cell Signaling Technology	Kalaitzidis et al., 2008
Stat4	pAb	-	Rab IgG	Hu, Ms	unconjugated	Invitrogen	Uzel et al., 2001

Table 1. continues on next page

Intracellular marker	Type	Clone name	Isotype	Reactivity	Labelling	Source	Ref.
Stat5	mAb	47	Ms IgG$_1$	Hu	AF-conjugated	BD Biosciences	Krutzik et al., 2003
Stat6	mAb	18	Ms IgG$_{2a}$	Hu	PE-conjugated	BD Biosciences	Krutzik et al., 2003
CD79a	mAb	HM57	Ms IgG$_1$	Hu	PE-conugated	Dako	Gerner et al., 2008
mTOR	mAb	D9C2	Rab IgG	Hu, Ms, Mk, (Rat)	unconjugated	Cell Signaling Technology	Kalaitzidis et al., 2008

Hu – human; Ms – mouse; Mk – monkey; Hm - hamster; Dm – drosophila melanogaster; Bov – bovine; Sc – saccharomyces cerevisiae; Mk – mink; Qua - quail

Table 1. Intracellular antigens for flow cytometry

A method for the direct labeling of antibodies that share the same subclass was tested. Because it was necessary to label relatively small amounts of antibodies, the Zenon-labeling technology was chosen.

The Zenon reagents are provided by Invitrogen. This labeling involves the binding of Fab fragments of the fluorochrome-labeled, subclass-specific secondary antibody to the primary antibody, prior to the labeling of the cells. The excess of the fluorochrome-labeled secondary antibody is neutralized by addition of irrelevant mouse antibody in excess, which is supplied within the Zenon kit. The disadvantage of these reagents is the relatively high price. The price is the same for all fluorochromes; however, the number of reactions that can be performed by the kit differ among fluorochromes. Packages with Alexa Fluor, FITC, Texas Red and Pacific dyes contain reagents for 50 rounds of labeling, packages with phycobiliprotein dyes such as R-PE or APC contain reagents for 25 rounds of labeling, and packages with tandem dyes contain reagents for only 10 rounds of labeling. Therefore, it is advantageous to choose Zenon reagents containing Alexa Fluor dyes.

The other disadvantage of the Zenon labeling is that the fluorescence yield of Zenon-labeled antibodies is slightly lower compared to classical indirect labeling. Therefore, antibodies against strongly expressed markers should be preferably stained with the Zenon reagents. (Ondrackova et al., 2010, 2011)

General protocol for cell surface staining followed by intracellular labeling

1. Place the samples of cell suspensions into a U-bottom 96-well plate, spin the plate, remove as much supernatant as possible, and vortex the plate.
   If viability staining with permanent dye combined with intracellular staining is performed, then:
2. Add the viability staining dye, incubate according the manufacturer's instructions, vortex the plate, and wash once.
   The cell surface staining is performed as follows:
3. Add the cocktail of primary antibodies against cell surface molecules in a total volume of 10 µl (dilute antibodies in CWS) + 10 µl of heat-inactivated, filtered goat serum, vortex the plate, incubate for 15 min at 4°C, and wash twice.
4. Add a secondary antibody cocktail in total volume of 25 µl (dilute antibodies in CWS), vortex the plate, and incubate for 20 min at 4°C.
   If the cell surface molecules are to be labeled with Zenon reagents or if intracellular staining follows, then:

5. Wash once and vortex the plate thoroughly because the plate in the next step cannot be vortexed due to the risk of the samples overflowing into the neighboring wells.
6. Add 100 µl of heat-inactivated, filtered mouse serum diluted 1:10 in CWS, incubate for 20 min at 4°C, and wash once.
    Labeling cell surface molecules with Zenon-labeled antibodies now follows, but this step can be omitted if it is not required:
7. Add the cocktail of Zenon-labeled antibodies against cell surface molecules in a total volume of 10 µl, incubate for 15 min at 4°C, vortex the plate, and wash –twice.
    Staining of intracellular molecules with directly-labeled, Zenon-labeled, or unlabeled antibodies follows, but these steps can be omitted if they not required:
8. Add 30 µl of Solution A from the IntraStain kit, vortex thoroughly to allow complete hemolysis of contaminating red blood cells, incubate for 15 min at room temperature, and wash –twice.
9. Add primary antibodies against intracellular molecules (directly-labeled, Zenon-labeled, or unlabeled) diluted in Solution B from the IntraStain kit in a total volume of 20 µl, vortex the plate, incubate for 20 min at room temperature, and wash twice.
10. Add secondary antibodies diluted in Solution B from the InraStain kit and CWS (ratio of Solution B and CWS 1:1) in total volume of 25 µl, incubate for 20 min at room temperature, wash twice.
11. Resuspend samples in CWS in the volume that is required for the subsequent measurement and measure by the appropriate method.

CWS solution: PBS containing 1.84 g/l EDTA, 1 g/l NaN3, 4 ml/l gelatin from cold water fish skin

Wash definition: Add as much CWS into each well as possible, spin 3 min at 500 g, remove much supernatant as possible, and vortex.

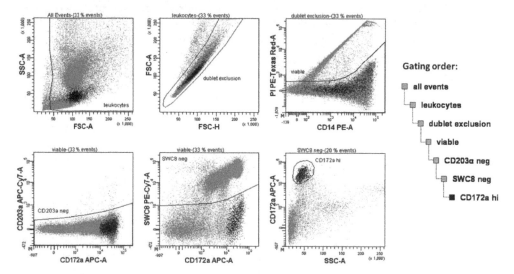

(a)

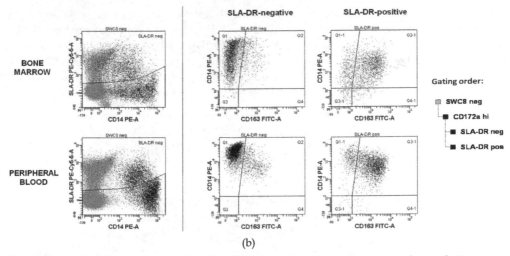

(b)

Fig. 3. Seven color flow cytometry for identification of porcine monocyte subpopulations using two Zenon-labeled antibodies

The red blood cells from the whole peripheral blood or from the bone marrow sample were lysed with ammonium chloride solution. The following fluorescent staining was performed:

Excitation wavelength	Primary antibody					Secondary antibody / Zenon reagent / viability stain				
	Target molecule	Clone	Manufacturer	Amount per well	Stained at point	Class/ subclass	Fluorochrome	Manufacturer	Dilution	Stained at point
488	CD163	2A10/11	VMRD, USA	0.1 µl	3)	IgG1	AlexaFluor488	Invitrogen, USA	1:750	4)
488	CD14	MIL-2	Serotec, UK	1 µl	3)	IgG2b	AlexaFluor647	Invitrogen, USA	1:750	4)
488						Propidium iodide				11)
488	SLA-DR	MSA3	VMRD, USA	0.05 µl	3)	IgG2a	PE-Cy5.5	Invitrogen, USA	1:100	4)
488	SWC8	MIL-3	Dr. J.K. Lunney *	1 µl	3)	IgM	DyLight405	GeneTex, USA	1:500	4)
640	CD172α	DH59B	VMRD, USA	0.1 µl	7)	Zenon IgG1 **	AlexaFluor647	Invitrogen, USA		
640	CD203α	PM 18-7	Serotec, UK	0.1 µl	7)	Zenon IgG1 ***	APC-AlexaFluor750	Invitrogen, USA		

* a generous gift from Dr. J.K. Lunney, Animal Parasitology Institute, Beltsville, MO, USA
** labeling with the Zenon® AlexaFluor647 Mouse IgG1 Labeling Kit perform as follows: 0.1 µl of anti-CD172α and 0.5 µl of Solution A of the Zenon Kit, mix well, incubate 10 min at 4°C, then add 0.5 µl of Solution B of the Zenon kit, mix well, incubate 10 min at 4°C
*** labeling with the Zenon® APC-AlexaFluor750 Mouse IgG1 Labeling Kit perform as follows: 0.1 µl of anti- CD203α and 0.5 µl of Solution A of the Zenon kit, mix well, incubate 10 min at 4°C, then add 0.5 µl of Solution B of the Zenon kit, mix well, incubate 10 min at 4°C
The measurement was performed by using BD FACSAria I flow cytometer (Becton Dickinson, USA).

The gating strategy for identification of monocytes in the bone marrow is depicted (A). Briefly, the leukocytes were gated according their light scatter properties (upper left dot-plot). The dublets of cells were excluded from the further analysis (upper middle dot-plot). The viable (propidium iodide-negative) cell were gated (upper right dot-plot). The CD203α-positive macrophages were excluded (lower left dot-plot). The monocytes were identified as SWC8-negative (lower middle dot-plot) CD172α-positive cells (lower right dot-plot).

Then monocyte subpopulations from the bone marrow and peripheral blood were identified based on expression of SLA-DR, CD14 and CD163 (B). The SLA-DR-positive and negative monocytes were gated (left dot-plots). Then SLA-DR-negative (middle dot-plots) and SLA-DR-positive (right dot-plots) monocyte subpopulations were depicted in CD163 vs. CD14 dot-plots.

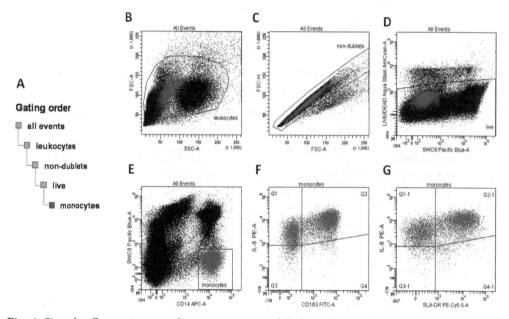

Fig. 4. Six color flow cytometry for measurement of IL-8 production by monocyte subpopulations using the Zenon-labeled anti-IL-8 antibody

The gating order for evaluation of IL-8 production by monocyte subpopulations is depicted (A). Briefly, the leukocytes were gated according their light scatter properties (B). The dublets of cells were excluded excluded from the further analysis (C). The viable (LIVE/DEAD® Fixable Aqua Dead Cell Stain-negative) cell were gated (D). The monocytes were identified as SWC8-negative CD14-positive cells (E). The IL-8 production by CD163-positive and negative monocytes (F) and by SLA-DR-positive and negative monocytes (G) was then evaluated.

The whole peripheral blood diluted 1:1 with RPMI 1640 was stimulated for 2 hours with LPS (1 µg/ml) in the presence of brefeldin A (10 µg/ml). The red blood cells were lysed with ammonium chloride solution. The following fluorescent staining was performed:

Excitation wavelength	Primary antibody					Secondary antibody / Zenon reagent / viability stain				
	Target molecule	Clone	Manufacturer	Amount per well	Stained at point	Class/subclass	Fluorochrome	Manufacturer	Dilution	Stained at point
405	SWC8	MIL-3	Dr. J.K. Lunney *	1 µl	3)	IgM	DyLight405	GeneTex, USA	1:500	4)
405						LIVE/DEAD® Fixable Aqua Dead Cell Stain Kit **				2)
488	CD163	2A10/11	VMRD, USA	0.1 µl	3)	IgG1	AlexaFluor488	Invitrogen, USA	1:750	4)
561	IL-8	8M6	Serotec, UK	0.2 µl	9)	Zenon IgG1 ***	R-PE	Invitrogen, USA		
561	SLA-DR	MSA3	VMRD, USA	0.05 µl	3)	IgG2a	PE-Cy5.5	Invitrogen, USA	1:100	4)
640	CD14	MIL-2	Serotec, UK	1 µl	3)	IgG2b	AlexaFluor647	Invitrogen, USA	1:750	4)

* a generous gift from Dr. J.K. Lunney, Animal Parasitology Institute, Beltsville, MO, USA
** diluted 1:1000 in PBS, 10 µl / well, incubation15 min at 4°C
*** labeling with the Zenon® R-Phycoerythrin Mouse IgG1 Labeling Kit perform as follows: 0.2 µl of anti-IL-8 and 4 µl of Solution A of the Zenon kit, mix well, incubate 10 min at 4°C, then add 4 µl of solution B of the Zenon kit, mix well, incubate 10 min at 4°C, then add 11.8 µl of Solution B of IntraStain kit
The measurement was performed by using BD LSRFortessa flow cytometer (Becton Dickinson, USA)

5. Intracellular measurement of the p-CrkL and CrkL levels using flow cytometry

5.1 Theoretical background

Chronic myeloid leukemia (CML) is a myeloproliferative disorder of hematopoietic stem cells that is characterized by the presence of the BCR-ABL fusion gene, which encodes the constitutively active BCR-ABL tyrosine kinase (Daley et al., 1990). Currently, the tyrosine kinase inhibitor imatinib (IM) (a potent inhibitor of BCR-ABL) is used as a first line therapy for CML patients (Baccarani et al., 2009).

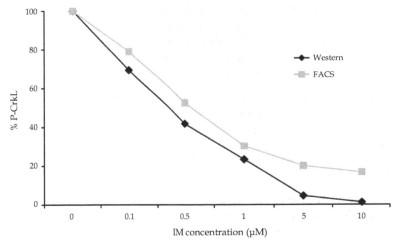

Fig. 5. The equivalence between flow cytometry and Western blot methods, in p-CrkL reduction in a BCR-ABL positive K562 cell line after 48 h treatment with imatinib (Hamilton et al., 2006)

The CrkL protein is a downstream signaling substrate of BCR-ABL, and its tyrosine phosphorylation (p-CrkL) serves as a specific indicator of BCR-ABL kinase activity in CML cells (Nichols et al., 1994; Patel et al., 2006). Recent studies have revealed that p-CrkL can act as a prognostic marker for imatinib treatment response of CML patients using either western blotting (White et al., 2005) or flow cytometry (Lucas et al., 2010). However, certain discrepancies have been found in the literature concerning the predictive value of p-CrkL in different cell types (mononuclear cells or CD34+ cells) used for analysis (Khorashad et al., 2009).

The technique for measuring p-CrkL levels using flow cytometry was originally described by Hamilton et al., (Hamilton et al., 2006) and the equivalence between flow cytometry and western blot methods was demonstrated (Figure 5).

5.2 The technique of intracellular p-CrkL and CrkL measurement by flow cytometry

Mononuclear cells (MNCs) were isolated from PB of newly diagnosed CML patients using Histopaque-1077 density gradient centrifugation (Sigma–Aldrich, St. Louis, MO, USA) and subsequently enriched for CD34+ cells using magnetic-activated cell sorting (MACS) with a CD34 MicroBead Kit (Miltenyi Biotec, Bergisch Gladbach, Germany) according to the manufacturer's instructions. CD34+ cells ($5x10^4$) were incubated for 16 h with 0, 0.5, 1.5, and 5 µM imatinib in 1 ml of serum-free medium (SFEM), supplemented with StemSpan CC100 cytokine mixture (StemCell Technologies, Köln, Germany) as previously described (Koutna et al., 2011). Then the cells were washed in phosphate buffered saline (PBS) and fixed with 4% formaldehyde for 10 min at 37°C, then washed in PBS and permeabilized by 90% methanol at 4°C for 30 min.

After permeabilization, the cells were washed in PBS and incubated with primary unlabeled antibody for 30 min at 4°C in 100 µl of FACS incubation buffer (0.5% bovine serum albumin in PBS). The concentration of primary antibodies was 12 µg/ml (p-CrkL, Cell Signaling Technology, Danvers, MA, USA; CrkL, Santa Cruz biotechnology, Santa Cruz, CA, USA; isotype control anti-normal-rabbit IgG G, R&D Systems, Minneapolis, MN, USA). The cells were washed in FACS incubation buffer and incubated with FITC-conjugated anti-rabbit IgG secondary antibody (Sigma-Aldrich) in a concentration of 10 µl/ml.

All samples were measured on a FACSCanto II Flow Cytometer (Becton Dickinson). For data analysis, BD FACSDiva (Becton-Dickinson) and FlowJo (Tree Star, Ashland, USA) software were used. The viable cell population was gated according to forward scatter and side scatter parameters. The level of p-CrkL and CrkL in the viable cells was determined as the geometric mean fluorescence intensity (MFI) of the p-CrkL- or CrkL-labeled sample minus the MFI of the isotype control (Figure 6).

The $IC50^{imatinib}$ was defined as the concentration of imatinib that caused a 50% decrease in the amount of p-CrkL compared to the untreated control (Figure 7) (White et al., 2005). The p-CrkL/CrkL ratio was calculated by dividing the concentrations of p-CrkL by those of CrkL and multiplying by 100 in untreated cells (Lucas et al., 2010). The p-CrkL ratio was assessed as a percentage of p-CrkL in the samples treated with a maximal imatinib concentration (5 µM) relative to the untreated control (Figure 7) (Khorashad et al., 2009).

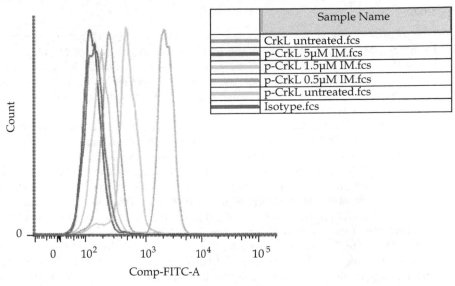

	Sample Name
	CrkL untreated.fcs
	p-CrkL 5µM IM.fcs
	p-CrkL 1.5µM IM.fcs
	p-CrkL 0.5µM IM.fcs
	p-CrkL untreated.fcs
	Isotype.fcs

Fig. 6. Representative FACS plot showing isotype, p-CrkL and CrkL MFI peaks in CD34+ cells isolated from peripheral blood of a newly diagnosed CML patient. Changes in p-CrkL MFI peaks following *in vitro* imatinib (IM) treatment are detectable and were visualized using FlowJo software

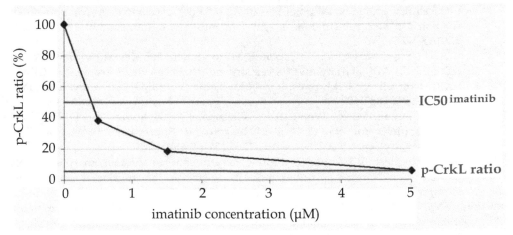

Fig. 7. The graph of p-CrkL decrease upon *in vitro* imatinib treatment. MFI peaks quantification was calculated in FlowJo software

6. Acknowledgment

This work was generously supported by grant of the Ministry of Education of the Czech Republic MSM 0021622430, grant and Ministryof Health NS-9681 and Ministry of Agriculture of the Czech Republic MZE0002716202

7. References

Baccarani, M., Cortes, J., Pane, F., Niederwieser, D., Saglio, G., Apperley, J., Cervantes, F., Deininger, M., Gratwohl, A., Guilhot, F., Hochhaus, A., Horowitz, M., Hughes, T., Kantarjian, H., Larson, R., Radich, J., Simonsson, B., Silver, R., Goldman, J., Hehlmann, R., & LeukemiaNet, E. (2009). Chronic myeloid leukemia: an update of concepts and management recommendations of European LeukemiaNet. J Clin Oncol, 27, 35, 6041-51, 1527-7755

Belloc, F., Belaud-Rotureau, M. A., Lavignolle, V., Bascans, E., Braz-Pereira, E., Durrieu, F., & Lacombe, F. (2000). Flow cytometry detection of caspase 3 activation in preapoptotic leukemic cells. Cytometry, 40, 2, 151-60, 0196-4763

Butts, C. L., Shukair, S.A., Duncan, K. M., Harris, C. W., Belyavskaya, E., & Sternberg, E.M. (2007). Evaluation of steroid hormone receptor protein expression in intact cells using flow cytometry. Nucl Recept Signal, 5, e007, 1550-7629

Cao, S., Hudnall, S.D., Kohen, F., & Lu, L. J. (2000). Measurement of estrogen receptors in intact cells by flow cytometry. Cytometry, 41, 2, 109-14, 0196-4763

Chow, S., Patel, H., & Hedley, D.W. (2001). Measurement of MAP kinase activation by flow cytometry using phospho-specific antibodies to MEK and ERK: potential for pharmacodynamic monitoring of signal transduction inhibitors. Cytometry, 46, 2, 72-8, 0196-4763

Daley, G. Q., Van Etten, R. A., & Baltimore, D. (1990). Induction of chronic myelogenous leukemia in mice by the P210bcr/abl gene of the Philadelphia chromosome. Science, 247, 4944, 824-30, 0036-8075

Gerner, W., Käser, T., Pintaric, M., Groiss, S., & Saalmüller, A. (2008). Detection of intracellular antigens in porcine PBMC by flow cytometry: A comparison of fixation and permeabilisation reagents. Vet Immunol Immunopathol, 121, 3-4, 251-9, 0165-2427

Hamilton, A., Elrick, L., Myssina, S., Copland, M., Jørgensen, H., Melo, J., & Holyoake, T. (2006). BCR-ABL activity and its response to drugs can be determined in CD34+ CML stem cells by CrkL phosphorylation status using flow cytometry. Leukemia, 20, 6, 1035-9, 0887-6924

Hulspas, R., O'Gorman, M. R., Wood, B. L., Gratama, J.W., & Sutherland, D. R. (2009). Considerations for the control of background fluorescence in clinical flow cytometry. Cytometry B Clin Cytom, 76, 6, 355-64, 1552-4957

Kalaitzidis, D., & Neel, B. G. (2008). Flow-cytometric phosphoprotein analysis reveals agonist and temporal differences in responses of murine hematopoietic stem/progenitor cells. PLoS One, 3, 11, e3776, 1932-6203

Karanikas, V., Lodding, J., Maino, V. C., & McKenzie, I. F. (2000). Flow cytometric measurement of intracellular cytokines detects immune responses in MUC1 immunotherapy. Clin Cancer Res, 6, 3, 829-37, 1078-0432

Khorashad, J., Wagner, S., Greener, L., Marin, D., Reid, A., Milojkovic, D., Patel, H., Willimott, S., Rezvani, K., Gerrard, G., Loaiza, S., Davis, J., Goldman, J., Melo, J., Apperley, J., & Foroni, L. (2009). The level of BCR-ABL1 kinase activity before treatment does not identify chronic myeloid leukemia patients who fail to achieve a complete cytogenetic response on imatinib. Haematologica, 94, 6, 861-4, 1592-8721

Koutna, I., Peterkova, M., Simara, P., Stejskal, S., Tesarova, L., & Kozubek, M. (2011). Proliferation and differentiation potential of CD133+ and CD34+ populations from the bone marrow and mobilized peripheral blood. Ann Hematol, 90, 2, 127-37, 1432-0584

Krutzik, P. O., & Nolan, G. P. (2003). Intracellular phospho-protein staining techniques for flow cytometry: monitoring single cell signaling events. Cytometry A, 55, 2, 61-70, 1552-4922

Lucas, C., Harris, R., Giannoudis, A., Knight, K., Watmough, S., & Clark, R. (2010). BCR-ABL1 tyrosine kinase activity at diagnosis, as determined via the pCrkL/CrkL ratio, is predictive of clinical outcome in chronic myeloid leukaemia. Br J Haematol, 149, 3, 458-60, 1365-2141

Maecker, H. T., & Trotter, J. (2006). Flow cytometry controls, instrument setup, and the determination of positivity. Cytometry A, 69, 9, 1037-42, 1552-4922

Mascola, J. R., Louder, M. K., Winter, C., Prabhakara, R., De Rosa, S. C., Douek, D. C., Hill, B. J., Gabuzda, D., & Roederer, M. (2002). Human immunodeficiency virus type 1 neutralization measured by flow cytometric quantitation of single-round infection of primary human T cells. J Virol, 76, 10, 4810-21, 0022-538X

Melan, M. A. (1999). Overview of cell fixatives and cell membrane permeants. Methods Mol Biol, 115, 45-55, 1064-3745

Millard, I., Degrave, E., Philippe, M., & Gala, J. L. (1998). Detection of intracellular antigens by flow cytometry: comparison of two chemical methods and microwave heating. Clin Chem, 44, 11, 2320-30, 0009-9147

Nichols, G., Raines, M., Vera, J., Lacomis, L., Tempst, P., & Golde, D. (1994). Identification of CRKL as the constitutively phosphorylated 39-kD tyrosine phosphoprotein in chronic myelogenous leukemia cells. Blood, 84, 9, 2912-8, 0006-4971

O'Gorman, M. R., & Thomas, J. (1999). Isotype controls--time to let go? Cytometry, 38, 2, 78-80, 0196-4763

Ondrackova, P., Nechvatalova, K., Kucerova, Z., Leva, L., Dominguez, J., & Faldyna, M. (2010). Porcine mononuclear phagocyte subpopulations in the lung, blood and bone marrow: dynamics during inflammation induced by Actinobacillus pleuropneumoniae. Vet Res, 41, 5, 64, 0928-4249

Ormerod, M. G. (2002). Investigating the relationship between the cell cycle and apoptosis using flow cytometry. J Immunol Methods, 265, 1-2, 73-80, 0022-1759

Pala, P., Hussell, T., & Openshaw, P. J. (2000). Flow cytometric measurement of intracellular cytokines. J Immunol Methods, 243, 1-2, 107-24, 0022-1759

Patel, H., Marley, S., & Gordon, M. (2006). Detection in primary chronic myeloid leukaemia cells of p210BCR-ABL1 in complexes with adaptor proteins CBL, CRKL, and GRB2. Genes Chromosomes Cancer, 45, 12, 1121-9, 1045-2257

Ruitenberg, J. J., & Waters, C. A. (2003). A rapid flow cytometric method for the detection of intracellular cyclooxygenases in human whole blood monocytes and a COX-2 inducible human cell line. J Immunol Methods, 274, 1-2, 93-104, 0022-1759

Tazzari, P. L., Cappellini, A., Bortul, R., Ricci, F., Billi, A. M., Tabellini, G., Conte, R., & Martelli, A. M. (2002). Flow cytometric detection of total and serine 473 phosphorylated Akt. J Cell Biochem, 86, 4, 704-15, 0730-2312

Uzel, G., Frucht, D. M., Fleisher, T. A., & Holland, S. M. (2001). Detection of intracellular phosphorylated STAT-4 by flow cytometry. Clin Immunol, 100, 3, 270-6, 1521-6616

van Stijn, A., Kok, A., van der Pol, M. A., Feller, N., Roemen, G. M., Westra, A. H., Ossenkoppele, G. J., & Schuurhuis, G. J. (2003). A flow cytometric method to detect apoptosis-related protein expression in minimal residual disease in acute myeloid leukemia. Leukemia, 17, 4, 780-6, 0887-6924

White, D., Saunders, V., Lyons, A., Branford, S., Grigg, A., To, L., & Hughes, T. (2005). In vitro sensitivity to imatinib-induced inhibition of ABL kinase activity is predictive of molecular response in patients with de novo CML. Blood, 106, 7, 2520-6, 0006-4971

Zelnickova, P., Faldyna, M., Stepanova, H., Ondracek, J., & Kovaru, F. (2007). Intracellular cytokine detection by flow cytometry in pigs: fixation, permeabilization and cell surface staining. J Immunol Methods, 327, 1-2, 18-29, 0022-1759

Zelnickova, P., Leva, L., Stepanova, H., Kovaru, F., & Faldyna, M. (2008). Age-dependent changes of proinflammatory cytokine production by porcine peripheral blood phagocytes. Vet Immunol Immunopathol, 124, 3-4, 367-78, 0165-2427

BD Biosciences; http://www.bdbiosciences.com

Beckman Coulter https://www.beckmancoulter.com

Cell Signaling Technology; http://www.cellsignal.com

Dako; http://www.dako.com

eBioscience (2011) http://www.ebioscience.com

R&D Systems http://www.rndsystems.com

10

Immunophenotypic Characterization of Normal Bone Marrow Stem Cells

Paula Laranjeira, Andreia Ribeiro, Sandrine Mendes,
Ana Henriques, M. Luísa Pais and Artur Paiva
Histocompatibility Center of Coimbra
Portugal

1. Introduction

Despite of being described more than one decade ago (Pittenger et al., 1999), the immunophenotypic profile of bone marrow mesenchymal stem cells (MSC) still not well documented. The difficulty in achieving a detailed phenotypic characterization is common in less-represented cell populations and/or populations lacking a specific known cell marker, like bone marrow MSC.

The recent advances in flow cytometry technology and the emergence of new high-speed flow cytometers have given a valuable contribute to diminish this problem in two different (but complementary) aspects: 1) by reducing dramatically the acquisition time period, making it more reasonable to study minor cell populations; and 2) by increasing the number of parameters that can be analyzed per cell at the same time, which is critical to improve the immunophenotypic characterization of those not-well characterized cell populations that lack a specific known marker.

A good example of the practical usefulness of such technical developments is the description of different cell compartments in the bone marrow CD34+ hematopoietic stem cell (HSC) population. Detailed studies on this minor bone marrow cell population demonstrated that each compartment is committed to a different hematopoietic cell lineage. An extensive immunophenotypic characterization of those CD34+ compartments allowed the development of protocols to easily and quickly identify, quantify and evaluate phenotypic aberrations and maturational blocks in those cells, which is decisive to the diagnosis, prognosis, or follow-up of a variety of hematological clonal diseases (del Cañizo et al., 2003; Lochem et al., 2004; Matarraz et al., 2008; Orfao et al. 2004).

2. Bone marrow mesenchymal stem cells

After the identification of a plastic-adherent bone marrow stromal cell population in 1976 by Friedenstein and colleagues and the first evidence of their multilineage potential (Pittenger et al., 1999) with subsequent confirmation of their stem cell nature, an increasing interest on these bone marrow MSC has emerged, mainly because of their promising therapeutic applications.

By definition, a stem cell is an undifferentiated cell with the potential ability of self-renewal and the capability of differentiation along different cell lineages (multipotency). MSC can be found on a great variety of adult tissues, where they play an important role in tissue regeneration, such as: bone marrow, adipose tissue, umbilical cord blood, umbilical cord matrix, menstrual blood, endometrium, placenta, dental pulp, skin and thymus, among others (Chamberlain et al., 2007; Ding et al. 2011; Kolf et al., 2007; Martins et al., 2009; Musina et al., 2005; Pittenger et al., 1999).

In addition to their presence in numerous adult tissues, MSC are relatively easy to isolate and have the capability to expand manyfold in culture without lose their stem cell properties. Moreover, when MSC are systemically transplanted, they are able to migrate to sites of injury and promote tissue repair, by producing growth factors or other soluble factors important to tissue regeneration, as well as by undergoing cellular differentiation (Chamberlain et al., 2007, Kolf et al., 2007; Mafi et al., 2011); such features explain the success of MSC transfusion therapy in genetic disorders affecting mesenchymal tissues (Horwitz et al., 2002; Undale et al., 2009). Furthermore, those cells have the ability of suppressing the immune response of a wide variety of immune cells, including T, B and NK lymphocytes, and antigen-presenting cells (Chamberlain et al., 2007; Stagg, 2007), and their importance in patients' clinical outcome has already been proven in severe acute graft-versus-host disease (Remberger et al., 2011; von Bahr et al., 2011). Moreover, the results achieved in animal models of autoimmune diseases are promising and encouraged the beginning of phase I clinical trials in multiple sclerosis (Constantin et al., 2009; Darlington et al., 2011; Siatskas et al., 2009).

2.1 Identification and quantification of bone marrow MSC

As referred previously, the study of minor cell populations with no known specific cell marker toke great advantage on the development of high-speed multi-parameter flow cytometers. The use of an 8-color FACSCanto II (Becton Dickinson Biosciences, BDB) flow cytometer allowed us to identify MSC in bone marrow, quantify them and further characterize their immunophenotypic profile. We employed a monoclonal antibody panel with a backbone of 3 common markers (CD13, CD45 and CD11b) for the identification of MSC (known to be CD13+CD45-CD11b-) in each tube that, at the same time, permitted the study of the expression of five more proteins on MSC per tube.

MSC are rare in bone marrow, being reported that they represent approximately 0,01% of all nucleated bone marrow cells (Chamberlain et al., 2007; Mafi et al., 2011), although is known that their number declines with aging (Caplan, 2007). Our data point to a percentage ranging between 0,01% and 0,03% of all nucleated bone marrow cells (Martins et al., 2009).

2.2 Immunophenotypic characterization of bone marrow MSC

2.2.1 Flow cytometer quality control, compensation setup strategies and other technical issues

According to the manufacturer's recommendations, it is done a daily quality control using the Rainbow Beads (BDB). In what concerns to cytometer's compensation setup, it is made once per month by setting up the Rainbow Beads (BDB) values according to the EuroFlow consortium's guidelines and then by doing a general compensation for stable fluorochromes and a specific compensation for each monoclonal antibody conjugated with tandem fluorochromes. Although the compensation is automatic, it is always revised by experienced staff at the end of the process.

In order to detect cellular autofluorescence, a negative control was made for each sample, where the bone marrow sample was only stained for CD45 PO and CD34 PerCPcy5.5.

SSC and FSC light dispersion properties allow a good discrimination between viable and dead cells and the doublets were excluded based on FSC-Area *versus* FSC-Height characteristics.

2.2.2 Material and methods

The immunophenotypic characterization of bone marrow MSC were performed in fresh EDTA-collected bone marrow samples from healthy individuals. After collection the samples were stored at 4 °C and processed within 24 hours.

Whole bone marrow samples were stained for surface cell markers using a stain-lyse-and-then-wash direct immunofluorescence technique. 200 µl of whole bone marrow were aliquoted in different tubes and stained with the following combinations of monoclonal antibodies in an 8-color staining protocol, detailed in table 1.

	FITC	PE	PerCPcy5.5	PEcy7	APC	APCH7	PB	PO
Tube 1	CD49e (SAM1) Beckman Coulter	CD73 (AD2) BD Pharmingen	CD34 (8G12) BDB	CD13 (Immu103.44) Beckman Coulter	CD90 (5E10) BD Pharmingen	HLA-DR (L243) BDB	CD11b (ICRF44) BD Pharmingen	CD45 (HI30) Invitrogen
Tube 2	CD31 (WM59) BD Pharmingen	NGFR (C40-1457) BD Pharmingen	CD14 (M5E2) BD Pharmingen	CD13	CD133 (293C3) Miltenyi Biotec	-	CD11b	CD45
Tube 3	CD15 (HI98) BDB	CD146 (P1H12) BD Pharmingen	CD24 (ALB9) Beckman Coulter	CD13	CD90	CD29 (TS2/16) BioLegend	CD11b	CD45
Tube 4	CD106 (51-10C9) BD Pharmingen	CD105 (1G2) Beckman Coulter	-	CD13	HLA-A, B, C (G46-2.6) BD Pharmingen	-	CD11b	CD45
Tube 5	-	CD73	CD24	CD13	CD90	-	CD11b	CD45

Table 1. Panel of monoclonal antibodies used for the bone marrow MSC characterization. FITC - fluorescein isothiocyanate; PE – phycoerythrin; PerCPcy5.5 - peridinin chlorophyll protein cyanine 5.5; PEcy7 - R-phycoerythrin cyanine 7; APC – allophycocyanin; APCH7 - allophycocyanin H 7; PB - pacific blue; PO - pacific orange

Data acquisition was performed in a FACSCanto II flow cytometer (BDB), using FACSDiva acquisition software (BDB). The total bone marrow cellularity of the whole sample was acquired (5 x 10⁶ events, minimum) for each tube. Bone marrow MSC were identified as CD13+/CD45-/CD11b-, as shown in Figure 1.

Data analysis was performed using Infinicyt software (Cytognos, Salamanca, Spain).

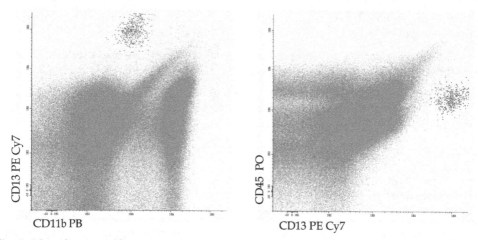

Fig. 1. Identification of bone marrow MSC (blue) present in a whole bone marrow sample, phenotypically characterized as CD13+CD45-CD11b-

2.2.3 Results and discussion

Bone marrow MSC showed to be uniformly positive to CD13, CD29, CD49e, CD90, CD106, CD146, CD73, NGFR, CD105 and HLA-A, B, C (Figure 1 and Figure 2); and negative to CD24, CD31, CD11b, CD14, CD15, CD34, CD45, CD133 and HLA-DR, which is in agreement with previous studies described in the literature (Chamberlain et al., 2007; Delorme et al., 2008; Ehninger & Trumpp, 2011; Fox et al., 2007; Jones & McGonagle, 2008; Kolf et al., 2007; Martins et al., 2009; Pittenger et al., 1999; Tormin et al., 2011). Based on the expression profile of these markers, bone marrow MSC behave as one sole cell population, as all the studied markers were homogeneously expressed inside the MSC population.

Several studies on adhesion molecules and chemokine receptors expression have been made in order to shed light on MSC migratory and homing ability. CD29 (integrin β_1-subunit) and CD106 (vascular cell adhesion molecule 1, VCAM-1) seem to be important in the adhesion of MSC to endothelial cells (Chamberlain et al., 2007; Kolf et al., 2007; Stagg, 2007) and CD29, which when dimerized with CD49e (integrin α_5-subunit) forms a receptor that binds to fibronectin and invasin, is likely to promote MSC-extracellular matrix interaction (Gu et al., 2009). CD146 (Muc18) plays an important role in cell-cell and cell-extracellular matrix adhesion and an increased expression of these marker on tumor cells is associated with an increased cell motility and invasiveness/ metastasis capability (Bardin et al., 2001; Zeng et al., 2011). The glycoprotein CD90 (Thy-1) regulates as well cell-cell and cell-extracellular matrix interactions, being involved in adhesion to endothelial cells, migration, metastasis and tissue regeneration (Jurisic et al., 2010; Rege & Hagood, 2006).

The enzyme CD73 is an ecto-5'-nucleotidase that produces extracellular adenosine. In animal tumor models, CD73-generated adenosine inhibits both homing and expansion of T cells via adenosine-receptor signaling. In fact, recent research shows that adenosine suppresses T cell immune response both in activation and effector phases, as well as NK cell immune activity (Wang et al., 2011; Zhang et al., 2010).

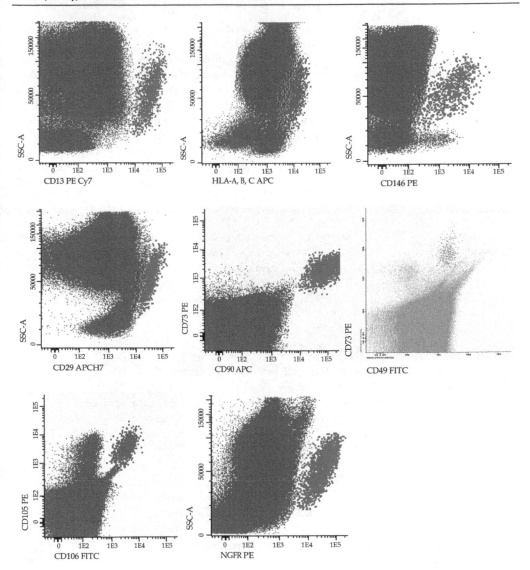

Fig. 2. Immunophenotypic characteristics of bone marrow MSC (blue). The remaining bone marrow nucleated cells are represented as grey events

In what concerns to growth factor receptors, NGFR (nerve growth factor receptor, CD271) is expressed in a wide variety of tissues and, depending on the cell type, signaling through this receptor regulates NF-kB activation, apoptosis, tissue regeneration, immune cell activation, proliferation and cell differentiation (Micera et al., 2007; Rogers et al., 2010). Finally, CD105 (endoglin) is one of the receptors for TGF-β, a growth factor involved in the regulation of development, maintenance and proliferation of MSC (Stagg, 2007), and also known to play an important role in tissue repair.

Some discrepancies described in the expression of adhesion molecules, chemokine receptors and other proteins, may be the reflex of the microenvironmental differences present in different studies. Although there are a great similitude in the phenotypic profile of MSC isolated from different tissues, differences do exist (Chamberlain et al., 2007; Kolf et al., 2007; Martins et al., 2009). As well as different cultures conditions can also change the MSC phenotype (Chamberlain et al., 2007; Halfon et al., 2011; Stagg, 2007; Tormin et al., 2011). This could be a clue of MSC highly sensitiveness to microenvironment alterations, and their potential to change their protein expression profile could be of great importance in giving an appropriate response to physiological or pathological challenges: by changing their migratory pattern, by initiating an immunomodulatory or immunosuppressive response, by modifying the production and release of soluble factors, or by undergoing cell differentiation.

As a minor bone marrow cell population easy to expand in vitro, it is attractive to characterize the MSC immunophenotype after culture cell expansion. Nevertheless, characterizing these cells directly (without previous culture) enables an analysis closest to their physiological conditions, excluding the phenotypic alterations induced by factors present in the culture medium. Moreover, this direct approach allows an accurate quantification of MSC in bone marrow. Also, this same strategy can be applied to MSC from other tissues.

3. Bone marrow hematopoietic stem cells

The multipotent hematopoietic stem cell is mainly located in the bone marrow of adult animals and has the ability to differentiate along all hematopoietic cell lineages. A number of studies based on in vitro cell culture, xeno-transplantation of hematopoietic human cells in immunodeficient mice and in pre-immune animal fetuses, were carried out to identify the human hematopoietic stem cell and unveil the hematopoietic precursors hierarchy (Nimer, 2008; Yin et al., 2007), becoming clear that CD34-positive cells were able to differentiate and give rise to all blood cells. There are evidences that, within this heterogeneous population, the more immature CD34+ HSC expresses CD133 and are CD38-negative/dim. It is also known that the CD34+CD133+ subpopulation can arise from the CD133+CD34-CD38-subset (Goussetis et al., 2006; Nimer, 2008; Yin et al., 1997).

3.1 Identification and quantification of the different bone marrow CD34+ HSC cell compartments

As already referred, CD34-positive cells are an heterogeneous bone marrow cell population, consisting in various cell compartments differing in immunophenotype, size and lineage commitment. The immunophenotypic pattern of each compartment is well described and, with a relatively low number of markers, the majority of those subsets can be accurately and easily identified.

Attending only to the immunophenotypic features, is possible to identify the following bone marrow CD34+ cell subsets by flow cytometry: uncommitted (more immature) precursors, neutrophil precursors, B cell precursors, monocytic precursors, plasmacytoid dendritic cells precursors, erythroid precursors, basophil precursors and mast cell precursors.

A detailed immunophenotypic description of human bone marrow CD34+ cells was published, few years ago, by Matarraz and colleagues (Matarraz et al., 2008) and Lochem and colleagues (Lochem et al., 2004), along with the frequency of each CD34+ cell subpopulation in normal hematopoiesis (Matarraz et al. 2008), presented on table 2.

Bone marrow CD34+ HSC compartments	Mean ± Standard deviation	Range
% Bone marrow CD34+ HSC (of total bone marrow)	0,9 ± 0,3	(0,2-1,6)
Immature CD34+ precursor (%) (within CD34+ cells)	52 ± 12	(19-66)
CD34+ neutrophil precursors (%) (within CD34+ cells)	34 ± 7	(15-47)
CD34+ B cell precursors (%) (within CD34+ cells)	14 ± 10	(1-36)
CD34+ monocytic precursors (%) (within CD34+ cells)	10 ± 7	(0-26)
CD34+ plasmacytoid dendritic cell precursors (%) (within CD34+ cells)	5 ± 2	(0-9)
CD34+ erythroid precursors (%) (within CD34+ cells)	18 ± 8	(1-36)
CD34+ basophil precursors (%) (within CD34+ cells)	0,7 ± 0,4	(0-1,5)
CD34+ mast cell precursors (%) (within CD34+ cells)	0 ± 0,005	(0-0,02)

Table 2. Distribution of the different cell compartments of bone marrow CD34+ HSC. The results are expressed as mean ± standard deviation (range). Adapted from Matarraz et al. Leukemia 2008

The most immature CD34+ subset can be identified based on CD133 expression (Goussetis et al., 2006; Pastore et al., 2008; Yin et al., 1997). When other markers are concerned, these cell are $CD34^{hi}/CD45^{int}/HLA\text{-}DR^{hi}/cyMPO^{-}/nTdT^{-}/CD117^{hi}$ and have intermediate side scatter (SSC) and forward scatter (FSC) light dispersion properties (Matarraz et al., 2008). As previously described by Matarraz and colleagues, the phenotypic profile of CD34+ B cell precursors is $CD34^{int}/CD45^{int/dim}/HLA\text{-}DR^{hi}/cyMPO^{-}/nTdT^{int}/CD117^{-}$ and these cells present the lowest SSC and FSC of all CD34+ subpopulations (Lochem et al., 2004; Matarraz et al., 2008); the CD34+ neutrophil precursors present $CD34^{hi}/CD45^{int/dim}/HLA\text{-}DR^{hi}/cyMPO^{int/hi}/nTdT^{-}/CD117^{hi}$, along with the highest values for SSC and FSC of all CD34+ subsets; the CD34+ plasmacytoid dendritic cell precursors are identified based on the expression of $CD34^{+}/CD123^{hi/int}/HLA\text{-}DR^{hi}$; CD34+ monocytic precursors display $CD34+/HLA\text{-}DR^{hi}/CD64^{hi}/CD45^{hi}/CD117^{-}$ immunophenotype; basophil precursors are described as being $CD34+/CD123^{int/hi}/HLA\text{-}DR^{-/+}$; and CD34+ mast cell precursors are $CD34+/CD117^{hi}/HLA\text{-}DR^{-/int}$ (Matarraz et al., 2008). Finally, CD34+ erythroid precursors are characterized by $CD34+/CD36+/CD64^{-}/CD45^{lo}$ immunophenotype (Matarraz et al., 2008) and by CD105 expression (Buhring et al., 1991; Rokhlin et al., 1995). As a matter of fact, CD105 and TGF-β_1 have a pivotal role in the regulation of the differentiation in the erythroid lineage (Fortunel et al., 2000; Moody et al., 2007).

3.2 A single-tube protocol to identify the different bone marrow CD34+ HSC compartments

Recently, we developed an 8-color single-tube protocol to identify the different bone marrow CD34+ HSC subsets by flow cytometry.

The single-tube protocol we propose here was constructed to allow an accurate, quick and easy identification and quantification of those cellular compartments. Attending to the monoclonal antibodies and fluorochrome-conjugation available on the market and to compensation issues, and based on our experience and knowledge on the hematopoietic maturation dynamics, we elected the best markers to identify with precision the cell populations of interest.

3.2.1 Material and methods

The immunophenotypic characterization of bone marrow CD34+ precursors were performed in fresh EDTA-collected bone marrow samples from healthy individuals. After collection, the samples were stored at 4 °C and processed within 24 hours. The quality control and compensation strategies are described in detail in section 2.2.1.

A stain-lyse-and-then-wash direct immunofluorescence protocol was used, and the monoclonal antibodies were combined as presented on table 3.

	FITC	PE	PerCPcy 5.5	PEcy7	APC	APCH7	PB	PO
Single Tube Protocol	CD35 (E11) BDB Pharmingem	CD123 (SSDCL Y107D2) Beckman Coulter	CD34 (8G12) BDB	CD117 (PN IM3698) Beckman Coulter	CD133 (293C3) Miltenyi Biotec	HLA-DR (L243) BDB	CD44 (IM7) Biolegend	CD45 (HI30) Invitrogen

FITC - fluorescein isothiocyanate; PE – phycoerythrin; PerCPcy5.5 - peridinin chlorophyll protein cyanine 5.5; PEcy7 - R-phycoerythrin cyanine 7; APC – allophycocyanin; APCH7 - allophycocyanin H 7; PB - pacific blue; PO - pacific orange.

Table 3. Panel of monoclonal antibodies used for the identification and quantification of the different subpopulations found in bone marrow CD34+ HSC

Data acquisition was performed on a FACSCanto II flow cytometer (BDB), using FACSDiva acquisition software (BDB). In a first step of acquisition, the whole bone marrow cellularity was stored (100.000 events). In a second step, only events within the CD34+ electronic gate were acquired (5.000 to 10.000 CD34+ events).

Data analysis was performed using Infinicyt software (Cytognos, Salamanca, Spain).

3.2.2 How to identify the different CD34+ HSC compartments with the single-tube protocol?

1. The most immature (uncommitted) compartment of bone marrow C34+ HSC
   The most immature compartment can be easily identified based on their positivity to CD133 marker (CD133hi). To differentiate this subset from CD34+ neutrophil precursors and CD34+ plasmacytoid dendritic cell precursors, also expressing CD133 (CD133int),

other important phenotypic characteristics have to be taken into account: CD35-/CD34hi/HLA-DRhi/CD117hi/FSCint/SSCint/CD123-. Figure 3 presents a detailed immunophenotype of this compartment considering all the markers used in this protocol.

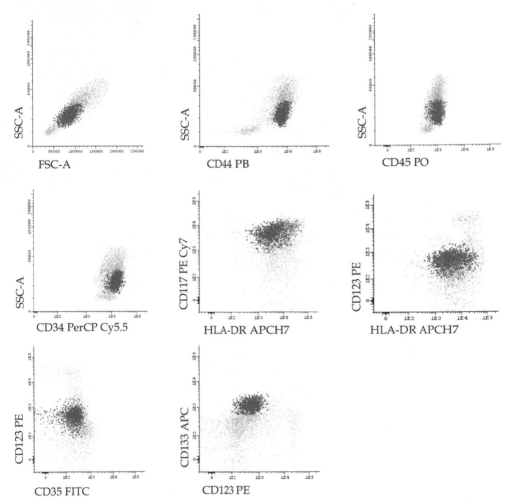

Fig. 3. Uncommitted bone marrow CD34+ HSC (red) immunophenotype. The remaining bone marrow CD34+ cell compartments are presented in grey

2.   Bone marrow CD34+ erythroid precursors
Both CD34+ erythroid precursors and monocytic precursors express CD35. The two CD34+ subpopulations can be distinguished in this protocol by the expression of CD117 and HLA-DR. The erythroid precursors are CD117+/HLA-DRint and the monocytic precursors are CD117dim/-/HLA-DRhi. Moreover, the erythroid precursors are characterized by a dim expression of CD34, CD45 and CD44 (Figure 4).

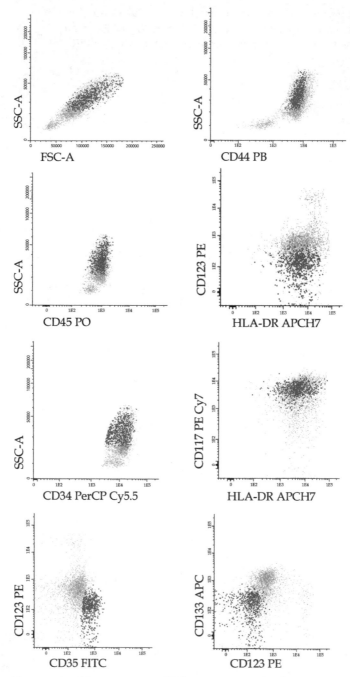

Fig. 4. Erythroid-committed bone marrow CD34+ precursors (red) immunophenotype. The remaining bone marrow CD34+ cell compartments correspond to the grey events

It is worth mentioning that our previous studies with simultaneous staining of CD105 and CD35 proved that the two markers were co-expressed in the same subset of CD34+ bone marrow cells and CD35 appears slightly before CD105 (Figure 5)[1].

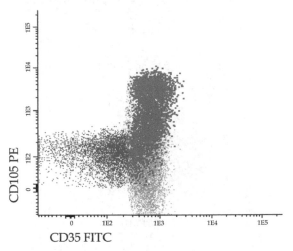

Fig. 5. Expression of CD105 and CD35 in bone marrow erythroid lineage: uncommitted CD34+ cells (red), CD34+erythroid precursors (blue) and CD34- erythroid precursors (grey)

3. Bone marrow CD34+ neutrophil precursors
    Neutrophil precursors show high reactivity to CD44 antigen, as the plasmacytoid dendritic cell precursors (CD44^{hi}), but in the absence of CD123 marker. Other important immunophenotypic features of this CD34+ compartment are: CD133^{int}/CD35-/HLA-DR^{hi}/CD117^{hi}/CD45$^{int/dim}$/FSC^{hi}/SSC^{hi} (Figure 6).

4. Bone marrow CD34+ monocyte precursors
    Using this single-tube approach, the monocyte precursors are primarily identified by exclusion of all the other myeloid CD34+ precursors. Is noteworthy that a large percentage of monocyte-committed CD34+ precursors express CD35, being discriminated from CD34+ erythroid precursors by their CD117$^{dim/-}$/HLA-DR+/CD45^{hi} phenotype. Although classically the identification of this CD34+ subset was made focusing on the expression of CD64, this marker seems to be also present on CD34+ plasmacytoid and myeloid dendritic cell precursors. In line with this, CD35 might be a good option to the identification of CD34+ monocyte precursors. The immunophenotype of this population is depicted in Figure 7.

5. Bone marrow CD34+ B cell precursors
    Even in the absence of an B-cell lineage specific marker, as CD19 or CD79a, CD34+ B cell precursors are clearly identified by the low expression of CD44 and CD45, along with low light scatter properties (Figure 8).

---

[1] According to our experience, CD35 seems to be expressed earlier than CD105 and CD36 on erythroid committed CD34+ precursors, allowing a more accurate quantification of this subset.

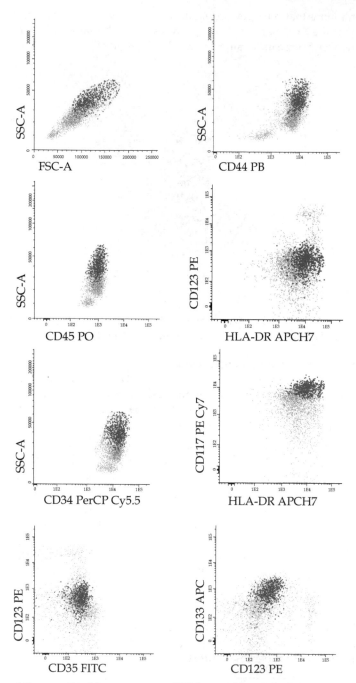

Fig. 6. Neutrophil-committed bone marrow CD34+ precursors (red) immunophenotype. The remaining bone marrow CD34+ cell compartments are presented in grey

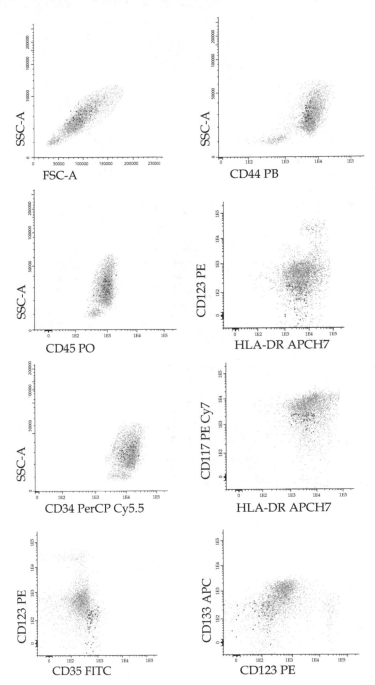

Fig. 7. Monocytic-committed bone marrow CD34+ precursors (red) immunophenotype. The remaining bone marrow CD34+ cell compartments are presented in grey

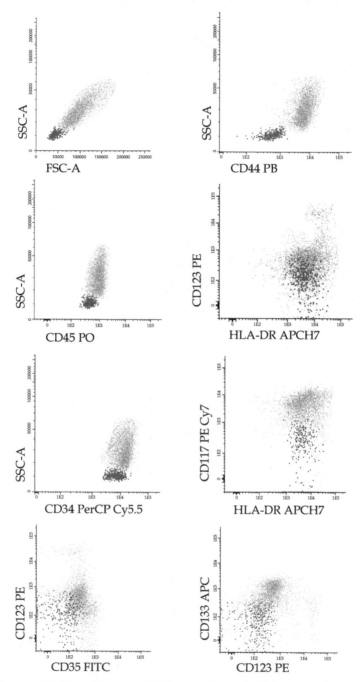

Fig. 8. B-cell-committed bone marrow CD34+ precursors (red) immunophenotype. The remaining bone marrow CD34+ cell compartments are presented in grey

6. Bone marrow CD34+ basophil precursors

Our protocol allows the identification of basophil precursors using the classical markers and attending to the immunophenotype HLA-DR$^{-/dim}$/CD123$^{int/hi}$. Of note, this CD34+ subset presents the lowest expression of CD44 among all bone marrow myeloid CD34+ cells, being easy to differentiate this precursors from all the other myeloid precursors by using CD44 marker (Figure 9).

7. Bone marrow CD34+ plasmacytoid dendritic cell precursors

The plasmacytoid dendritic cell precursors are identified using the classical markers, as being HLA-DR^{hi}/CD123$^{hi/int}$. The most immature forms of this precursor express CD133 (CD133^{int}). The immunophenotypic characteristics of this population are represented on Figure 10.

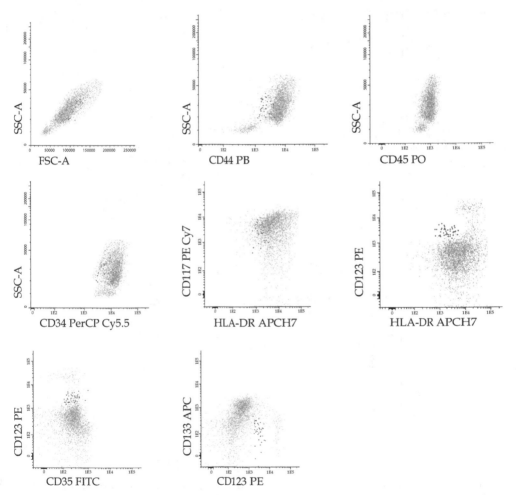

Fig. 9. Basophil-committed bone marrow CD34+ precursors (red) immunophenotype. The remaining bone marrow CD34+ cell compartments are presented in grey

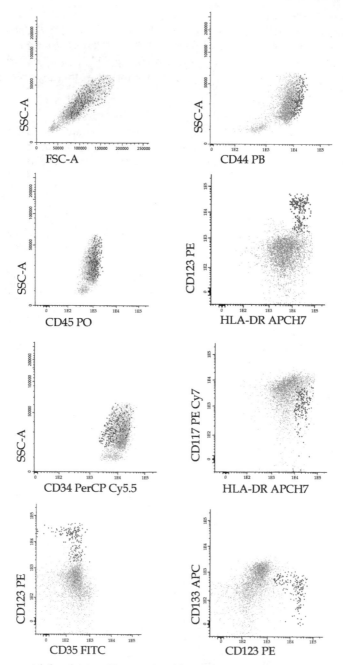

Fig. 10. Plasmacytoid dendritic cell-committed bone marrow CD34+ precursors (red) immunophenotype. The remaining bone marrow CD34+ cell compartments are presented in grey

8. Bone marrow CD34+ mast cell precursors

The classical markers for the identification of CD34+ mast cell precursors are included in our protocol, and these cells are CD117^{hi}/HLA-DR$^{-/int}$. This subset expresses high levels of CD44. Other immunophenotypic characteristics of this subset are illustrated in Figure 11.

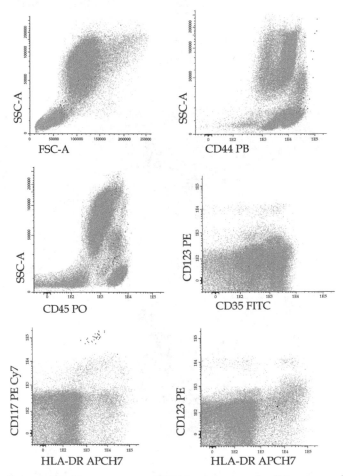

Fig. 11. Mast cell-committed bone marrow CD34+ precursors (red) immunophenotype. The events in grey correspond to remaining whole bone marrow nucleated cells

3.3 The maturation dynamic of bone marrow CD34+ hematopoietic stem cell

The possibility of a multiparameter analysis in a single cell basis conduct to a broader knowledge on the immunophenotypic characteristics of bone marrow CD34+ compartments and how it varies along the differentiation through different hematological cell lineages. Figure 12 depicts the dynamic of the maturation of different bone marrow CD34+ cell compartments.

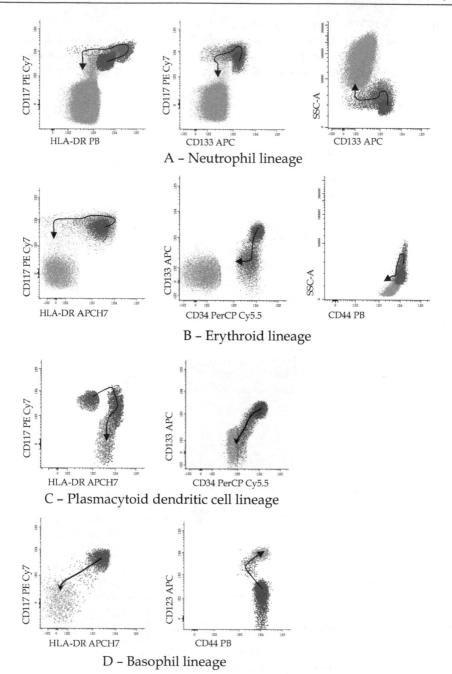

A – Neutrophil lineage

B – Erythroid lineage

C – Plasmacytoid dendritic cell lineage

D – Basophil lineage

Fig. 12. Maturational dynamic of bone marrow CD34+ HSC. Uncommitted CD34+ cells are presented in red, lineage committed CD34+ cells are presented in blue and the lineage committed CD34- cells correspond to grey events

4. Conclusion

The emergence of high-speed multi-parameter flow cytometers have given an important contribute to unveil the phenotypic characteristics of minor cell populations and/or populations without a known specific cell marker.

Using flow cytometry to characterize bone marrow MSC directly (without in vitro cell culture) represents a great advantage by enabling an analysis closest to the physiologic conditions of the cells, excluding all the phenotypic alterations induced by factors present in the culture medium. Moreover, this direct analysis allows an accurate quantification of these cells in bone marrow. In addition, the strategy used for bone marrow can also be applied in MSC from other tissues, allowing their direct quantification and characterization.

A broader knowledge about the immunophenotypic characteristics of the different compartments of bone marrow HSC could improve their identification, allow a more accurate quantification of those compartments, as well as shed light on the protein expression patterns in the earliest stages of maturation of each hematological cell lineage. Furthermore, a better knowledge of those protein expression patterns might contribute to the development of new strategies to identify aberrant phenotypes in hematological diseases affecting the more immature bone marrow cells compartments, which can be helpful in the classification of acute leukemias, diagnosis of myelodysplastic syndromes and detection of minimal residual disease. A more extensive understanding of the phenotype of CD34+ hematopoietic stem cells in the different maturational stages could also be useful to monitoring and investigate if different mobilization regimens have the capability of mobilizing distinct CD34+ hematopoietic stem cells subpopulations.

Here, we presented a simple, quick and economic approach to identify and quantify the different bone marrow CD34+ HSC compartments.

5. References

Bardin N, Anfosso F, Massé J, Cramer E, Sabatier F, Le Bivic A, Sampol J & Dignat-George F. (2001). Identification of CD146 as a component of the endothelial junction involved in the control of cell-cell cohesion. *Blood*, Vol.98, No.13, (December 2001), pp. 3677-3684, ISSN 0006-4971.

Bühring HJ, Müller CA, Letarte M, Gougos A, Saalmüller A, van Agthoven AJ, Busch FW. (1991). Endoglin is expressed on a subpopulation of immature erythroid cells of normal human bone marrow. *Leukemia*, Vol.5, No.10, (October 1991), pp. 841-847, ISSN 0887-6924.

Caplan A. (2007). Adult mesenchymal stem cells for tissue engineering versus regenerative medicine. *J Cell Phisiol*, Vol.213, No.2, (November 2007), pp. 341-347, ISSN 1097-4652.

Chamberlain G, Fox J, Ashton B & Middleton J. (2007). Concise review: mesenchymal stem cells: their phenotype, differentiation capacity, immunological features, and potential for homing. *Stem Cells*, Vol.25, No.11 (November 2007), pp. 2739-2749, ISSN 1549-4918.

Constantin G, Marconi S, Rossi B, Angiari S, Calderan L, Anghileri E, Gini B, Bach D, Martinello M, Bifari F, Galie M, Turano E, Budui S, Sbarabti A, Krampera M &

Bonetti B. (2009). *Stem Cells*, Vol.27, No.10, (October 2009), pp. 2624-2635, ISSN 1549-4918.

Darlington PJ, Boivin MN, Bar-Or A. (2011). Harnessing the therapeutic potential of mesenchymal stem cells in multiple sclerosis. *Expert Rev Neurother*, Vol.11, no.9, (September 2011), pp. 1295-303, ISSN 1473-7175.

Del Cañizo MC, Fernández ME, López A, Vidriales B, Villarón E, Arroyo JL, Ortuño F, Orfao A & San Miguel JF. (2003). Immunophenotypic analysis of myelodysplastic syndromes. *Haematologica*, Vol.88, No.4, (April 2003), pp. 402-407, ISSN 0390-6078.

Delorme B, Ringe J, Gallay N, Le Vern Y, Kerboeuf D, Jorgensen C, Rosset P, Sensebé L, Layrolle P, Häupl T & Charbord P. (2008).Specific plasma membrane protein phenotype of culture-amplified and native human bone marrow mesenchymal stem cells. *Blood*, Vol.111, No.5, (March 2008), pp. 2631-2635, ISSN 0006-4971.

Ding DC, Shyu WC & Lin S. (2011). Mesenchymal stem cells. *Cell Transplant*, Vol.20, No.1, (2011), pp. 5–14, ISSN 1555-3892.

Ehninger A & Trumpp A. (2011). The bone marrow stem cell niche grows up: mesenchymal stem cells and macrophages move in. *J Exp Med*, Vol.208, No.3, (March 2011), pp. 421-428, ISSN 0022-1007.

Fortunel N, Hatzfeld A, & Hatzfeld J. (2000). Transforming growth factor-b: pleiotropic role in the regulation of hematopoiesis. *Blood*, Vol.96, No.6, (September 2000), pp. 2022-2036, ISSN 0006-4971.

Fox JM, Chamberlain G, Ashton BA & Middleton J. (2007). Recent advances into the understanding of mesenchymal stem cell trafficking. *Br J Haematol*, Vol.137, No.6, (June 2007), pp. 491-502, ISSN 1365-2141.

Goussetis E, Theodosaki M, Paterakis G, Tsecoura C & Graphakos S. (2006). In vitro identification of a cord blood CD133+CD34-Lin+ cell subset that gives rise to myeloid dendritic precursors. *Stem Cells*, Vol.24, No.4, (April 2006), pp. 1137-1140, ISSN 1549-4918.

Gu J, Isaji T, Sato Y, Kariya Y& Fukuda T. (2009).Importance of N-glycosylation on alpha5beta1 integrin for its biological functions. *Biol Pharm Bull*, Vol.32, No.5, (May 2009), pp. 780-785, ISSN 0918-6158.

Halfon S, Abramov N, Grinblat B & Ginis I. (2011). Markers distinguishing mesenchymal stem cells from fibroblasts are downregulated with passaging. *Stem Cell Dev*, Vol.20, No.1, (January 2011), ISSN 1547-3287.

Horwitz EM, Gordon PL, Koo WK, Marx JC, Neel MD, McNall RY, Muul L & Hofmann T. (2002). Isolated allogeneic bone marrow-derived mesenchymal cells engraft and stimulate growth in children with osteogenesis imperfecta: Implications for cell therapy of bone. *Proc Natl Acad Sci USA*, Vol.99, No.13, (June 2002), pp. 8932-8937, ISSN 0027-8424.

Jones E & McGonagle D. (2008).Human bone marrow mesenchymal stem cells in vivo. *Rheumatology (Oxford)*. Vol.47, No.2, (February 2008), pp. 126-131, ISSN 1462-0324.

Jurisic G, Iolyeva M, Proulx ST, Halin C & Detmar M. (2010). Thymus cell antigen 1 (Thy1, CD90) is expressed by lymphatic vessels and mediates cell adhesion to lymphatic endothelium. *Exp Cell Res*, Vol.316, No.17, (October 2010) pp. 2982-2992, ISSN 0014-4827.

Kolf CM, Cho E & Tuan RS. (2007). Mesenchymal stromal cells. Biology of adult mesenchymal stem cells: regulation of niche, self-renewal and differentiation. *Arthritis Res Ther*, Vol.9, No.1, (February 2007), pp. 204-214, ISSN 1478-6354.

Martins AA, Paiva A, Morgado JM, Gomes A & Pais ML. (2009). Quantification and immunophenotypic characterization of bone marrow and umbilical cord blood mesenchymal stem cells by multicolor flow cytometry. *Transplant Proc*, Vol.41, No.3, (April 2009), pp. 943-946, ISSN 0041-1345.

Matarraz S, López A, Barrena S, Fernandez C, Jensen E, Flores J, Bárcena P, Rasillo A, Sayagues JM, Sánchez ML, Hernandez-Campo P, Hernandez Rivas JM, Salvador C, Fernandez-Mosteirín N, Giralt M, Perdiguer L & Orfao A. (2008). The immunophenotype of different immature, myeloid and B-cell lineage-committed CD34+ hematopoietic cells allows discrimination between normal/reactive and myelodysplastic syndrome precursors. *Leukemia*, Vol.22, No.6, (June 2008), pp. 1175-1183, ISSN 0887-6924.

Micera A, Lambiase A, Stampachiacchiere B, Bonini S, Bonini S & Levi-Schaffer F. (2007). Nerve growth factor and tissue repair remodeling: trkA(NGFR) and p75(NTR), two receptors one fate. *Cytokine Growth Factor Rev*, Vol.18, No.3-4, (June-August 2007), pp. 245-256, ISSN 1359-6101.

Moody JL, Singbrant S, Karlsson G, Blank U, Aspling M, Flygare J, Bryder D & Karlsson S. (2007). Endoglin is not critical for hematopoietic stem cell engraftment and reconstitution but regulates adult erythroid development. *Stem Cells*, Vol.25, No.11, (November 2007), pp. 2809-2819, ISSN 1549-4918.

Musina RA, Bekchanova ES & Sukhikh GT. (2005). Comparison of mesenchymal stem cells obtained from different human tissues. *Bull Exp Biol Med*, Vol.139, No.4, (April 2005), pp. 504-9, ISSN 0007-4888.

Nimer S. (2008). MDS: a stem cell disorder--but what exactly is wrong with the primitive hematopoietic cells in this disease? *Hematology Am Soc Hematol Educ Program*, (2008), pp. 43-51. ISSN 1520-4391.

Orfao A, Ortuño F, Santiago M, Lopez A & San Miguel J. (2004). Immunophenotyping of Acute Leukemias and Myelodysplastic Syndromes. *Cytometry A*, Vol.58A, No.1, (March 2004), pp. 62-71. ISSN 1552-4930.

Pastore D, Mestice A, Perrone T, Gaudio F, Delia M, Albano F, Russo Rossi A, Carluiccio P, Leo M, Liso V& Specchia G. (2008). Subsets of CD34+ and early engraftment kinetics in allogeneic peripheral SCT for AML. *Bone Marrow Transplant*, Vol.41, No.11, (June 2008), pp. 977-981, ISSN 0268-3369.

Pittenger MF, Mackay AM, Beck SC, Jaiswal RK, Douglas R, Mosca JD, Moorman MA, Simonetti DW, Craig S & Marshak DR. (1999). Multilineage potential of adult human mesenchymal stem cells. *Science*, Vol.284, No.5411, (April 1999), pp. 143-7, ISSN 0036-8075.

Rege TA & Hagood JS. (2006). Thy-1 as a regulator of cell-cell and cell-matrix interactions in axon regeneration, apoptosis, adhesion, migration, cancer, and fibrosis. *FASEB J*, Vol.20, No.8, (June 2006), pp. 1045-54, ISSN: 0892-6638.

Remberger M, Uhlin M, Karlsson H, Omazic B, Svahn BM & Mattsson J. (2011). Treatment with mesenchymal stromal cells does not improve long-term survival in patients with severe acute GVHD. *Transpl Immunol*, 2011 Sep 10 [Epub ahead of print], ISSN 0966-3274.

Rogers ML, Bailey S, Matusica D, Nicholson I, Muyderman H, Pagadala PC, Neet KE, Zola H, Macardle P & Rush RA. ProNGF mediates death of Natural Killer cells through activation of the p75NTR-sortilin complex. *J Neuroimmunol*, Vol.226, No.1-2, (September 2010), pp. 93-103, ISSN: 0165-5728.

Rokhlin OW, Cohen MB, Kubagawa H, Letarte M & Cooper MD. (2005). Differential expression of endoglin on fetal and adult hematopoietic cells in human bone marrow. *J Immunol*, Vol.154, No.9, (May 1995), pp. 4456-4465, ISSN: 0022-1767.

Siatskas C, Payne NL, Short MA & Bernard CC. (2010). A consensus statement addressing mesenchymal stem cell transplantation for multiple sclerosis: it's time! *Stem Cell Rev*, Vol.6, No,4, (December 2010), pp. 500-506, ISSN 1550-8943.

Stagg J. (2007). Immune regulation by mesenchymal stem cells: two sides to the coin. *Tissue Antigens*, Vol.69, No.1, (January 2007), pp. 1-9, ISSN 0001-2815.

Tormin A, Li O, Brune JC, Walsh S, Schütz B, Ehinger M, Ditzel N, Kassem M & Scheding S. (2011). CD146 expression on primary nonhematopoietic bone marrow stem cells is correlated with in situ localization. *Blood*, Vol.117, No.19, (May 2011), pp. 5067-77, ISSN 0006-4971.

Undale AH, Westendorf JJ, Yaszemski MJ & Khosla S. (2009). Mesenchymal stem cells for bone repair and metabolic bone diseases. *Mayo Clin Proc*, Vol.84, No.10, (October 2009), pp. 893-902, ISSN: 0025-6196.

van Lochem EG, van der Velden VH, Wind HK, te Marvelde JG, Westerdaal NA & van Dongen JJ. (2004). Immunophenotypic differentiation patterns of normal hematopoiesis in human bone marrow: reference patterns for age-related changes and disease-induced shifts. *Cytometry B Clin Cytom*, Vol.60, No.1, (July 2004), pp. 1-13, ISSN 1552-4957.

von Bahr L, Sundberg B, Lönnies L, Sander B, Karbach H, Hägglund H, Ljungman P, Gustafsson B, Karlsson H, Le Blanc K & Ringdén O. (2011). Long-Term Complications, Immunologic Effects, and Role of Passage for Outcome in Mesenchymal Stromal Cell Therapy. *Biol Blood Marrow Transplant*, 2011 Aug 4 [Epub ahead of print]. ISSN 1083-8791.

Wang L, Fan J, Thompson LF, Zhang Y, Shin T, Curiel TJ & Zhang B. (2011). CD73 has distinct roles in nonhematopoietic and hematopoietic cells to promote tumor growth in mice. *J Clin Invest*, Vol.121, No.6, (June 2011), pp. 2371-2382, ISSN 0021-9738.

Yin AH, Miraglia S, Zanjani ED, Almeida-Porada G, Ogawa M, Leary AG, Olweus J, Kearney J & Buck DW. (1997). AC133, a novel marker for human hematopoietic stem and progenitor cells. *Blood*, Vol.90, No.12, (December 1997), pp. 5002-5012, ISSN 0006-4971.

Zeng GF, Cai SX, Wu GJ. (2011). Up-regulation of METCAM/MUC18 promotes motility, invasion, and tumorigenesis of human breast cancer cells. *BMC Cancer*, Vol. 11, (March 2011), pp. 113-126, ISSN 1471-2407 .

Zhang B. (2010). CD73: A novel target for cancer immunotherapy. *Cancer Res*, Vol.70, No.16, (August 2010), pp. 6407–6411, ISSN 0008-5472.

11

Biological Effects Induced by Ultraviolet Radiation in Human Fibroblasts

Silvana Gaiba et al*,
Universidade Federal de São Paulo – Unifesp,
Universidade Estadual de Santa Cruz – Uesc
Universidade Nove de Julho – Uninove
Brazil

1. Introduction

As the most superficial body organ, skin plays an important role in protecting the body from environmental damage. The skin is composed of three layers: the epidermis, dermis and subcutaneous tissue. The epidermis, the outermost layer, has as main functions to protect the body against harmful environmental stimuli and to reduce fluid loss. It is a stratified squamous epithelium with several layers and its major cell type is the keratinocyte. This tissue is constantly being renewed by keratinization, a process of detachment of cornified cells (Blumenberg & Tomic-Canic, 1997). Located under the epidermis are the dermis and the dermal connective tissue, with extracellular matrix proteins such as collagen, elastic fibers, fibronectin, glycosaminoglycans and proteoglycans, which are produced and secreted into the extracellular space by fibroblasts, the major cell type found in this tissue (Makrantonaki & Zouboulis, 2007). The extracellular matrix proteins in the dermal connective tissue contribute for maintaining skin preservation and integrity (Hwang *et al.*, 2011). Stromal fibroblasts play an important role in tissue homeostasis regulation and wound repair via protein synthesis and secretion of growth factors or cytokines of paracrine action with direct effect on proliferation and differentiation of adjacent epithelial tissues (Andriani *et al.*, 2011). Solar ultraviolet (UV) radiation is a predictable epidemiologic risk factor for melanoma and non-melanoma skin cancers. (Katiyar *et al.*, 2011). UV irradiation can impair cellular functions by directly damaging DNA to induce apoptosis (Wäster & Ollinger, 2009). Among other things, longer UV wavelengths (UVB, UVA) induce oxidative stress and protein denaturation whereas short wavelength UV radiation (UVC) causes predominantly DNA damage to cells in the form of pyrimidine dimers, 6-4 photoproducts and apoptosis (Armstrong & Kricker, 2001; Gruijl *et al.*, 2001). UVB irradiation damages skin cells by the formation of ROS (Reactive Oxygen Species) resulting in oxidative stress, an important mediator of damage to cell structures, including lipids and membranes, proteins, and DNA (Wäster & Ollinger, 2009). However, it has less penetrating power than UVA and acts mainly on the epidermal basal layer of the skin. UVC, on the other hand, is extremely damaging to the skin because its wavelengths have enormous energy and induce genotoxic

* Vanina M. Tucci-Viegas, Lucimar P. França, Fernanda Lasakosvitsch, Fernanda L. A. Azevedo, Andrea A. F. S. Moraes, Alice T. Ferreira and Jerônimo P. França

stress. Fortunately, UVC is prevented from reaching the earth, as it is largely absorbed by atmospheric ozone layer (Afag, 2011). It has already been proposed that programmed cell death (apoptosis) can be induced by UV light in various cell types (reviewed in Schwarz, 1998). The cellular responses to injuries or stresses are important in determining cell fate (Aylon & Oren, 2007). Many signaling pathways participate in this process, with the mitogen-activated protein kinase (MAPK) cascades and p53 pathway being two of the major pathways implicated (Aylon & Oren, 2007; Li et al., 2009). The cellular response to DNA damage is focused on p53, which can induce the cell to apoptosis by the protein PUMA (p53 up-regulated modulator of apoptosis), a member of the Bcl-2 homology (BH)3-only Bcl-2 family proteins. Recent studies suggest that Bcl-2 family members play an essential role in regulating apoptosis initiation through the mitochondria (Zhang et al., 2009). UV irradiation induces permeabilization of the lysosomal membrane with release of cathepsin B and D to the cytosol, translocation of the proapoptotic Bcl-2 proteins Bax and Bid to mitochondrial-like structures. Subsequently, there is cytochrome c release and activation of caspase-3 (Bivik et al., 2006). p38 MAPK, one of the four MAPK subfamilies in mammalian cells, is activated by proinflammatory cytokines and environmental stress (Brown & Benchimol, 2006; Johnson & Lapadat, 2002). p38 is not only reported to be phosphorylated and activated to mediate cell apoptosis and the differentiation process (Thornton & Rincon, 2009), but also to have cell protective effects under certain circumstances (Chouinard et al., 2002). MAPK pathways mediate cellular responses to many different extracellular signaling molecules such as the ones involved in differentiation, gene expression, regulation of proliferation, apoptosis, development, motility or metabolism. The typical MAPK pathways, characterized by the ERK1/2, ERK5, JNK, and p38[MAPK] components, comprise a cascade of three successive phosphorylation events exerted by a MAPK kinase kinase (MAPKKK), a MAPK kinase (MAPKK), and a MAPK (Kostenko et al., 2011).

Ultraviolet UVA light absorption after solar exposure is responsible for photoactivation of DNA and other biomolecules. Additionally, UVA radiation (320-400 nm) induces photoaddition, oxidative stress and DNA damage, which may be continuous. The cell is also unable to replicate in case of severe DNA damage. This way, DNA repair must be considered essential for genetic information preservation and transmission in any life form. UVA light generates mutagenic DNA lesions in the skin. Exposure to solar UVB radiation is responsible for skin inflammation and tumorigenesis.

Besides that, oxidative stress induced by solar radiation could be responsible, as well, for the increased frequency of DNA mutations in photoaged human skin. Genomic DNA damage triggers the activation of a network of pathways that rapidly modulate several cellular activities. ROS and hydrogen peroxide can damage DNA. Furthermore, it has been recently shown that increased oxidative stress is correlated to DNA alterations. ROS are deleterious to DNA, membranes and proteins although their exact role in mutagenesis and lethality is still unclear in the many skin cell types. In addition, repair ability and defense mechanisms may differ a lot from one cellular type to another.

Epidermal and dermal cells are targets for UVA oxidative stress and their antioxidant defenses can be defeated. Keratinocytes and fibroblasts may respond differently to UV radiation depending on their localization in the body or their functional and metabolic characteristics. Cell culture models have helped to describe the cytotoxic action of UVA and the role of ROS in UVA-induced cellular damage (Tyrrell, 1990). p53 stabilization and activation, an essential outcome of the DNA damage response pathway, leads to cell cycle arrest and DNA repair, apoptosis, or cellular senescence. More specifically, initial genomic insults lead to p53 stabilization and nuclear localization where transient cell cycle arrest can be quickly activated, allowing damaged DNA repair prior to replication. A signaling cascade can be activated by p53 in case of extreme and irreversible DNA damage to induce programmed cell death through transcription of different proapoptotic factors.

UV radiation induces phospholipids peroxidation in cellular membrane. Lipid peroxidation is a consequence of free primary radicals (ROS). This, in turn, leads to the generation of polar products and increase the membrane dielectric constant and capacitance. An important consequence of this phenomenon is the alteration of transport particles across the membrane (Strässle *et al.*, 1991)

During the Fenton reaction, singlet oxygen directly initiates lipid peroxides and hydrogen peroxides indirectly initiate hydroxyl radicals (Halliwell & Gutteridge, 1999).

Cellular responses that lead to cell cycle arrest, DNA repair, apoptosis or senescence are induced by the p53 tumor suppressor pathway upon activation by genotoxic stress. This pathway works mostly through transactivation of different downstream targets, for example, p21 cell cycle inhibitor, required for short-term cell cycle arrest or long-term cellular senescence, or other proapoptotic genes such as p53 upregulated modulator of apoptosis (PUMA) (Tavana *et al.*, 2010). Yet, the mechanism that regulates the switching from cell cycle arrest to apoptosis is still unknown. In case of extreme or irreversible damage, p53 can additionally activate a signaling cascade to induce apoptosis through transcription of pro-apoptotic genes, most particularly p53-upregulated modulator of apoptosis (PUMA) and trans-repression of anti-apoptotic genes including Bcl-2. Programmed cell death directly protects cells against the accumulation of genomic instability that could lead to tumorigenesis.

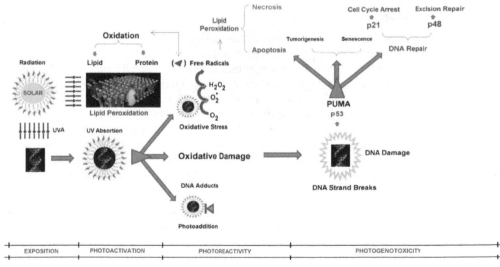

Fig. 1. Schematic representation. Photobiological effects of ultraviolet radiation on human skin cells

Senescence, an irreversible cell cycle arrest, can also be induced by DNA damage. p21, the cyclin-dependent kinase inhibitor, plays an important role in cell cycle checkpoint regulation and induction of cellular senescence, thus being one key p53 target. After DNA damage, p21 is commonly transactivated and induces G1arrest by inhibiting the cyclinE/CDK2 complex. (Campisi, 2009). Many different stimuli can induce cellular senescence including telomere shortening (replicative senescence), oncogenic signaling (oncogene-induced senescence), or stress/DNA damage irrespectively of the two previous signaling pathways (premature senescence) (Campisi, 2009). Despite the stimuli, cellular senescence and apoptosis are somewhat equivalent in preventing genomic instability and consequently inhibiting tumor formation (Van Nguyen, 2007). Upon UV exposure, p48 mRNA levels strongly depended on basal p53 expression and increased even more after DNA damage in a p53-dependent manner thus pointing as the link between p53 and the nucleotide excision repair apparatus (Hwang et al., 1999).

2. Objective

The objective of this study was to investigate modifications in cytoskeleton through the formation of blebs and apoptosis in cultured human fibroblasts by confocal microscopy and flow cytometry.

3. Methods

This study was performed in accordance with the ethical standards laid down in the updated version of the 1964 Declaration of Helsinki and was approved by the Research Ethics Committee of the Federal University of São Paulo. All patients signed a free and informed consent form. Samples of normal adult human skin (6 women, 18-50 years, skin phototype Fitzpatrick class. III-IV) were obtained as discarded tissue from trunk cosmetic surgery.

3.1 Fibroblast culture

Primary human skin fibroblast culture was done by explant. Fragments were placed in 15 ml conic tubes and exhaustively rinsed (six times) with 10 ml PBS (Phosphate-Buffered Saline, Cultilab, Campinas, SP, Brazil) containing penicillin (100 Ul/ml, Gibco, Carlsbad, CA, USA) and streptomycin (100µg/ml, Gibco) under vigorous agitation, changing tubes and PBS at each repetition. Then, fragments were transferred to 60 mm^2 diameter Petri dishes, in grid areas scratched with a scalpel. Dishes were left semi-opened in the laminar flow for 20 min, for the fragments to adhere to its surface. Then, 6 ml of DMEM (Dulbecco's Modified Eagle's Medium, Cultilab) supplemented with 10% FBS (Fetal Bovine Serum, Cultilab), 1% glutamine, penicillin (100 UI/ml, Gibco) and streptomycin (100 µg/ml, Gibco) were carefully added to each plate. Plates were kept in humidified incubator (37°C, 95% O_2, 5%CO_2).

Culture medium was changed every two days and a few days after establishing the primary culture, spindle-like cells were seen proliferating from the edges of the explanted tissue, regarded as culturing fibroblasts. Fibroblast satisfactory proliferation was observed in approximately 7-14 days and subculturing (passage) was performed when cellular confluence reached approximately 80% at the Petri dish. For all experiments, cells from passages one to five (Figure 2) were used after harvesting by trypsinization [0.025% trypsin,

0.02% ethylene diamine tetra acetic acid (EDTA; Sigma Chemical Co., Saint Louis, MO, USA) in PBS].

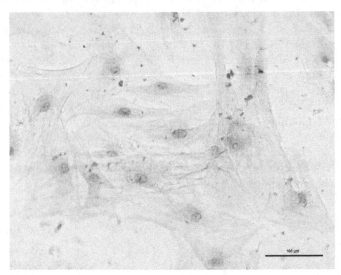

Fig. 2. Optical microscopy. Primary human skin fibroblast culture. Hematoxilin and eosin staining

3.2 Ultraviolet irradiation

Cells were rinsed in PBS. The PBS was then removed and a thin layer of buffer was left on top of the coverslip. Fibroblasts were irradiated in culture dishes in a 10cm^2 field using a UV chamber (with 6 UV F40 Philips lamps) in exposure times of 30 and 60 minutes.

3.3 Immunofluorescence labeling

Primary human skin fibroblast culture were used after harvesting by trypsinization [0.025% trypsin, 0.02% ethylene diamine tetra acetic acid (EDTA; Sigma Chemical Co., Saint Louis, MO, USA) in PBS]. The cells were washed 3 times with phosphate-buffered saline (PBS). Human fibroblasts were plated on glass coverslips, fixed in 2% paraformaldehyde for 10 minutes at 4°C, washed 3 times in PBS, and washed twice in PBS with 50 mmol/L NH$_4$Cl. Cells were permeabilized with 0.1% saponin in PBS containing 10% normal bovine serum for 30 minutes at 22°C and stained with a combination of fluorescent dyes. Filaments of cytoskeleton immunostained with phalloidin conjugated fluorescent with Alexa Fluor 594 (red) - Molecular Probe, were used to identify actin filaments F inside the cells. Phalloidin (1:500) incubation was performed in PBS containing 10% normal bovine serum and 0.1% saponin. Nuclei were counter stained with blue - fluorescent DNA stain DAPI (4_6-diamidino-2-phenylindole) 1:10000 (catalog #D1036; Molecular Probes, Invitrogen, Carlsbad, CA), and excited using a 750nm multiphoton source (two simultaneous photon excitations at 375nm). The images are a composite of three images acquired using filter sets appropriate for blue and red fluorescence, on a Zeiss confocal microscope (LSM 510, Germany).

3.4 Determination of MDA-TBA levels

Taking the 1h time-point, which proved to be optimal for the determination of MDA increase, we then studied dose kinetics. Fibroblasts were exposed to a series of single doses UV irradiation in exposure times of 30 and 60 minutes. Markedly elevated MDA concentrations in the UV and TBARs–MDA complex concentrations were determined by high-performance liquid chromatography (HPLC) as described by Gueguen *et al.*, 2002. The MDA-TBA test, which is the colorimetric reaction of malondialdehyde and thiobarbituric acid in acid solution, was used to determine the MDA levels. HPLC was used after the formation of the MDA-TBA complex (Figure 3) to assess the concentration of the complex based on a known standard curve. After heating at 95 °C for 60 min, the MDA-TBA chromogen was fluorometrically analyzed using a reversed-phase C18 column HPLC and a wavelength of 532 nm. The MDA-TBA method was previously described by Chirico *et al.* (1993). MDA levels were expressed in relation to the total cellular lysate protein amount, which was assessed using Bradford's method (Bradford, 1976).

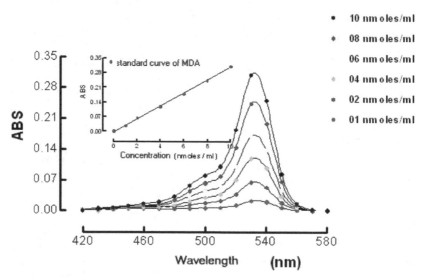

Fig. 3. Absorbance spectra for MDA – TBA chromogen complex standards in Thiobarbituric Acid Reactions (TBARs). Malondialdehyde (MDA) is a very effective method for determining lipid peroxidation levels in fibroblasts exposed to ultraviolet radiation. The standards absorption peaks of the inserted curve were highly linear in the range of 0 to 10nmoles/mL with maximum absorption at 532nm

3.5 Apoptosis assay

Flow cytometry technique, using propidium iodide, was used to detect apoptosis in fibroblast culture of human skin exposed to UV radiation (Nicoletti *et al.*, 1991).

Human fibroblasts were labeled with annexin V-FITC (Roche), which bind to phosphatidylserine at the cell surface of apoptotic cells, and propidium iodide (PI; Sigma Aldrich), was used as a marker of cell membrane permeability according to manufacturer's directions. Samples were examined by fluorescence-activated cell sorter (FACS) analysis, and the results were analyzed using Cell-Quest software (Becton Dickinson, San Jose, CA) (Vermes *et al.*, 1995).

3.6 Flow cytometric analysis of caspase 3 and p53

Briefly, normal human fibroblast cells from cultures with increasing passage number were collected and re-suspended in a buffer saline (PBS) containing 0.1% sodium azide (Sigma) containing 20 mM HEPES (pH 7.5), the cells were homogenized and centrifuged at 10,000 x g for 5 min. For analysis of caspase 3 and p53 expression, cells were fixed in 2% paraformaldehyde for 10 minutes at 4°C, washed 3 times in PBS, then washed twice in PBS with 50 mmol/L NH$_4$Cl. Cells were permeabilized with 0.1% saponin in PBS containing 10% normal bovine serum for 30 minutes at 22°C. The first primary antibody incubation (anti-p53 (SER 15) or anti–cleaved caspase 3) was performed in PBS containing 10% normal bovine serum and 0.1% saponin. Aliquots were then incubated for 60 minutes with anti-caspase 3 and p53 antibodies (Santa Cruz Biotechnology, Santa Cruz, CA), final dilution 1:800, or rabbit IgG as a control, followed by washing in PBS containing 0.1% saponin 3 times for 5 minutes each at 22°C. Cells were then incubated with the first fluorochrome-conjugated secondary antibodies Alexa 488 and 594 diluted 1:1600, and incubation was performed for 40 minutes at 37°C in the dark (Danova *et al.*, 1990).

3.7 Statistical analysis

The results obtained were analyzed using a one-way analysis of variance (ANOVA) followed by the Student–Newman–Keuls Multiple Range Test. Data were analyzed by GraphPad Prism v.3.0 software.

4. Results and discussion

Skin cells exposure to solar radiation may result in biological consequences, one of the most important being skin DNA photodamage due to sunlight ultraviolet (UV) radiation. Wavelengths in the UVB range are absorbed by DNA and can induce mutagenesis. It has been suggested that p53-independent mechanisms of killing tumor cells may not involve programmed cell death and could be a result of induced mechanical damage, rather than apoptosis (Funkel, 1999).

Ultraviolet A radiation (UVA, 320–400 nm), an oxidizing component of sunlight, exerts its biological effect mainly by producing reactive oxygen species (ROS) which cause biological damage in exposed tissues, including the lipid bilayer, via iron-catalyzed oxidative reactions (Halliwell & Gutteridge, 1999; Tyrrell, 1990). Membrane alterations induced by UV irradiation were determined, such as MDA concentration increase, which indicates lipid peroxidation levels (methods previously described -Figure 3). The UV radiation effects in the cellular production of ROS were indirectly determined by the ratio: (MDA concentration / total amount of fibroblasts) at the sample. Analyses of the lipid peroxidation by measuring the products that react with Tiobarbituric Acids (TBARs) normalizing the obtained MDA (malondialdehyde) results by the number of cells in the sample (Figure 4). A significant MDA increase was observed, of about 45.0 % after 30 min of UV exposure and 130% after 60 min of UV exposure.

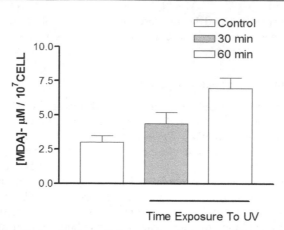

Time Exposure To UV

Fig. 4. Effect of UV radiation in the cellular production of oxygen reactive species measured by ratio: MDA concentration by total number of fibroblast in the sample. Histograms values differ significantly from each other. *Data analyzed with one-way ANOVA followed by Newman Keuls (significance level $p < 0.05$). Values represent the mean ± SEM of at least four different experiments

Similar results are also described by several studies demonstrating that low UVA radiation doses can induce lipid peroxidation in membranes of both human fibroblasts and keratinocytes via pathways involving singlet oxygen and iron (Morliere *et al*, 1991).

Looking from a different angle, cells also have repair mechanisms to respond to DNA damage, and at least two different mechanisms are responsible for UVA-induced DNA damage repair. The primary process that removes bulky damage is the nucleotide excision repair pathway. Small lesions induced by ROS are mostly processed by base excision repair pathway. On the other hand, highly damaged cells may undergo cell cycle arrest, apoptosis and senescence (Hazane *et al.*, 2006). Our results are consistent with those of Shindo *et al.* (1994) who investigated antioxidant molecules in crude extracts of human epidermis and dermis. In addition, Moysan *et al.* (1995), using cells from the same biopsy, found no link between UVA cytotoxicity and antioxidant capacity since SOD, catalase and GSH were identical in both cells and GSH-Px was higher in fibroblasts (Degterev *et al.*, 2008). Other authors, however, have found more antioxidant molecules in fibroblasts than in keratinocytes. Yohn *et al.* (1991), using cells from different donors, found increased GSH-Px, SOD and catalase in fibroblasts compared to keratinocytes, and in keratinocytes compared to melanocytes (Huang *et al.*, 2008).

Several *in vitro* and *in vivo* studies on skin cells have demonstrated that UV radiation can damage many molecules and structures (Matsumura & Ananthaswamy, 2004). Corroborating these results, morphological analysis by confocal fluorescence microscopy of fibroblasts group control showed characteristics of nuclear and cytoskeleton integrity. High cellularity was also observed (Figure 5). In contrast, exposed to UV for 30 and 60 minutes showed changes in the actin filaments arrangement of the cellular cytoskeleton. Groups irradiated for 30 and 60 min presented disruption of the actin filaments, with the formation of blebbing and nuclear fragmentation as a consequence of the ultraviolet radiation (Figure 6).

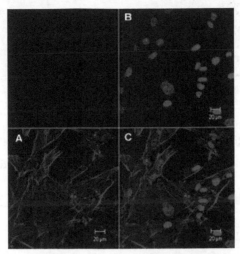

Fig. 5. Confocal microscopy. Cultured human skin fibroblasts. Control group. Cellular localization of actin filaments and nuclei. A) Actin filaments immunostained with phalloidin conjugated with Alexa Fluor 594 (red). B) Cell nuclei stained with DAPI (blue), showing characteristics of nuclear and cytoskeleton integrity. High cellularity was also observed. C) Overlapped images A and B

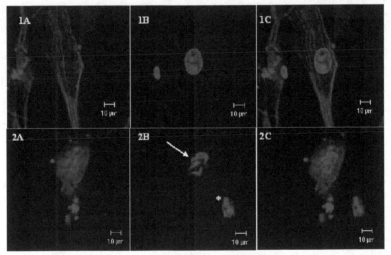

Fig. 6. Confocal microscopy. Cultured human skin fibroblasts. Cellular localization of actin filaments and nuclei. Cells exposed to UV radiation for 30 min (1A, 1B, 1C) or 60 min (2A, 2B, 2C). 1A) Actin filaments immunostained with phalloidin conjugated with Alexa Fluor 594 (red). The occurrence of blebbing can be observed. 1B) Cell nuclei stained with DAPI (blue), showing characteristics of nuclear and cytoskeleton integrity. 1C) Overlapped images A and B. 2A) Actin filaments immunostained with phalloidin conjugated with Alexa Fluor 594 (red). 2B) Pyknotic nuclei (*) and nuclear fragmentation (arrow) were observed. 2C) Overlapped images A and B

In addition, skin fibroblasts viability, stained by propidium iodide (PI), was analyzed by flow cytometry. Viable cells were characterized by a structurally intact cell membrane and no PI uptake. In contrast, dead cells (necrosis or late apoptotic cells) were characterized by loss of the integrity of their membranes and were stained by PI. At all UV radiation tested doses, the amount of viable cells was reduced, as verified by PI staining. The amount of viable fibroblasts was dramatically reduced by UV radiation at all tested doses/exposure times, about 80% after 30 min of exposure and 30% after 60 min of exposure (Figures 7A and 8A).

There are strong evidences that skin cancer can be developed as a result of ultraviolet radiation, which is directly associated to the TP53-gene tumor mutation.

To further investigate whether p53 is involved in the apoptosis induced by UV, cells were first stained for membrane-exposed phosphatidylserine using annexin-V conjugated to fluorescein (FITC). There was a significant increase of the number of apoptotic cells: about 21.0 % (30 min) and 50% (60 min) after irradiation (Figures 7B and 8B) and (Figures 7C and 8C), respectively.

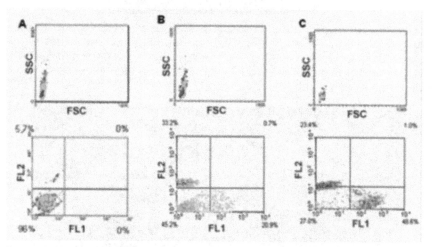

Fig. 7. Contour diagram of PI flow cytometry of cultured fibroblasts for groups: A) Control; B) UV irradiated for 30 min and C) UV irradiated for 60 min. The lower left quadrant of the cytograms shows the viable cells, which excluded PI. The upper right quadrants represent the apoptotic cells showing PI uptake. Panel (B) shows cells number (%) for apoptosis and necrosis 30 minutes after exposure to ultraviolet radiation. Panel (C) shows cells number (%) for apoptosis and necrosis 60 minutes after exposure to ultraviolet radiation. Data are representative of 04 independent experiments

Ultraviolet radiation is a carcinogenic agent for the skin. Even though being a tumor suppressor gene, details are still needed in order to understand the signaling mechanisms of skin cell death induced by UV radiations, which can lead to cancer and/or cell aging.

DNA alteration can ultimately lead to the development of skin cancer, so DNA itself is a critical target (Matsumura et al., 2004). Skin DNA photodamage activates the signaling pathway of cell death by apoptosis. Apoptosis is a crucial mechanism in eliminating cells with unrepaired DNA damage and preventing carcinogenesis.

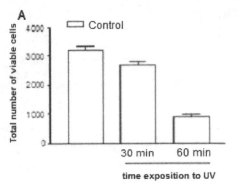
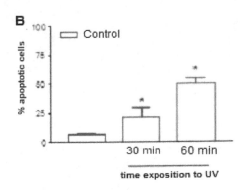

Fig. 8. Mean percentage ± of apoptotic cells in groups control and 30min and 60min after exposure to UV radiation. Data are the means of triplicate assays of one experiment representative of three that gave similar results. A) Total number of viable cells and B) Percentage of apoptotic cells. *Data analyzed with one-way **ANOVA** followed by Newman Keuls (significance level p < 0.05). Values represent the mean ± SEM of at least four different experiments

DNA is a critical target because its alteration can ultimately lead to the development of skin cancer (Matsumura *et al.*, 2004). In addition, Skin DNA photodamage activates the signaling pathway of cell death by apoptosis. Apoptosis is a crucial mechanism in eliminating cells with unrepaired DNA damage and preventing carcinogenesis (or preventing the formation of malignant tumors).

Apoptosis is characterized by a p53-dependent induction of pro-apoptotic proteins, leading to permeabilization of the outer mitochondrial membrane, release of apoptogenic factors into the cytoplasm, activation of caspases (cysteine-aspartic proteases) and subsequent cleavage of various cellular proteins. Apoptogenic effects include chromatin condensation and exposure of phosphatidylserine on the cell membrane surface (Meier *et al.*, 2007).

p53 levels increased about 40% after 30 min of UV exposure and about 60% after 60 min of UV exposure (Figure 9).

Previous studies indicated that BimL was involved in UV-induced apoptosis, but it remains unclear whether Bim directly activates Bax or if this activation occurs via the release of pro-survival factors (antiapoptotic) such as Bcl-xL. In recent studies, Wang *et al.* (2009) determined the interactions between BimL and Bax/Bcl-xL during UV-induced apoptosis.

Caspases have a major role in apoptosis. They are synthesized as inactive proenzymes that become activated by cleavage. Procaspase 3 is a constitutive proenzyme activated by cleavage during apoptosis. (Cohen, 1997). Caspase-3 is the most important protease in the caspase-dependent apoptosis pathway, as it is required for chromatin condensation and fragmentation (Porter & Jänicke, 1999). Poly-ADP ribose polymerase (PARP-1) is a major target of caspase-3, since cleavage-mediated inactivation of PARP-1 preserves cellular ATP that is required for apoptosis (Bouchard *et al.*, 2003).

Regarding the caspases, the resulting enzyme is able to cleave several aspartate residues of many target proteins, after a DEVD sequence common to all caspases 3 and 7 substrates

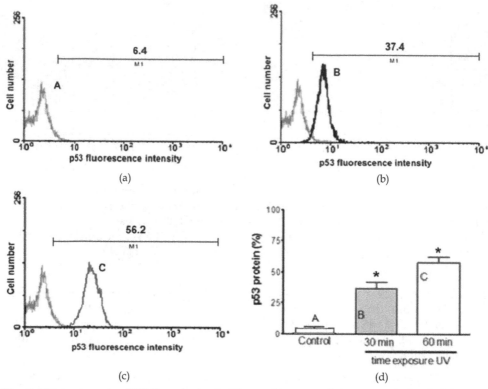

Fig. 9. Flow cytometry (FCM) analysis of p53 protein accumulation control (upper left set of panel – figure 9A – green line) and / or activation by UV can be followed of cultured fibroblasts for groups: UV irradiated for 30 min (right set of figure 9B – Black line) and UV irradiated for 60 min (right set of figure 9C – red line). The fibroblasts treated in 2% paraformaldehyde are the same as those shown in Figure 05 (control group) and figure 06 (UV irradiated groups). Cells were permeabilized with 0.1% saponin in PBS containing 10% normal bovine serum for 30 minutes at 22°C and stained with anti-p53 (SER 15) antibodies at figures (9A), (9B), and (9C) after the beginning of the experiment and analyzed by FCM. A control performed with an irrelevant antibody is shown figure 9A. The percentage of cells exhibiting active p53 conjugated with FITC is indicated on each histogram. The results from one representative experiment of four experiments performed are shown. The numbers indicate the percentages of positive cells and fluorescence intensity. Histogram overlays show the FL1 (green fluorescence) intensity corresponding to a given p53 (black line – UV irradiated for 30min and red line – UV irradiated for 60min) compared to the intensity for the control (green line). 9D Mean percentage ± of cells exhibiting active p53 in groups control (figure 9A – green line) and groups UV irradiate for 30 min (figure 9B – Black line) and UV irradiated for 60 min (figure 9C – red line). Data are the means of triplicate assays of one experiment representative of three that gave similar results. A) Total number cells fibroblasts exhibiting active p53 antibodies at figures (9A), (9B), and (9C). Histograms values differ significantly from each other. *Data analyzed with one-way ANOVA followed by Newman Keuls (significance level $p < 0.05$). Values represent the mean ± SEM of at least four different experiments

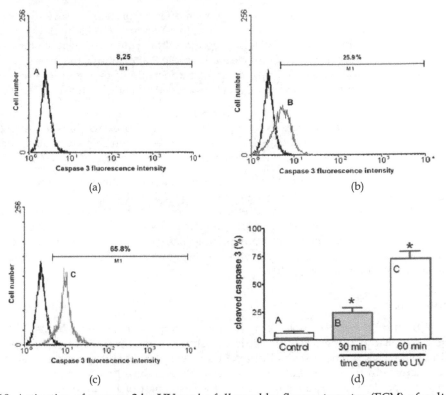

Fig. 10. Activation of caspase 3 by UV can be followed by flow cytometry (FCM) of cultured fibroblasts for groups: A) control (upper left set of panel – figure 10A - Black line) and UV irradiated for 30 min (right set of figure 10B – Blue line) and UV irradiated for 60 min (right set of 10C – orange line). The fibroblasts treated in 2% paraformaldehyde are the same as those shown in figure 05 (control group) and figure 06 (UV irradiated groups). Cells were permeabilized with 0.1% saponin in PBS containing 10% normal bovine serum for 30 minutes at 22°C and stained with anti–cleaved caspase 3 antibodies at figures (10A), (10B), and (10C) after the beginning of the experiment and analyzed by FCM. A control performed with an irrelevant antibody is shown 10A. The percentage of cells exhibiting active caspase 3 conjugated with FITC is indicated on each histogram. The results from one representative experiment of four experiments performed are shown. The numbers indicate the percentages of positive cells and fluorescence intensity. Histogram overlays show the FL1 (green fluorescence) intensity corresponding to a given caspase 3:(blue line – UV irradiated for 30min and red line – UV irradiated for 60min compared to the intensity for the control (black line). 10D Mean percentage ± of cells exhibiting active caspase 3 in groups control (figure 9A – green line) and groups UV irradiated for 30 min (figure 9B – Black line) and UV irradiated for 60 min (figure 9C – red line). Data are the means of triplicate assays of one experiment representative of three that gave similar results. A) Total number cells fibroblasts exhibiting active caspase 3 antibodies at figures (10A), (10B), and (10C). Histograms values differ significantly from each other. *Data analyzed with one-way ANOVA followed by Newman Keuls (significance level $p < 0.05$). Values represent the mean ± SEM of at least four different experiments

(DEVDase). Thus, active caspase 3 is a common effector protein in several apoptotic pathways, and it may be a good marker to detect (pre-) apoptotic cells by flow cytometry (Porter & Jänicke, 1999). Taking this into consideration, apoptosis was confirmed by determining the increased expression of cleaved caspase 3 after fibroblasts exposure to UV radiation. In this work we could verify a significant increase of cleaved caspase 3 levels, about 25.0 % after 30 min of UV exposure and 75% after 60 min of UV exposure (Figure 10).

Although caspases represent a significant component of the apoptotic pathway, there is indication that a caspase-independent apoptosis pathway also exists (Broker *et al.*, 2005). This pathway involves the Apoptosis-Inducing Factor (AIF), which translocates from the mitochondria to the nucleus to cause chromatin condensation (Daugas *et al.*, 2000).

Then again, genotoxic effects of solar UVA are mediated essentially by the activation of endogenous photosensitizers which generate a local oxidative stress. Depending on the dose and duration of exposure, UV-induced effects may occur, and DNA damage can lead to mutations and genetic instability. This is one of the reasons why sunlight overexposure increases the risk of skin cancer and DNA photolesions can also be involved in other skin-specific responses to UV radiation: erythema, immunosuppression, and melanogenesis (Matsumura & Ananthaswamy, 2004).

5. Conclusion

Damages occurring on DNA molecules not always induce mutagenesis. We should take in consideration many strong scientific evidences showing that specific activation molecular signaling pathways promote several different answers. Both the prolonged exposure time and the increase in the UV radiation dose were able to induce lipid peroxidation and cell death by apoptosis. Our results suggest that the major part of UV induced apoptosis cell death is caspase-dependent, although a minority of cells may die by a caspase-independent pathway, presumably apoptotic. In this work we also showed that p53 levels increased after UV exposure. In these circumstances, the action of UV radiation on skin cells still involves many issues depending on the cell type and on different cellular response pathways induced by phototoxic stress. Skin fibroblasts are surely sensitive to UV radiation, thus, from a better understanding of the molecular mechanisms triggered by the action of UV radiation on skin cells, it will be possible to work on improving skin radioprotection and attenuating the effects of sunlight exposure.

6. Acknowledgment

We would like to thank UNIFESP and UESC (collaborators) for their help in experiments with fibroblast cell culture. The authors gratefully acknowledge the financial support from FAPESP, FAPESB and CNPq grants.

7. References

Abbas, T & Dutta, A. (2009). p21 in cancer: intricate networks and multiple activities. *Nature reviews. Cancer*, Vol.9, pp.400-414, ISSN 1759-4782
Afag, F. (2011). Natural agents: cellular and molecular mechanisms of photoprotection. *Archives of Biochemistry and Biophysics*, Vol.508, pp. 144-151, ISSN 0003-9861

Andriani, F.; Marquelis, A.; Lin, N.; Griffey, S. & Garlick, J. A. (2003). Analysis of
    microenvironmental factors contributing to basement membrane assembly and
    normalized epidermal phenotype. *Journal of Investigative Dermatology*, Vol.120, pp.
    923-931, ISSN 0022-202X

Armstrong, B. K. & Kricker, A. (2001). The epidemiology of UV induced skin cancer. *Journal
    of Photochemistry and Photobiology B*, Vol.63, pp.8-18, ISSN 1011-1344

Aylon & Oren. (2007). Living with p53, dying of p53. *Cell*, Vol. 130, No.4, pp. 597-600, ISSN
    1097-4172

Bivik, C. A.; Larsson P. K.; Kadegal, K. M.; Rosdah, l. K. & Ollinger, K. M. (2006). UVA/B –
    induced apoptosis in human melanocytes involves translocation of cathepsins and
    Bcl-2 family members. *Journal of Investigative Dermatology*, Vol.126, pp. 1119-1127,
    ISSN 0022-202X .

Blumenberg, M. & Tomić-Canić M. (1997). Human epidermal keratinocyte: keratinization
    processes. *EXS*, Vol.78, pp.1-29, ISSN 1023-294X

Bouchard, V. J.; Rouleau, M.; Poirier, G. G. (2003). PARP-1, a determinant of cell survival in
    response to DNA damage. *Experimental Hematology*, Vol.31, pp.446–454, ISSN 0301-
    472X

Bradford, M. M. (1976). A rapid and sensitive method for the quantification of microgram
    quantities of protein utilizing the principle of protein-dye binding. *Analytical
    Biochemistry*, Vol.72, pp.248–254, ISSN0003-2697

Broker, L. E.; Kruyt, F. A. E & Giaccone, G. (2005). Cell death independent of caspases: a
    review. *Clinical Cancer Research,* Vol.11, pp.3155–3162, ISSN: 1078-0432

Brown, L & Benchimol, S. (2006). The involvement of MAPK signaling pathways in
    determining the cellular response to p53 activation: cell cycle arrest or apoptosis.
    The Journal Of Biological Chemistry, Vol.281, No.7, pp.:3832-3840, ISSN 1083-351X

Campisi, J. (2005). Senescent cells, tumor suppression and organismal aging: good citizens,
    bad neighbors. *Cell*, Vol.120, pp.513-522, ISSN 1097-4172

Canman, C. E. & Kastan, M. B. (1996). Signal transduction. Three paths to stress relief.
    Nature, Vol.384, pp.213-214, ISSN 0028-0836

Chirico, S.; Smith, C.; Marchant, C.; Mitchinson, M. J. & Halliwell, B. (1993). Lipid
    peroxidation in hyperlipidaemic patients. A study of plasma using an HPLC-based
    thiobarbituric acid test. *Free radical research communications*, Vol.19, pp.51-57, ISSN
    8755-0199

Chouinard, N.; Valerie, K.; Rouabhia, M & Huot J. (2002). UVB-mediated activation of p38
    mitogen-activated protein kinase enhances resistance of normal human
    keratinocytes to apoptosis by stabilizing cytoplasmic p53. *Biochemical Jounal*,
    Vol.365, pp.133-134, ISSN · 0264-6021

Chung, J. H.; Hanft, V. N. & Kang, S. (2003). Aging and photoaging. *Journal of the American
    Academy of Dermatology.* Vol.49, pp.690–697, ISSN 0190-9622

Chung, J. H.; Seo, J. Y.; Choi, H. R.; Lee, M. K.; Youn, C. S.; Rhie, G.; Cho, K. H.; Kim, K. H.;
    Park, K. C. & Eun, H. C. (2001). Modulation of skin collagen metabolism in aged
    and photo-aged human skin in vivo. *Journal of Investigative Dermatology*, Vol.117,
    pp. 1218-1224, ISSN 0022-202X

Chung, J. H.; Seo, J. Y.; Lee, M. K.; Eun, H. C.; Lee, J. H.; Kang, S.; Fisher, G. J. & Voorhees, J.
    J. (2002). Ultraviolet modulation of human macrophage metalloelastase in human

skin in vivo. *Journal of Investigative Dermatology*, Vol.119, pp.507-512, ISSN 0022-202X

Cohen, G. (1997). Caspases: the executioners of apoptosis. *Biochemical Jounal*, Vol.326, pp.1–16, ISSN 0264-6021

Danova, M.; Giordano, M; Mazzini, G. & Riccardi, A. (1990). Expression of p53 protein during the cell cycle measured by flow cytometry in human leukemia. *Leukemia Research*, Vol.14, pp.417–422, ISSN 0145-2126

Daugas, E.; Nochy, D.; Ravagnan, L.; Loeffler, M.; Susin, S. A.; Zamzami, N. & Kroemer, G. (2000). Apoptosis inducing factor (AIF): a ubiquitous mitochondrial oxidoreductase involved in apoptosis. *FEBS Letters*, Vol.476, pp.118–123, ISSN 0014-5793

Degterev, A.; Hitomi, J.; Germscheid, M.; Chen, I. L.; Korkina, O. *et al.* (2008). Identification of RIP1 kinase as a specific cellular target of necrostatins. *Nature Chemical Biology*, Vol.5, pp.313–321, ISSN 1552-4450

Florence, H.; Sauvaigo, S.; Douki, T.; Favier, A. & Beani, J-C. (2006). Age-dependent DNA repair and cell cycle distribution of human skin fibroblasts in response to UVA irradiation. Journal of *Photochemistry and Photobiology* B, Vol.82, pp.214–223, ISSN 0031-8655

Funkel, E. (1999). Does cancer therapy trigger cell suicide? *Science*, Vol.286, pp.2256–2258, ISSN 0036-8075

Gruijl, F. R. (1999). Skin cancer and solar UV radiation. *European Journal of Cancer*, Vol.35, pp.2003-2009, ISSN 1359-6349

Gruijl, F. R.; Van Kranen, H. J. & Mullenders, L. H. (2001). UV-induced DNA damage, repair, mutations and oncogenic pathways in skin cancer. *Journal of Photochemistry and Photobiology B*, Vol.63, pp.19-27, ISSN 1011-1344

Gueguen, S.; Herbeth, B.; Siest, G. & Leroy P. (2002). An isocratic liquid chromatographic method with diode-array detection for the simultaneous determination of alpha-tocopherol, retinol, and five carotenoids in human serum. *Journal of Chromatographic Science*, Vol.40, No.2, pp.69-76, ISSN 1570-0232

Halliwell, B. & Gutteridge, J. M. (1999). Measurement of reactive species, In: *Free Radicals in Biology and Medicine*, pp.284 -313, Oxford University Press, Inc., ISBN 019-8568-69-X New York

Huang, H.; Fletcher, L.; Beeharry, N.; Daniel, R.; Kao, G.; Yen, T. J. & Ruth J. Muschel, R. J. (2008). Abnormal cytokinesis after X-irradiation in tumor cells that override the G2 DNA damage checkpoint. *Cancer Research*, Vol.68, pp.3724–3732, ISSN 0008-5472

Hwang, B. J; Ford, J.M.; Hanawallt, P.C.; Chu, G. (1999). Expression of the p48 xeroderma pigmentosum gene is p53-dependent and is involved in global genomic repair. *Proceedings of the National Academy of Sciences of the United States of America. Biochemistry*, Vol.96, pp.424–428, ISSN 0027-8424

Hwang, K. A.; Yi, B. R. & Choi, K. C. (2011). Molecular mechanisms and in vivo mouse models of skin aging associated with dermal matrix alteration. *Laboratory Animal Research*, Vol.27, pp.1-8, ISSN 1738-6055

Janicke, R. U.; Sprengart, M. L.; Wati, M. R. & Porter, A. G. (1998). Caspase-3 is required for DNA fragmentation and morphological changes associated with apoptosis. *Journal of Biological Chemistry*, Vol.273, pp.9357–9360, ISSN 0021-9258

Johnson, G. L. & Lapadat, R. Mitogen-activated protein kinase pathways mediated by ERK, JNK, and p38 protein kinases. (2002). *Science*, Vol.298, pp.1911-1912, ISSN 0036-8075

Katiyar, S. K.; Mantena, S. K. & Meiran, S. M. (2011). Silymarin protects epidermal keratinocytes from ultraviolet radiation-induced apoptosis and DNA damage by nucleotide excision repair mechanism. *PLoS One*, Vol.6, pp.1-11, ISSN 1932-6203

Kostenko, S.; Shiryaev, A.; Gerits, N.; Dumitriu, G.; Klenow, H.; Johannessen, M. & Moens, U. (2011). Serine residue 115 of MAPK- activated protein kinase MK5 is crucial for its PKA- regulated nuclear export and biological function. *Cellular and Molecular Life Sciences*, Vol.68, pp.847-862, ISSN 1420-682X

Li, G.; Mongillo, M.; Chin, K. T.; Harding. H.; Ron, D.; Marks, A. R. & Tabas, I. (2009). Role of ERO1-α–mediated stimulation of inositol 1,4,5-triphosphate receptor activity in endoplasmic reticulum stress–induced apoptosis. The Journal of Cell Biology, Vol.186, pp.783–792, ISSN 0021-9525

Makrantonaki, E. & Zouboulis, C. C. (2007). The skin as a mirror of the aging process in the human organism-state of the art and results of the aging research in the German National Genoma Research Network 2 (NGFN-2) *Experimental Gerontology*, Vol.42, pp.879-886, ISSN 0531-5565

Matsumura, Y. & Ananthaswamy, H. N. (2004). Toxic effects of ultraviolet radiation on skin. *Toxicology and Applied Pharmacology*, Vol.195, pp.298-308, ISSN 0041-008X

Meier, P. & Vousden, K. H. (2007). Lucifer's labyrinth - ten years of path finding in cell death. *Molecular Cell*, V.28, pp. 746–754, ISSN 1097-2765

Morliere, P.; Moysan, A.; Santus, R.; Huppe, G.; Maziere, J. C. & Dubertret, L. (1991). UVA-induced lipid peroxidation in cultured human fibroblasts. *Biochimica et Biophysica Acta*, Vol.084, pp.261–268, ISSN 0006-3002

Moysan, A.; Clement-Lacroix, P. M.; Dubertret. L. & Morliere, P. (1995). Effects of ultraviolet A and antioxidant defense in cultured fibroblasts and keratinocytes. *Photodermatology, photoimmunology & photomedicine*, Vol.11, pp.192-197, ISSN 0905-4383

Nicoletti, I.; Migliorati, G.; Pagliaci, M. C.; Grignani, F. & Riccardi, C. (1991). A rapid and simple method for measuring thymocyte apoptosis by propidium iodide staining and flow cytometry. *Journal of Immunological Methods*, Vol.139, pp.271-279, ISSN 1872-7905

Porter, A. G. & Janicke, R. U. (1999). Emerging roles of caspase-3 in apoptosis. *Cell Death & Differentiation*,Vol.6, pp.99–104, ISSN 1350-9047

Schwarz, T. (1998). UV light affects cell membrane and cytoplasmic targets. *Journal of Photochemistry and Photobiology B*, Vol.44, pp.91-96, ISSN 1011-1344

Shindo, Y.; Witt, E.; Han, D.; Epstein, W. & Packer, L. (1994). Enzymic and non-enzymic antioxidants in epidermis and dermis of human skin. *Journal of Investigative Dermatology*, Vol.102, pp.122-124, ISSN 0022-202X

Shreeram, S.; Hee, W. K. & Bulavin, D. V. (2008). Cdc25A serine 123 phosphorylation couples centrosome duplication with DNA replication and regulates tumorigenesis. *Molecular and Cell Biology*. Vol.28, No.24, pp.7442-7450, ISSN 0270-7306

Strässle, M.; Wilhelm, M. & Stark, G. (1991). The Increase of Membrane Capacitance as a Consequence of Radiation-induced Lipid Peroxidation. *International Journal of Radiation Biology*, Vol. 59, No.1, pp.71-83, INSS 0360-3016

Tavana, O.; Benjamin, C. L.; Puebla-Osorio, N.; Sang, M.; Ullrich, S. E.; Ananthaswamy, H. N. & Zhu, C. (2010). Absence of p53-dependent apoptosis leads to UV radiation hypersensitivity, enhanced immunosuppression and cellular senescence. *Cell Cycle,* Vol.9, pp.3328-3336, ISSN 1551-4005

Tyrrell, R. M. & Keyse, S. M. (1990). New trends in photobiology. The interactions of UV-A radiation with cultured cells *Journal of Photochemistry and Photobiology B,* Vol.4, pp.349-361, ISSN 1011-1344

Thornton, T.M. & Rincon, M. (2009). Non-Classical P38 Map Kinase Functions: Cell Cycle Checkpoints and Survival. *International Journal of Biological Scienses,* Vol.1, pp.44-52, ISSN 1449-2288

Van Nguyen, T.; Puebla-Osorio, N.; Pang, H.; Dujka, M. E. & Zhu, C. (2007. DNA damage-induced cellular senescence is sufficient to suppress tumorigenesis: a mouse model. *The Journal of experimental medicine,* Vol.204 pp.1453-1461, ISSN 1540-9538

Varani, J.; Perone, P.; Fligiel, S. E.; Fisher, G. J. & Voorhees, J. J. (2002). Inhibition of type I procollagen production in photodamage: correlation between presence of high molecular weight collagen fragments and reduced procollagen synthesis. *Journal of Investigative Dermatology,* Vol.119, pp. 122-129, ISSN 0022-202X

Vermes, I.; Haanen, C.; Steffens-Nakken, H. & Reutelingsperger, C. (1995). A novel assay for apoptosis. Flow cytometric detection of phosphatidylserine expression on early apoptotic cells using fluorescein labelled Annexin V. *Journal of Immunological Methods,* Vol.184, pp.39–51, ISSN 0022-1759

Wang, X.; Xing, D.; Liu, L. & Chen, W. R. (2009). BimL directly neutralizes Bcl-xL to promote Bax activation during UV-induced apoptosis. *FEBS Letters,* Vol.583, pp.1873-1879, ISSN

Wäster, P. K. & Ollinger, K. M. (2009). Redox-dependent translocation of p53 to mitochondria or nucleus in human melanocytes after UVA- an UVB- induced apoptosis. *Journal of Investigative Dermatology,* Vol.129, pp.1769-1781, ISSN 0022-202X

Wäster, P. K. & Ollinger, K. M. (2009). Redox-dependent translocation of p53 to mitochondria or nucleus in human melanocytes after UVA- an UVB- induced apoptosis. *Journal of Investigative Dermatology,* Vol.129, pp. 1769-1781, ISSN 0022-202X

Yohn, J. J.; Norris, D. A.; Yrastorza, D. G.; Buno, I. J,. Leff, J. A.; Hake, S. S. & Repine, J. E. (1991). Disparate antioxidant enzyme activities in cultured human cutaneous fibroblasts, keratinocytes and melanocytes. *Journal of Investigative Dermatology,* Vol.97, pp.405-409. ISSN, 0022-202X

Zhang, Y.; Xing, D. & Liu, L. (2009). PUMA promotes Bax translocation by both directly interacting with Bax and by competitive binding to Bcl-XL during UV- induced apoptosis. *Molecular Biology of the Cell,* Vol.20, pp.3077-3087, ISSN 1936-4586

12

Ethanol Extract of *Tripterygium wilfordii* Hook. f. Induces G0/G1 Phase Arrest and Apoptosis in Human Leukemia HL-60 Cells Through c-Myc and Mitochondria-Dependent Caspase Signaling Pathways

Chung-Jen Chiang et al.*
Department of Medical Laboratory Science and Biotechnology,
China Medical University, Taichung,
Taiwan

1. Introduction

Tripterygium wilfordii Hook. f. is a traditional Chinese herb (Murphy, 2006; Qiu et al., 2003). The extract of *Tripterygium wilfordii* Hook. f. has been widely applied to the treatment of immune-related diseases, such as rheumatoid arthritis (RA), nephritis, and systemic lupus erythematosus (SLE) (Chang et al., 1999; Wang et al., 2000). Extracts of *Tripterygium wilfordii* Hook. f. have been shown to inhibit lymphocyte proliferation induced by mitogentic stimulation *in-vitro* (Wu et al., 2003). Triptolide (PG490, one of the most active components in *Tripterygium wilfordii* Hook. f. extract, possesses immunosuppressive, anti-inflammatory and anti-fertility actions *in vivo* and *in vitro* (Zhao et al., 2005; Leuenroth et al., 2005). Many reports have demonstrated that triptolide has anti-proliferate activity against L1210, U937, K562, HL60, and P388 leukemia cells (Lou et al., 2004; Chan et al., 2001; Wei et al., 1991). However, the cellular and molecular mechanisms underlying mediating *Tripterygium wilfordii* Hook. f.-induced differentiation and/or apoptosis in leukemia cells have not been well studied.

Leukemia is a malignant disease characterized by uncontrolled cellular growth and disrupted differentiation of hematopoietic stem cells (Lichtman et al., 2005; O'Hare et al., 2006). Chemotherapy can be effective in certain types of leukemia, but in cases in which it

* Jai-Sing Yang[1], Yun-Peng Chao[2], Li-Jen Lin[3] Wen-Wen Huang[4], Jing-Gung Chung[4], Shu-Fen Peng[4], Chi-Cheng Lu[5], Jo-Hua Chiang[5], Shu-Ren Pai[4] and Minoru Tsuzuki[6,7]
[1]*Department of Pharmacology, China Medical University, Taichung, Taiwan*
[2]*Department of Chemical Engineering, Feng Chia University, Taichung, Taiwan*
[3]*School of Chinese Medicine, China Medical University, Taichung, Taiwan*
[4]*Department of Biological Science and Technology, China Medical University, Taichung, Taiwan*
[5]*Department of Life Sciences, National Chung Hsing University, Taichung, Taiwan*
[6]*Nihon Pharmaceutical University, Saitama, Japan*
[7]*Tsuzuki Institute for Traditional Medicine China Medical University, Taichung, Taiwan*

is not effective additional therapeutic strategies are needed. (Faderl et al., 2005; Frankfurt et al., 2006; ter Bals et al., 2005). Several compounds are capable of inducing the differentiation of leukemia cells into mature cells *in vitro*, and differentiation therapy has been shown to be an effective approach for treating leukemia (Altucci et al., 2004; Altucci et al., 2005; Takahashi et al., 2002). Human promyelocytic leukemia HL-60 cells and mouse monocytic leukemia WEHI-3 cells are commonly used to study various properties of leukemia cell proliferation and differentiation *in vitro* (Lin et al., 2006; Abe et al., 1987). Differentiation of HL-60 is induced into granulocytes by dimethyl sulfoxide (DMSO) and all-trans retinoic acid (ATRA), and into monocytic-like cells by phorbol ester (TPA) and 1,25-dihydroxy-vitamin D3 (Tsiftsoglou et al., 2003). In HL-60 cells, differentiation is induce-specific and is characterized by agents to differentiated is a marked increase in the proportion of G0/G1 cells (Yen et al., 2006), and the modulation of cyclin/CDK (Horie et al., 2004; Wang et al., 1996; Barrera et al., 2004; Pizzimenti et al., 1999; Kumakura et al., 1996).

In hematopoietic cells, apoptosis can be coupled to terminal differentiation of myeloid progenitor (Yazdanparast et al., 2005; Samudio et al., 2005). Cells undergoing apoptosis have observable morphology changes expressed as nuclear condensation, DNA fragmentation, and compact packaging of the cellular debris into apoptotic bodies (Fleischer et al., 2006; Bohm et al., 2006). The delivery and performance of apoptotic signals requires a coordinated cascade of caspase activation and action. The initiator caspases include caspase-8 in Fas-induced apoptosis, and caspase-9, the activation of which is triggered by cytochrome *c* release from mitochondria in response to various stimuli. Those caspases can directly activate downstream effectors of caspase-3, -6, and -7, which cleave death substrates, such as poly(ADP-ribose) polymerase (PARP) (Christophe et al., 2006; Lucken et al., 2005; Lockshin et al., 2005).

In this study, we investigated the cytotoxic effects of ethanol extract of *Tripterygium wilfordii* Hook. f. (ETW) on the promotion of cell cycle arrest and apoptosis in HL-60 cells. Our results indicated that ETW effectively induces both G0/G1 phase arrest and apoptosis of HL-60 cells *in vitro*. The mechanisms governing ETW-induced G0/G1 phase arrest included down regulation of cyclin E, Bcl-2 and Bax, and -triggered apoptosis through caspase-9, caspase-8 and caspase-3-dependent pathways.

2. Materials and methods

2.1 Chemicals and reagents

EDTA, Propidium iodide (PI), RNase A, Tris-HCl, Tritox X-100, Tween 20 and Proteinase K were obtained from Sigma Chemical Co. (St.Louis, MO, USA). RPMI-1640 medium, fetal bovine serum (FBS), and L-glutamine, penicillin/streptomycin were obtained from Gibco BRL Co. (Grand Island, NY, USA). The caspase-3, caspase-8 and caspase-9 activity assay kits were bought from R&D Systems, Inc. (Minneapolis, MN, USA)

2.2 Ethanol *Tripterygium wilfordii* Hook. f. (ETW) extraction

Dried and powdered plant materials were subjected to continuous ethanol extraction in a Soxhlet extractor with absolute ethanol for 72 h. The ethanol extract was collected and

Ethanol Extract of Tripterygium wilfordii Hook. f. Induces G0/G1 Phase Arrest and Apoptosis in Human Leukemia
HL-60 Cells Through c-Myc and Mitochondria-Dependent Caspase Signaling Pathways

221

concentrated by vacuum distillation. The extract was evaporated to dryness and reconstituted in ethanol before experiment.

2.3 Cell culture and viability assay

The human promyelocytic leukemia cell line (HL-60) was obtained from the Culture Collection and Research Center (CCRC, Taiwan, R.O.C.), originally from the American Type Culture Collection (ATCC, USA). Cells were cultured in RPMI-1640 culture medium (Gibco/Life Technologies, Taipei, Taiwan) supplemented with 10% heated-inactive fetal bovine serum (Gibco/Life Technologies), 2 mM L-glutamine, penicillin (100 units/ml), and streptomycin (100 µg/ml) (Gibco/Life Technologies) and incubated at 37°C in humidified 5% CO_2 atmosphere.

For viability analysis, 2.5 X 10^5 cells/well were seeded in 24-well culture plates. ETW was added to each well and the plates were incubated at 37°C for 24, 48 and 72 h. Cell viability was estimated by a propidium iodide (PI) incorporation assay and flow cytometry (FACS CaliburTM, Becton Dickinson) analysis (Aouacheria et al., 2002).

2.4 Cell cycle analysis

Cells were incubated with 50, 100 or 200 µg/mL of ETW for 0, 24 or 48 h. After treatment, cells were washed with phosphate-buffered saline (PBS) twice. The cells were re-suspended in hypotonic PI solution (0.1% sodium citrate, 0.1% Triton X-100, and 50 µg/ml propidium iodide), and then cellular DNA content was determined by flow cytometry (Kamikubo et al., 2003).

2.5 Western blotting analysis

Total protein was prepared with protein lysising buffer (PRO-PREP™ protein extraction solution, iNtRON Biotechnology, Seongnam, Gyeonggi-Do, Korea). The concentration of protein was determined by the Bradford method using the Bio-Rad protein assay dye reagent. The lysates containing 40 µg of protein were separated by SDS-PAGE and transferred onto PVDF membrane. Nonspecific binding sites were blocked with 5% non-fat milk in PBST buffer (0.05 % Triton X-100 in PBS) for 1 h. The PVDF membrane was incubated overnight at 4°C with specific primary antibodies against cyclin D1, cyclin E, Bcl-2, and α-tubulin (Santa Cruz Biotechnology, Inc., Santa Cruz, CA, USA). After being washed with PBST buffer, the membrane was incubated with horseradish peroxidase (HRP)-conjugated secondary antibodies (Santa Cruz). Immunoreactive proteins were detected using a Western Blotting Chemiluminescence Reagent Plus kit (NENTM Life Science) and exposed to Chemiluminescence films (Choi et al., 2003).

2.6 Caspase activities assays

Cells were collected in lysis buffer (50 mM Tris-HCl, 1 mM EDTA, 10 mM EGTA, 10 mM digitonin and 2 mM DTT) and placed on ice for 10 min. The lysates were centrifuged at 15,800g at 4°C for 10 min. Cell lysates (50 µg of protein) were incubated with caspase -3, -9, and -8 specific substrates (Ac-DEVD-pNA, Ac-LEHD-pNA, and Ac-IETD-pNA) with reaction buffer in a 96-well plate at 37°C for 1 h. The caspase activity was determined by measuring OD405 of the released pNA (An et al., 2004).

2.7 Statistical analysis

Results are presented as mean ± S.D. Differences between the different treatment groups, which consisted of matched samples, were assessed by the Student's *t*-test. A p value of less than 0.05 was considered to be significant.

3. Results

3.1 Effects of ETW on cell viability in HL-60 and WEHI-3 cells

We treated HL-60 cells with ETW at the concentrations of 0, 50, 100 and 200 µg/ml. The number of viable cells was counted by a PI exclusion method 0, 24 and 48 h later. As shown in Fig. 1, ETW exerted a dose- and time-dependent loss of cell membrane integrity and viability in HL-60 cells.

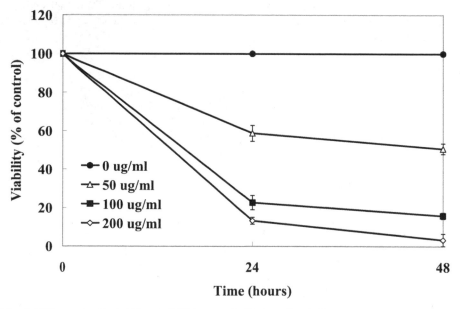

Fig. 1. Effects of cell viability in ETW treated HL-60 cells. Cells were treated with various concentrations of ETW for indicated duration. Viable cells were measured by PI exclusion and immediately analyzed by flow cytometry. The percentage of cell viability was calculated as a ratio between drug-treated cells and control cells. Each value represents mean±S.D. from three independent experiments

3.2 Effects of ETW on cell cycle progression in HL-60 cells

To investigate the mechanisms by which ETW induced cytotoxicity effect in HL-60 cells, we cultured cells for various time periods with 100 µg/ml ETW and analyzed DNA content by flow cytometry. Cell cycle analysis showed that ETW induced a prominent G0/G1 population arrest in HL-60 cells (Fig. 2.). In addition, 100 µg/ml of ETW increased the sub-G0/G1 nuclei population in HL-60 cells in a time-dependent manner (Fig. 2.).

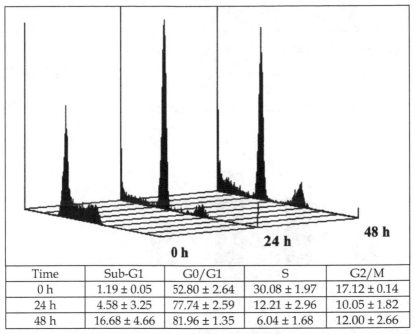

Time	Sub-G1	G0/G1	S	G2/M
0 h	1.19 ± 0.05	52.80 ± 2.64	30.08 ± 1.97	17.12 ± 0.14
24 h	4.58 ± 3.25	77.74 ± 2.59	12.21 ± 2.96	10.05 ± 1.82
48 h	16.68 ± 4.66	81.96 ± 1.35	6.04 ± 1.68	12.00 ± 2.66

Fig. 2. Cell cycle progression on HL-60 cells after treated with ETW. Cells were treated with ETW for the indicated incubation times, then stained for DNA with PI, and analyzed for cell cycle progression or apoptosis by flow cytometry. Cell cycle analysis showed that ETW induced a prominent G0/G1 population arrest and apoptosis in HL-60 cells. Each value represents mean±S.D. from three independent experiments

3.3 Effects of ETW on cyclin D1, cyclin E, Bcl-2 and c-Myc proteins expression in HL-60 cells

To better understand how ETW induces G0/G1 arrest, we investigated the protein expressions of cyclin D1 and cyclin E. After treatment with 100 µg/ml of ETW, there was a marked increase in protein levels of cyclin D1 and a marked decrease in cyclin E (Fig. 3A and 3B.) We also examined the expression levels of Bcl-2 and c-Myc protein. As shown in Fig. 3C and 3D, Bcl-2 and c-Myc protein levels decreased in HL-60 cells relative to controls. Our results suggest that ETW induces G0/G1 arrest and apoptosis in HL-60 cells by regulating cyclin D1, cyclin E, Bcl-2 and c-Myc protein expression.

3.4 ETW induced apoptosis is mediated by the activations of caspase-9, caspase-8 and caspase-3

Activation of caspase plays a key role in the induction of apoptosis. We used a fluorogenic enzymatic assay to detect activated caspase-9, caspase-8 and caspase-3 in ETW-treated HL-60 cells. Both caspase-9 and caspase-3 activities increased 24 h after ETW treatment and caspase-8 activities increased 48 h after ETW treatment (Fig. 4). Our results suggest that ETW-induced apoptosis is mediated through the activation of caspase-9, caspase-3 and then caspase-8.

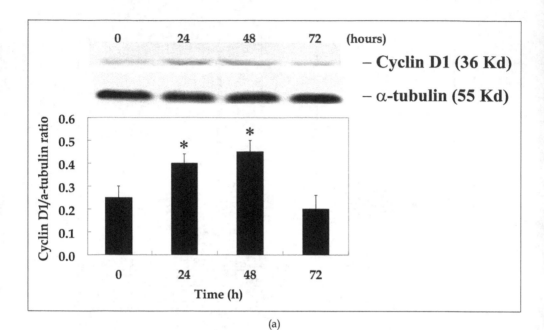

(a)

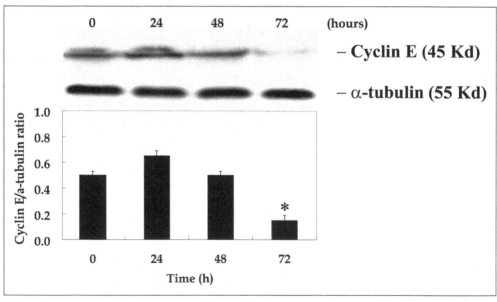

(b)

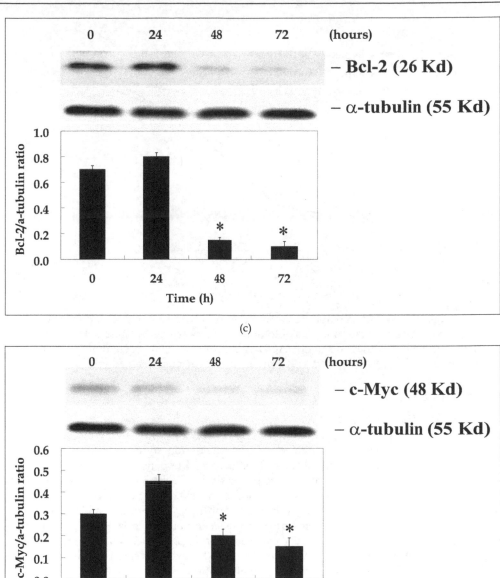

Fig. 3. Representative Western blotting showing changes on the levels of (A) cyclin D1,
(B) cyclin E, (C) Bcl-2 and (D) c-Myc in HL-60 cells after exposure to ETW (100 μg/ml). Cells
were treated with ETW for the indicated incubation times then total protein were prepared
and determined as described in Materials and Methods

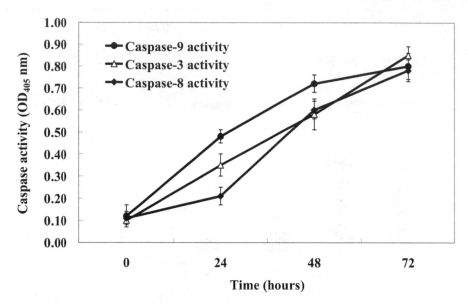

Fig. 4. Effects of ETW induced apoptosis on HL-60 cells by caspases-9, -8 and -3 activities. For caspase activity analysis, aliquots of total cell extracts were incubated with caspases-3, -9 and -8 specific substrates, respectively (Ac-DEVD-pNA, Ac-LEHD-pNA and Ac-IETD-pNA). The release of pNA was measured at 405 nm by a spectrophotometer

4. Discussion

Tripterygium wilfordii Hook. f. (TWHF) is used to treat inflammatory and immune-related diseases. Triptolide, a diterpenoid triepoxide extracted from the TWHF, exerts anti-tumorigenic actions against leukemia cells. In Differentiation -inducing activity study, some triterpene aglycones and Betulinic acid (pentacyclic triterpene) showed differentiation-inducing activity and against human acute promyelocytic leukemia HL-60 cells (Poon, 2004; Umehara et al., 1992). The preclinical laboratory work of identification and testing of potential anti-leukemia agents is designed for three categories: inhibition of cell proliferation, promotion of cell cycle arrest and induction of apoptosis. In the present study, we demonstrated that ETW induces cell cycle arrest and apoptosis in HL-60 cells. Hence, we suggest that ETW is a potent Chinese herb in HL-60 leukemia cells. However, it remains unclear whether ETW effectively induces the elimination of premalignant cells apoptosis *in vivo*.

The effects of ETW on HL-60 cells were associated with a specific disruption of cell cycle events and an induction of G0/G1 arrest. Our results show that ETW led to a loss of cell viability in a dose- and time-dependent manner (Fig. 1). Our study demonstrated that G1 cyclins (cyclin D1 and E) were regulated of HL-60 cells induced by ETW. A recent investigation of leukemia cell differentiation agent-induced differentiation of HL-60 leukemia cells has suggested that TPA to differentiate along the monocyte/macrophage lineage up-regulated of cyclin D1, and all-trans retinoic acid (ATRA) to differentiate along the Granulocyte lineage down-regulated cyclin E expression (Wang et al., 1996; Barrera et

Ethanol Extract of Tripterygium wilfordii Hook. f. Induces G0/G1 Phase Arrest and Apoptosis in Human Leukemia
HL-60 Cells Through c-Myc and Mitochondria-Dependent Caspase Signaling Pathways

227

al., 2004; Pizzimenti et al., 1999). Thus, it could be suggested that the regulation of cyclin D1 and E as well as CDK2 might anticipate in part the early events in differentiation in ETW-treated HL-60 cells. Our studies found that ETW reduced the level of Bcl-2 and c-Myc in a time-dependent manner. Regulation of the relative levels of Bcl-2 and c-Myc may play an important role in modulating the susceptibility of cells to differentiation (Li et al., 2004; Wu et al., 2002). Previous studies have demonstrated that HL-60 cells exhibited an over-expression of Bcl-2 and c-Myc proto-oncogene and that alteration of cellular oncogenes occur during the differentiation of HL-60 cells (Kumakura et al., 1996). Within myeloid lineage, Bcl-2 is over-expressed in early myeloid precursors but under-expressed or absent in matured myeloid cells and neutrophils (Gazitt et al., 2001; Blagosklonny et al., 1996).

Apoptosis is an evolutionarily conserved process that regulates development and homeostasis, and defects in the mechanisms that regulate cell death are implicated in both tumor genesis and multidrug resistance. Two distinct pathways for apoptosis have been defined, namely the death-receptor pathway and mitochondria pathway (Bohm et al., 2006; Christophe et al., 2006; Lucken et al., 2005; Lockshin et al., 2005). The signal transmitted to the mitochondria pathway causes the release of cytochrome c into cytosol. We analyzed apoptosis induction in ETW-treated HL-60 cells by measuring the accumulation of sub-G1 nuclei overtime. We observed the induction of caspase-9 and caspase-8 at 24 h of treatment, and caspase-3 activities at 48 h of treatment before the onset of DNA fragmentation at 72 h treatment by at ETW (Fig. 4). Furthermore, we detected loss of mitochondria membrane potential ($\Delta\Psi$m) in ETW-treated HL-60 cells and release of mitochondrial cytochrome c to cytosol after 18 h of treatment (data not shown). Recent investigation of triptolide-induced apoptosis of U937 cells has suggested that induced caspase-3 activation and down-regulation of the caspase inhibitory protein, XIAP, are involved in this apoptotic process (Choi et al., 2003). Recent reports suggest that DNA damage results in onset of mitochondrial permeability transition, which plays a major role in the apoptotic processes (Choi et al., 2003). A common step in apoptosis involves the loss of mitochondrial membrane potential resulting in increased generation of reactive oxygen species (ROS) from the mitochondrial respiratory chain. Our results suggest that ETW-induced apoptosis is mediated through the loss of mitochondria membrane potential and activation of caspase cascades by activated caspase 9, 8 and caspase-3 in a cytochrome c-dependent manner.

In summary, our results show that ETW induced G0/G1 arrest of HL-60 leukemia cells by regulating the protein expression of cyclin D1, cyclin E, Bcl-2 and c-Myc, and that it induced apoptosis in HL-60 cells by activating caspase-9, caspase-8 and caspase-3. ETW might, therefore, be an alternative cancer therapy in treatment of leukemia patients.

5. References

Abe J., Morikawa M., Miyamoto K., Kaiho S., Fukushima M., Miyaura C., Abe E., Suda T., Nishii Y., 1987. Synthetic analogues of vitamin D3 with an oxygen atom in the side chain skeleton. A trial of the development of vitamin D compounds which exhibit potent differentiation-inducing activity without inducing hypercalcemia. FEBS Letters 226, 58-62.

Altucci L., Rossin A., Hirsch O., Nebbioso A., Vitoux D., Wilhelm E., Guidez F., De Simone M., Schiavone EM., Grimwade D., Zelent A., de The H., Gronemeyer H., 2005. Rexinoid-triggered differentiation and tumor-selective apoptosis of acute myeloid leukemia by protein kinase A-mediated desubordination of retinoid X receptor. Cancer Research 65, 8754-8765.

Altucci L., Wilhelm E., Gronemeyer H., 2004. Leukemia: beneficial actions of retinoids and rexinoids. International Journal of Biochemistry & Cell Biology 36, 178-182.

An WW., Wang MW., Tashiro S., Onodera S., Ikejima T., 2004. Norcantharidin induces human melanoma A375-S2 cell apoptosis through mitochondrial and caspase pathways. Journal of Korean Medical Science 19, 560-566.

Aouacheria A., Neel B., Bouaziz Z., Dominique R., Walchshofer N., Paris J., Fillion H., Gillet G., 2002. Carbazolequinone induction of caspase-dependent cell death in Src-overexpressing cells. Biochemical Pharmacology 64, 1605-1616.

Barrera G., Pizzimenti S., Dianzani MU., 2004. 4-hydroxynonenal and regulation of cell cycle: effects on the pRb/E2F pathway. Free Radical Biology & Medicine 37, 597-606.

Blagosklonny MV., Alvarez M., Fojo A., Neckers LM., 1996. bcl-2 protein downregulation is not required for differentiation of multidrug resistant HL60 leukemia cells. Leukemia Research 20, 101-7.

Bohm I., Traber F., Block W., Schild H., 2006. Molecular imaging of apoptosis and necrosis – basic principles of cell biology and use in oncology. Rofo: Fortschritte auf dem Gebiete der Rontgenstrahlen und der Nuklearmedizin 178, 263-271.

Brown G., Drayson MT., Durham J., Toellner KM., Hughes PJ., Choudhry MA., Taylor DR., Bird R., Michell RH., 2002. HL60 cells halted in G1 or S phase differentiate normally. Experimental Cell Research 281, 28-38.

Chan EW., Cheng SC., Sin FW., Xie Y., 2001. Triptolide induced cytotoxic effects on human promyelocytic leukemia, T cell lymphoma and human hepatocellular carcinoma cell lines. Toxicology Letters. 122, 81-7.

Chang DM., Kuo SY., Lai JH., Chang ML., 1999. Effects of anti-rheumatic herbal medicines on cellular adhesion molecules. Annals of the Rheumatic Diseases 58, 366-371.

Choi YJ., Kim TG., Kim YH., Lee SH., Kwon YK., Suh SI,. Park JW., Kwon TK., 2003. Immunosuppressant PG490 (triptolide) induces apoptosis through the activation of caspase-3 and down-regulation of XIAP in U937 cells. Biochemical Pharmacology 66, 273-80.

Christophe M., Nicolas S., 2006. Mitochondria: a target for neuroprotective interventions in cerebral ischemia-reperfusion. Current Pharmaceutical Design 12, 739-57.

Faderl SJ., Keating MJ., 2005. Treatment of chronic lymphocytic leukemia. Current Hematology Reports 4, 31-38.

Fleischer A., Ghadiri A., Dessauge F., Duhamel M., Rebollo MP., Alvarez-Franco F., Rebollo A., 2006. Modulating apoptosis as a target for effective therapy. Molecular Immunology 43, 1065-1079.

Frankfurt O., Tallman MS., 2006. Strategies for the treatment of acute promyelocytic leukemia. Journal of the National Comprehensive Cancer Network 4, 37-50.

Gazitt Y., Reddy SV., Alcantara O., Yang J., Boldt DH., 2001. A new molecular role for iron in regulation of cell cycling and differentiation of HL-60 human leukemia cells: iron is required for transcription of p21(WAF1/CIP1) in cells induced by phorbol myristate acetate. Journal of Cellular Physiology 187, 124-135.

Hickey EJ., Raje RR., Reid VE., Gross SM., Ray SD., 2001. Diclofenac induced in vivo nephrotoxicity may involve oxidative stress-mediated massive genomic DNA

Ethanol Extract of Tripterygium wilfordii Hook. f. Induces G0/G1 Phase Arrest and Apoptosis in Human Leukemia
HL-60 Cells Through c-Myc and Mitochondria-Dependent Caspase Signaling Pathways

229

fragmentation and apoptotic cell death. Free Radical Biology & Medicine 31, 139-152.

Ho LJ., Lai JH., 2004. Chinese herbs as immunomodulators and potential disease-modifying antirheumatic drugs in autoimmune disorders. Current Drug Metabolism 5, 181-192,

Horie N., Mori T., Asada H., Ishikawa A., Johnston PG., Takeishi K., 2004. Implication of CDK inhibitors p21 and p27 in the differentiation of HL-60 cells. Biological & Pharmaceutical Bulletin 27, 992-997.

Kamikubo Y., Takaori-Kondo A., Uchiyama T., Hori T., 2003. Inhibition of cell growth by conditional expression of kpm, a human homologue of Drosophila warts/lats tumor suppressor. Journal of Biological Chemistry 278, 17609-17614.

Kumakura S., Ishikura H., Tsumura H., Iwata Y., Endo J., Kobayashi S., 1996. c-Myc and Bcl-2 protein expression during the induction of apoptosis and differentiation in TNF alpha-treated HL-60 cells. Leukemia & Lymphoma 23, 383-394.

Leuenroth SJ., Crews CM., 2005. Studies on calcium dependence reveal multiple modes of action for triptolide. Chemistry & Biology 12, 1259-1268.

Li CY., Zhan YQ., Xu CW., Xu WX., Wang SY., Lv J., Zhou Y., Yue PB., Chen B., Yang XM., 2004. EDAG regulates the proliferation and differentiation of hematopoietic cells and resists cell apoptosis through the activation of nuclear factor-kappa B. Cell Death & Differentiation 11, 1299-1308.

Lichtman MA., Segel GB., 2005. Uncommon phenotypes of acute myelogenous leukemia: basophilic, mast cell, eosinophilic, and myeloid dendritic cell subtypes: a review. Blood Cells Molecules & Diseases 35, 370-383.

Lin CC., Kao ST., Chen GW., Ho HC., Chung JG., 2006. Apoptosis of human leukemia HL-60 cells and murine leukemia WEHI-3 cells induced by berberine through the activation of caspase-3. Anticancer Research 26, 227-242.

Lockshin RA., 2005. Programmed cell death: history and future of a concept. Journal de la Societe de Biologie 199, 169-173.

Lou YJ., Jin J., 2004. Triptolide down-regulates bcr-abl expression and induces apoptosis in chronic myelogenous leukemia cells. Leukemia & Lymphoma 45, 373-376.

Lucken Ardjomande S., Montessuit S., Martinou JC., 2005. Changes in the outer mitochondrial membranes during apoptosis. Journal de la Societe de Biologie 199, 207-210.

Murphy, 2006. Rachel Chinese antirheumatic remedy for the treatment of SLE. Nature Clinical Practice Rheumatology 2, 180-181.

O'Hare T., Corbin AS., Druker BJ., 2006. Targeted CML therapy: controlling drug resistance, seeking cure. Current Opinion in Genetics & Development 16, 92-99.

Peng X., Zhao Y., Liang X., Wu L., Cui S., Guo A., Wang W., 2006. Assessing the quality of RCTs on the effect of beta-elemene, one ingredient of a Chinese herb, against malignant tumors. Contemporary Clinical Trials 27, 70-82.

Pizzimenti S., Barrera G., Dianzani MU., Brusselbach S., 1999. Inhibition of D1, D2, and A-cyclin expression in HL-60 cells by the lipid peroxydation product 4-hydroxynonenal. Free Radical Biology & Medicine 26, 1578-1586.

Poon KH., Zhang J., Wang C., Tse AK., Wan CK., Fong WF., 2004. Betulinic acid enhances 1,25-dihydroxyvitamin D3-induced differentiation in human HL-60 promyelocytic leukemia cells. Anti-Cancer Drugs 15, 619-624.

Qian SZ., 1987. *Tripterygium wilfordii*, a Chinese herb effective in male fertility regulation. Contraception 36, 335-345.

Qiu Daoming, Kao PN., 2003. Immunosuppressive and Anti-Inflammatory Mechanisms of Triptolide, the Principal Active Diterpenoid from the Chinese Medicinal Herb *Tripterygium wilfordii* Hook. f. Drugs in R & D 4, 1-18.

Samudio I., Konopleva M., Safe S., McQueen T., Andreeff M., 2005. Guggulsterones induce apoptosis and differentiation in acute myeloid leukemia: identification of isomer-specific antileukemic activities of the pregnadienedione structure. Molecular Cancer Therapeutics 4, 1982-1992.

Takahashi N., 2002. Induction of cell differentiation and development of new anticancer drugs. Journal of the Pharmaceutical Society of Japan. 122, 547-563.

Ter BE., Kaspers GJ., 2005. Treatment of childhood acute myeloid leukemia. Expert Review of Anticancer Therapy 5, 917-929.

Tsiftsoglou AS., Pappas IS., Vizirianakis IS., 2003. Mechanisms involved in the induced differentiation of leukemia cells. Pharmacology & Therapeutics 100, 257-290.

Umehara K., Takagi R., Kuroyanagi M., Ueno A., Taki T., Chen YJ., 1992. Studies on differentiation-inducing activities of triterpenes. Chemical & Pharmaceutical Bulletin. 40, 401-405.

Veselska R., Zitterbart K., Auer J., Neradil J., 2004. Differentiation of HL-60 myeloid leukemia cells induced by all-trans retinoic acid is enhanced in combination with caffeic acid. International Journal of Molecular Medicine 14, 305-310.

Wang J., Xu R., Jin RL., Chen ZQ., Fidler JM., 2000. Immuno-supperssive activity of the Chinese medicinal plant *Tripterygium wilfordii* Hook.: I. prolongation of rat cardiac and renal allograft survival by the PG27 extract and Immunosuppressive Synergy in Combination Therapy with Cyclosporine. Transplantation 70, 447-455.

Wang QM., Jones JB., Studzinski GP., 1996. Cyclin-dependent kinase inhibitor p27 as a mediator of the G1-S phase block induced by 1,25-dihydroxyvitamin D3 in HL60 cells. Cancer Research 56, 264-267.

Wei YS., Adachi I., 1991. Inhibitory effect of triptolide on colony formation of breast and stomach cancer cell lines. Zhongguo Yao Li Xue Bao/Acta Pharmacologica Sinica 12, 406-410.

Wu LD., Chen YZ., Li NN., Wu Y., 2002. Study on telomerase activity and expression of hTERT, c-myc and bcl-2 during terminal differentiation of HL-60 cells induced by retinoic acid. Zhongguo Shi Yan Xue Ye Xue Za Zhi 10, 395-399.

Wu Y., Wang Y., Zhong C., Li Y., Li X., Sun B., 2003. The suppressive effect of triptolide on experimental autoimmune uveoretinitis by down-regulating Th1-type response. International Immunopharmacology 3, 1457-1465.

Yazdanparast R., Moosavi MA., Mahdavi M., Sanati MH., 2005. 3-Hydrogenkwadaphnin from Dendrostellera lessertii induces differentiation and apoptosis in HL-60 cells. Planta Medica 71, 1112-1117.

Yen A., Varvayanis S., Smith JL., Lamkin TJ., 2006. Retinoic acid induces expression of SLP-76: expression with c-FMS enhances ERK activation and retinoic acid-induced differentiation G0/G1 arrest of HL-60 cells. European Journal of Cell Biology 85, 117-132.

Zhao YF., Zhai WL., Zhang SJ., Chen XP., 2005. Protection effect of triptolide to liver injury in rats with severe acute pancreatitis. Hepatobiliary & Pancreatic Diseases International 4, 604-608.

13

Applications of Quantum Dots in Flow Cytometry

Dimitrios Kirmizis[1], Fani Chatzopoulou[2],
Eleni Gavriilaki[2] and Dimitrios Chatzidimitriou[2]
[1]Medical School, Aristotle University, Thessaloniki,
[2]Laboratory of Microbiology, Aristotle University, Thessaloniki,
Greece

1. Introduction

Among several applications of flow cytometry is the identification of cell populations, which is a demanding and often daunting task, given the multitude and the often intercalating pattern of protein expression between different cell types. This complexity nescessitated the use of multicolor flow cytometry, a technique that has been given new perspectives with the emergence of Quandum Dot (QD) technology, which permitted overcoming obstacles, such as limited fluorochrome availability or limited sensitivity of combining multiple organic fluorochromes. The first systematic studies of size-dependent optical properties of semiconductor crystals in colloidal solutions were performed at early 1980s (Henglein, 1982; Brus, 1983). Later, Spanhel et al (1987) performed one of the first core–shell syntheses, a major advance in increasing the quantum yield. Major improvements leading to highly fluorescent QDs were made in the mid-1990s (Hines MA & Guyot-Sionnest P, 1996; Dabbousi RO et al, 1997; Peng et al, 1997). Subsequently, CdSe crystals with silane-modified hydrophilic surfaces were introduced for biological applications (Bruchez et al, 1998; Chan, WCW & Nie S, 1998). Even more recent developments include the encapsulation of CdSe/ZnS core–hell nanocrystals into carboxylated polymer, followed by chemical modification of the surface with long-chain polyethylene glycol (PEG) (Quantum Dot Corporation, Hayward, CA).

2. Properties of Quantum Dots

QDs are inorganic fluorochromes manufactured with the use of semiconductor materials (cadmium selenide for QDs emitting light in the 525- to 655-nm range or cadmium telluride for QDs emitting higher wavelength light) that assemble into nanometer-scale crystals (Chan et al, 2002; Bruchez, 2005). The tiny size of the QD nanocrystals gives these materials unique physical properties which seem tailor-made for multicolour flow cytometry compared to typical semiconductors or other fluorochromes. The most important of them is their broad excitation spectra (Bruchez M, 2005). Actually, QDs can be excited over the entire visual wavelength range as well as far into the ultraviolet. Because of their exceptionally large Stokes shifts (up to 400 nm), QDs can potentially be used for the multicolor detection even by a single laser flow cytometer, whereas organic dyes require multiple lasers for excitation in order to be used in multiplexed analysis (Bruchez M, 2005; Perfetto et al, 2004; Chattopadhyay et al, 2006). The light that the flow cytometer detects is the light which the electrons in QDs, after having been excited (excitons) by light absorption, emit as they return back from their conduction to their valence bands. What differentiates QDs from the typical fluorochromes is the so called "quantum confinement" phenomenon (Andersen et al, 2002), i.e. the phenomenon whereby, in contrast with the typical semiconductor materials where the distances between the bands are intimisimal (continuous), the excitons in QDs jump a discrete distance (known as the band gap) between bands, as a result of the very small size of the QD nanocrystal core. The narrow emission spectra of QDs usually overcomes the need for compensation, a standard process used in organic fluorochromes, which subtracts spillover fluorescence by estimating its magnitude as a fraction of the measured fluorescence in the primary detector (Roederer M, 2001) (see Fig. 1). In practice, it is reported that most QDs can be used simultaneously with only minimal (<10%) compensation between channels (Roederer et al, 2004). Moreover, when QD reagents are used with common fluorochromes excited by 488, 532, or 633 lasers (e.g., fluoroscein isothiacyanate, phycoerythrin [PE], or allophycocyanin), almost no spillover signal from other fluorochromes in the QD channels is found. Thus, in instruments with two or more lasers, QDs can be multiplexed with other fluorochromes to successfully measure even more colors (Chattopadhyay et al, 2007; 2010). In addition, the signals produced can be extremely bright, such as when an ultraviolet (350 nm) or violet (408 nm) laser is used to excite longer wavelength QDs (like QD605 andQD655), because of their high absorbance and low background levels (Hotz CZ, 2005; Wu et al, 2003). Finally, the emission properties of QDs also offer advantages over organic fluorochromes, albeit to a lesser extent.

Whereas organic fluorochromes of different colors come from a wide variety of source materials, each with distinct (and complex) physical, chemical, and biological properties which may not be compatible with each other or with staining conditions, this becomes less of a concern in QDs since QDs of different colors can be synthesized from the same starting materials (Chan et al, 2002), and thus multiplexed analysis is easier. However, the large surface-to-volume ratio in a nanosized crystal (about 50% of all atoms are on the surface) affects the emission of photons. Photochemical oxidation and surface defects in a crystal with no shell may lead to a broad emission and lower quantum yields. Indeed, early QD nanocrystals did not give stable or bright signals, exhibited poor solubility, and could not be attached to biologic probes (Riegler J & Nann T, 2004). These challenges were overcome by coating the QD nanocrystal with various materials such as inorganic zinc-sulfide, which is in turn coated with organic polymers (Bruchez et al, 1998). These organic polymers increase solubility and provide a platform of functional groups (such as amines, NH3) for conjugation to antibodies, streptavidin, and nucleic acids. Because they have similar coatings, QDs of various colors share uniform biophysical properties and a common conjugation procedure. The final QD product is about the size of PE, and can be linked to antibodies using a very similar conjugation chemistry. The shell helps to confine the excitation to the CdSe core and prevent the non-radiative relaxation.

The fluorescent properties of QDs are derived from their nanocrystal cores and not from the overall size of QD, which is actually similar in all QDs as a result of the fact that the cores are coated with various materials as mentioned above. Each QD has its characteristic emission peak, as long as the excitonic energy levels and quantum yields of fluorescence depend on exciton–photon interaction in the crystal and the size of the crystalline core. The primary determinant of the emission spectrum of each particle is its size so that smaller nanocrystals have different quantum confinement properties than bigger ones, as a result of the fact that the distance jumped by the exciton differs (the band gap is larger), and light is emitted at a different wavelength upon return to resting state. Increasing crystal size (from 2–3 to 10–12 nm) results in shift of the emission maximum from 500 to 800 nm. However, although they fluoresce at different wavelengths, they are excited at the same wavelength, allowing detection of multiple QD colors from just one laser. Thus, QDs with the smallest nanocrystal cores (3 nm) emit light in the blue region of the spectrum, whereas QDs with the largest cores (~6 nm) emit far red light (Bruchez et al, 1998). The most commonly used QDs in multicolor flow cytometry emit light at 525 (referred to as QD525), 545, 565, 585, 605, 655, 705, and 800 nm (Perfetto et al, 2004); their nanocrystal cores range in size from 2 to 6 nm (Biju et al, 2008). Other QDs emitting light at intermediate wavelengths (like 625 nm) are also commercially available. Thus, by 'tuning' the size of the nanocrystal core with various procedures (Peng et al, 1998; Smet et al, 1999), the description of which is beyond the scope of this chapter, QDs of different colors can be produced from the same starting material (*see* Fig. 1).

The core–shell nanocrystals have large extinction coefficients and high quantum yields. These parameters describe the capacity of the system to capture and subsequently rerelease light. Although quantum yields of QD conjugates in aqueous buffers (20–50%) are comparable with those of conventional fluorophores, the excitation efficiency of QD conjugates is much higher, making them about two orders of magnitude more efficient at absorbing excitation light than organic dyes and fluorescent proteins. QDs have a fluorescence lifetime of 20–30 ns — about 10 times longer than the background autofluorescence of proteins. Thus, fluorescence from single CdSe crystals has been observed much longer than from other fluorophores, resulting in high turnover rates and a large number of emitted photons (Doose, 2003).

The procedure for conjugation of antibodies to QDs is similar to conjugation of antibodies to PE, with slight variations in the reagents used and ratio of antibodies to fluorescent molecules. Successful conjugation relies on the coupling of malemide groups on the QDs to thiol groups on the antibody. These groups are generated during the initial steps of the procedure, as amine groups on the QDs are activated with a heterobifunctional crosslinker (sulfosuccinimidyl 4-[N-maleimidomethyl]cyclohexane-1-carboxylate, sulfo-SMCC) to generate the malemide moieties, and disulfide bonds in the antibody are reduced to thiol groups using dithiothreitol (DTT). Before conjugation, the DTT-reduced antibody is then mixed with two dye-labeled markers, Cyanin-3 (Cy3) and dextran blue, which track the monomeric fraction of antibodies as it passes through purification columns. Activated QDs and reduced antibody are subsequently purified over columns and mixed for conjugation.

A number of laser choices are available to excite QDs. Low wavelength ultraviolet (UV) and violet lasers are typically employed, since they induce maximal fluorescence emission. In theory, QD fluorescence arising from UV excitation is greater than that resulting from violet excitation; however, in practice, UV lasers induce much higher autofluorescence of cells, thereby negating the benefit of higher signal intensity. Still, users who rely on UV-excited probes (like DAPI and Hoechst) should note that QDs are compatible with their systems (Telford WG, 2004). Wheremultiplexed analysis of QDs is important, UV or violet excitation systems can be coupled to as many as eight photomultiplier tubes, allowing simultaneous measurement of QD545, QD565, QD585, QD605, QD655, QD705, QD800, and/or a violet-excitable organic fluorochrome (Chattopadhyay et al, 2010). To detect QD signals, we employ a filter strategy that first selects light sharply with a dichroic mirror, allowing only light above a certain wavelength to pass (long-pass filter). A second filter (known as a band pass filter) is stationed in front of the PMT, in order to collect a broad band of wavelengths for maximal signal. The light reflected by the first long-pass filter is passed to the next detector where it is queried in a similar fashion.

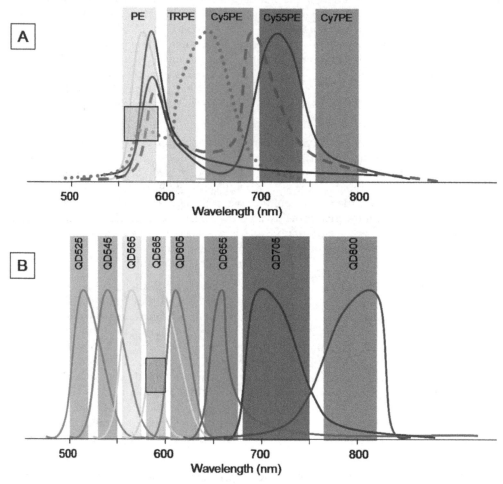

Fig. 1. Emission spectra of fluorochromes. The colored bars represent the wavelength range of filters used for the detection of each fluorochrome. The grey squares note the overlap of neighboring fluorochromes and signify the need for compensation. A significant overlap of phycoerythrin (PE) and PE-based tandems is apparent, which necessitates high compensation values, whereas their long tails of emission can induce significant spreading error (A). In contrast, such an overlap is avoided in QDs, whose emission spectra are narrow and symmetrical, their sensitivity better and the need for compensation less (B)

3. Quantum Dot applications

Multicolor Flow Cytometry: The utility of QDs in multicolor flow cytometry has been documented by several studies. Chattopadhyay et al (2006) in their interesting study analyzed the maturity of various antigenspecific T-cell populations using a 17-color staining panel. This panel consisted of 7 QDs and 10 organic fluorochromes, which were measured simultaneously in the same sample. The QD reagents used were conjugates with conventional antibodies (against CD4, CD45RA, and CD57), as well as peptide MHC Class I (pMHCI) multimers designed to detect those antigen-specific T-cells directed against HIV, EBV, and CMV epitopes. By identifying multiple phenotypically distinct subsets within each antigenspecific T-cell population, the remarkable intricacy of T-cell immunity as well as the power of a multiplexed approach was shown. QDs also allowed the reasearchers to measure many antigen-specific populations simultaneously, an important factor when sample availability is limited.

Markers of interest for use in multicolor flow cytometry are assigned to three categories: primary, secondary, and tertiary (Chattopadhyay et al, 2006, 2010; Perfetto et al, 2004; Mahnke YD & Roederer M, 2007). Primary markers are those that are highly expressed on cells, without intermediate fluorescence (i.e., they exhibit on/off expression). Secondary markers alike are expressed brightly and are well-characterized, but can be expressed at intermediate levels, and therefore resolution of dimly staining populations may be important. Thus, the fluorochromes assigned to secondary markers should be those with the less spreading error. Finally, tertiary markers are particularly dim, poorly characterized, or expressed by only a small proportion of cells. For the latter, bright fluorochromes are necessary. In practice, tertiary markers must be considered first. If these markers are particularly dim, they are assigned to fluorochrome channels that receive very little spreading error. QDs are particularly useful in this regard. However, some QDs are dim (QD 525) (Chattopadhyay et al, 2006), and therefore are not suitable for the measurement of dim cell populations. Among QDs, the brightest choices for tertiary markers are QD655, QD605, and QD585, in order of signal intensity. Secondary markers are ideal candidates for conjugation to QDs, especially for slightly dimmer channels, such as QD545, QD565, or QD800, as long as these are often brightly expressed. Finally, primary markers can be assigned to dim channels or to fluorochrome pairs with significant spectral overlap and spreading error.

Intracellular staining: Although QDs are not always compatible with intracellular staining, there have been recent advances in the ability to stain intracellularly with QDs. One approach, designed to avoid steric issues or intracellular degradation, is to target the QD (with or without conjugated antibody) into a cell using enzymes, such as matrix metalloproteinases (Zhang et al, 2006; Tekle et al, 2008), or nuclear or mitochondrial signal peptides (Hoshino et al, 2004). When coupled to antibodies, QDs bound to delivery molecules might allow organelle directed, specific intracellular staining without fixation/permeabilization.

Tetramer production: In the past, only FITC-, PE-, APC-tetramers were available, which limited panel design because many novel or dimly staining antibodies are only found on these fluorochromes. QDs with SA groups can be used to produce pMHCI multimers (commonly called 'tetramers') (Chattopadhyay et al, 2006), displaying higher valency than PE or APC and, thus, allowing brighter signals and better staining resolution (Chattopadhyay et al, 2008).

Pathogen detection: Efforts to detect whole pathogens have been considerably more successful with the introduction of QDs. When applied to a mixture of pathogenic and harmless *Escherichia coli* strains, QDs conjugated to antibodies against *E. coli* can detect one pathogenic bacterium among 99 harmless ones (Hahn et al, 2008). These detection limits are comparable to current assays that use FITC, but QDs are 10-fold brighter and give more accurate results.

Fluorescence resonance energy transfer (FRET) assays: Another interesting potential application of QDs is for new QD-based FRET assays. In particular, recent studies have reported on efforts to achieve FRET with QDs as the donor or acceptor fluorochrome (Willard DM & VanOrden A, 2003). This might not be so difficult to do, since QDs may be available in a wide variety of colors but share similar biochemistry and, thus, it is easy to find an acceptor dye that emits fluorescence at the desired wavelength. Furthermore, signal from the acceptor and the donor are well discrete and easily recognised, because of their narrow emission spectra. Similarly, donors and acceptors can be chosen such that spectral overlap is minimized; this reduces background emission and increases sensitivity. These advantages are not available in traditional FRET assays using organic fluorochromes.

4. Caveats, safety & toxicity

Although QDs are emerging as useful tools in multicolor flow cytometry, they are not fully characterized and occasionally exhibit peculiar properties. As mentioned above, not all antibodies will successfully conjugate to QDs. In particular, markers for intracellular flow cytometry (e.g., reagents for intracellular cytokine detection) have been problematic to conjugate in our facility, owing in part to the presence of excessive quantities of unconjugated QDs, to limited access to intracellular compartments due to QD size-related steric problems, to uneven dispersion of QDs throughout the intracellular environment, or to high sensitivity of QDs to chemicals used in the fixation and permeabilization process associated with intracellular staining (Riegler J & Nann T, 2004; Jaiswal et al, 2004b; Tekle et al, 2008). Variation within the QDs themselves occasionally might also be considerable, due to difficulties in the control of their production process. Thus, subtle differences in incubation time or injection of precursor solutions can cause differences in size distribution, shape, and surface defects among QDs (Dabbousi et al, 1997). These can potentially impact basic properties like fluorescence. As a rule of thumb, when using QDs in multicolor flow cytometry it might be useful to engage compensation controls using exactly the same reagent as the experimental panel. Another matter of potential concern with QDs is storage method and stability, as long as QDs are prone to form aggregates or precipitate out of solution, albeit the organic coating surrounding QDs has significantly improved solubility (Jaiswal J & Simon S, 2004) and any precipitation does not actually result in loss of activity, nor does it affect staining patterns (since these aggregates stain very brightly in all channels and are easily gated out of analyses). Manufacturers typically recommend storage in glass vials or in specially coated, non-adherent plastic tubes, since in standard microcentrifuge tubes, QDs may bind plastic, precipitate, and lose activity, especially at low volumes.

Two important obstacles to biological applications of commercially available QDs until recently are low quantum yields in aqueous buffers and strong aggregation of conjugates, both determined by the surface chemistry. For the use of QDs as antibody labeled probes, their outer layer must insulate the CdSe/ZnS core structure from the aqueous environment, prevent the nonspecific adsorption of QDs to cells, as well as provide the functional groups necessary for covalent attachment of antibodies. Lately, improvements on both nanocrystal core and shell technologies have enabled production of QD conjugates with exceptional brightness and low nonspecific adsorption (Larson DR et al, 2003). Recently, a new generation of QD nanocrystals was introduced with the application of a novel surface chemistry with the use of polymeric shell modified with long-chain, amino-functionalized PEGs. This new generation of QD nanocrystals has low nonspecific binding to cells and can be directly conjugated to antibodies through the introduced amino groups, using bisfunctional cross-linkers.

Since QDs are a new technology, their safety and toxicity are still a matter of concern. Although preliminary data from literature employing QDs for *in vivo* imaging of mice suggested that QDs are both safe and nontoxic (Voura et al, 2004; Shiohara et al, 2004; Bruchez, 2005; Gao et al, 2007), recent *in vitro* toxicology studies have questioned this assumption (Shiohara et al, 2004; Male et al, 2008). It appears that QDs coated with organic shells are relatively nontoxic for short incubation periods, but their degradation products (in particular Cd and Se), principally as a result of their oxidation and photolysis, may be toxic. Since QD size, charge, and composition of the outer shell are the main factors determinig oxidation and photolysis, toxicity likely differs by QD color and preparation. Regarding their risk on human health, data suggest that this is rather minimal, as long as QDs cannot enter the skin. However, this might not be true upon inhalation or ingestion as well, where there seems to be some potential for toxicity (Oberdorster et al, 2005; Hoet et al, 2004)

5. Conclusion

Although specific applications for QDs are still emerging, the basic technology has matured to the point that it can be relatively easily employed in multicolour flow cytometry. Unfortunately, just as applications for QDs are still nascent, so too is the commercial market for QD reagents. Therefore, to maximize the utility of QDs researchers must turn to in-house conjugations. Once implemented, the powerful potential of QD technology becomes evident. The remarkable spectral properties of QDs allow easy multiplexing, and therefore more information can be acquired from fewer samples. These properties make QDs useful in studying complex biologic systems.

6. References

Andersen KE, Fong C, Pickett W. Quantum confinement in CdSe nanocrystallites. *J Non-Cryst Solids* 2008, 2002:1105–1110.

Bentzen EL, Tomlinson ID, Mason J, Gresch P,Warnement MR, et al. Surface modification to reduce nonspecific binding of quantum dots in live cell assays. *Bioconjug Chem* 2005, 16:1488–1494.

Biju V, Itoh T, Anas A, Sujith A, Ishikawa M. Semiconductor quantum dots and metal nanoparticles: syntheses, optical properties, and biological applications. *Anal Bioanal Chem* 2008, 391:2469–2495.

Brus, L. E. (1983) A simple model for the ionization potential, electron affinity, and aqueous redox potentials of small semiconductor crystallites. *J. Chem. Phys.* 79, 5566–5571.

Bruchez, M., Moronne, M., Gin, P., Weiss, S., and Alivisatos, A. P. (1998) Semiconductor nanocrystals as fluorescent biological labels. *Science* 281, 2013–2016.

Bruchez M. Turning all the lights on: quantum dots in cellular assays. *Curr Opin Chem Biol* 2005, 9:533–537.

Chan, W. C. W. and Nie, S. (1998) Quantum dots bioconjugates for ultrasensitive nonisotopic detection. *Science* 281, 2016–2018.

Chan WC, Maxwell DJ, Gao X, Bailey RE, Han M, et al. Luminescent quantum dots for multiplexed biological detection and imaging. *Curr Opin Biotechnol* 2002, 13:40–46.

Chattopadhyay PK, Price DA,Harper TF, Betts MR, Yu J, et al. Quantum dot semiconductor nanocrystals for immunophenotyping by polychromatic flow cytometry. *Nat Med* 2006, 12:972–977.

Chattopadhyay PK, Perfetto SP, Yu J, et al. Application of Quantum Dots to Multicolor Flow Cytometry. In: Quantum dots : biological applications (editors: Bruchez M, Hotz CZ), 2007, Humana Press Inc., New Jersey

Chattopadhyay PK, Yu J, Roederer M. Application of quantum dots to multicolor flow cytometry. *Methods Mol Biol (Clifton,NJ)* 2007, 374:175–184.

Chattopadhyay PK,Melenhorst JJ, Ladell K, Gostick E, Scheinberg P, et al. Techniques to improve the direct ex vivo detection of low frequency antigen-specific CD8+ T cells with peptide-major histocompatibility complex class I tetramers. *Cytometry A* 2008, 73:1001–1009.

Chattopadhyay PK, Perfetto SP, Yu J, et al. The use of quantum dot nanocrystals in multicolor flow cytometry. *WIREs Nanomed Nanobiotechnol* 2010, 2: 334–348.

Clapp AR, Medintz IL, Mauro JM, Fisher BR, Bawendi MG, et al. Fluorescence resonance energy transfer between quantum dot donors and dye-labeled protein acceptors. *J Am Chem Soc* 2004, 126:301–310.

Dabbousi, R. O., Rodriguez-Viejo, J., Mikulec, F. V., et al. (1997) (CdSe)ZnS coreshell quantum dots: synthesis and characterization of a size series of highly luminescent nanocrystallites. *J. Phys. Chem. B* 101, 9463–9475.

Doose, S. (2003) Single molecule characterization of photophysical and colloidal properties of biocompatible quantum dots. Dissertation, Ruprecht-Karls University, Heidelberg, Germany.

Dwarakanath S, Bruno JG, Shastry A, Phillips T, John AA, et al. Quantum dot-antibody and aptamer conjugates shift fluorescence upon binding bacteria. *Biochem Biophys Res Commun* 2004, 325:739–743.

GaoX, ChanWC, Nie S.Quantum-dot nanocrystals for ultrasensitive biological labeling and multicolor optical encoding. *J Biomed Opt* 2002, 7:532–537.

Gao X, Nie S. Quantum dot-encoded beads. *Methods Mol Biol (Clifton, NJ)* 2005, 303:61–71.

Gao X, Chung LWK, Nie S. Quantum dots for in vivo molecular and cellular imaging. *Methods Mol Biol* 2007, 374:135–145.

Grabolle M, Ziegler J, Merkulov A, Nann T, Resch-Genger U. Stability and fluorescence quantum yield of CdSe-ZnS quantum dots — influence of the thickness of the ZnS shell. *Ann NY Acad Sci* 2008, 1130:235–241.

Hahn MA, Keng PC, Krauss TD. Flow cytometric analysis to detect pathogens in bacterial cell mixtures using semiconductor quantum dots. *Anal Chem* 2008, 80:864–872.

Henglein, A. (1982) Photochemistry of colloidal cadmium sulfide. 2. Effects of adsorbed methyl viologen and of colloidal platinum. *J. Phys. Chem.* 86, 2291–2293.

Hines, M. A. and Guyot-Sionnest, P. (1996) Synthesis and characterization of strongly luminescing ZnS-capped CdSe nanocrystals. *J. Phys. Chem.* 100, 468–471.

Hoet PH, Broske-Hohlfeld I, Salata OV. Nanoparticles— known and unknown health risks. *J Nanobiotechnol* 2004, 2:12–27.

Hoshino A, Fujioka K, Oku T, Nakamura S, Suga M, et al. Quantum dots targeted to the assigned organelle in living cells. *Microbiol Immunol* 2004, 48:985–994.

Hotz CZ. Applications of quantum dots in biology: an overview. *Methods Mol Biol (Clifton, NJ)* 2005, 303:1–17.

Jaiswal JK, Mattoussi H, Mauro JM, Simon SM. Longterm multiple color imaging of live cells using quantum dot bioconjugates. *Nat Biotechnol* 2003, 21:47–51.

Jaiswal J, Simon S. Potentials and pitfalls of fluorescent quantum dots for biological imaging. *Trends Cell Biol* 2004, 14:497–504.

Jaiswal JK, Goldman ER, Mattoussi H, Simon SM. Use of quantum dots for live cell imaging. *Nat Methods* 2004, 1:73.

Larson, D. R., Zipfel,W. R.,Williams, R. M., et al. (2003) Water-soluble quantum dots for multiphoton fluorescence imaging *in vivo. Science* 300, 1434–1436.

Medintz IL, Clapp AR, Mattoussi H, Goldman ER, Fisher B, et al. Self-assembled nanoscale biosensors based on quantum dot FRET donors. *Nat Mater* 2003, 2:630–638.

Mahnke YD, Roederer M. Optimizing a multicolor immunophenotyping assay. *Clin Lab Med* 2007, 27:469–485.

Male KB, Lachance B, Hrapovic S, Sunahara G, Luong JH. Assessment of cytotoxicity of quantum dots and gold nanoparticles using cell-based impedance spectroscopy. *Anal Chem* 2008, 80:5487–5493.

Mattoussi H, Mauro JM, Goldman ER, Anderson GP, Sundar VC, et al. Self-assembly of CdSe-ZnS quantum dot bioconjugates using an engineered recombinant protein. *J Am Chem Soc* 2000, 122:12142–12150.

Oberdorster G, Oberdorster E, Oberdorster J. NANOTOXICOLOGY: an emerging discipline evolving from studies of ultrafine particles. *Environ Health Perspect* 2005, 17:823–839.

Peng, X., Schlamp, M. C., Kadavanich, A. V., and Alivisatos, A. P. (1997) Epitaxial growth of highly luminescent CdSe/CdS core/shell nanocrystals with photostability and electronic accessibility. *J. Amer. Chem. Soc.* 119, 7019–7029.

Peng X, Wickham J, Alivisatos AP. Kinetics of II-VI and III-V colloidal semiconductor nanocrystal growth: focusing of size distributions. *J Am Chem Soc* 1998, 120:5343–5349.

Perfetto SP, Chattopadhyay PK, Roederer M. Seventeen-colour flow cytometry: unravelling the immune system. *Nat Rev Immunol* 2004, 4:648–655.

Reiss P, Bleuse J, Pron A. Highly Luminescent CdSe ZnSe core/shell nanocrystals of low size dispersion 781–784. *Nano Lett* 2002, 2:781–784.

Riegler J, Nann T. Application of luminescent nanocrystals as labels for biological molecules. *Anal Bioanal Chem* 2004, 379:913–919.

Roederer M. Spectral compensation for flow cytometry: visualization artifacts, limitations, and caveats. *Cytometry* 2001, 45:194–205.

Roederer, M., Perfetto, S. P., Chattopadhyay, P. K., Harper, T., and Bruchez, M. (2004) Quantum dots for multicolor flow cytometry. Poster, International Society for Analytical Cytology Conference; May 24–28, Montpelier, France.

Shahzi SI,Michael WM, John GB, Burt VB, Carl AB, et al. A review of molecular recognition technologies fordetection of biological threat agents. *Biosens Bioelectron* 2000, 15:549–578.

Shiohara A, Hoshino A, Hanaki K, Suzuki K, Yamamoto K. On the cyto-toxicity caused by quantum dots. *Microbiol Immunol* 2004, 48:669–675.

Smet YD, Deriemaeker L, Parloo E, Finsy R. On the determination of ostwald ripening rates from dynamic light scattering measurements. *Langmuir* 1999, 15:2327–2332.

Spanhel, L., Haase, M.,Weller, H., and Henglein, A. (1987) Photochemistry of colloidal semiconductors. Surface modification and stability of strong luminescing CdS particles. *J. Amer. Chem. Soc.* 109, 5649–5662.

Tekle C, Deurs B, Sandvig K, Iversen T. Cellular trafficking of quantum dot-ligand bioconjugates and their induction of changes in normal routing of unconjugated ligands. *Nano Lett* 2008, 8:1858–1865.

Telford WG. Analysis of UV-excited fluorochromes by flow cytometry using near-ultraviolet laser diodes. *Cytometry A* 2004, 61:9–17.

Voura EB, Jaiswal JK, Mattoussi H, Simon SM. Tracking metastatic tumor cell extravasation with quantum dot nanocrystals and fluorescence emission-scanning microscopy. *Nat Med* 2004, 10:993–998.

Wang HQ, Liu TC, Cao YC, Huang ZL, Wang JH, et al. A flow cytometric assay technology based on quantum dots-encoded beads. *Anal Chim Acta* 2006, 580:18–23.

Willard DM, Carillo LL, Jung J, Van Orden A. CdSe-ZnS quantum dots as resonance energy transfer donors in a model protein-protein binding assay. *Nano Lett* 2001, 1:469–474.

Willard DM, Van Orden A. Quantum dots: resonant energy-transfer sensor. *Nat Mater* 2003, 2:575–576.

Wu X, Liu H, Liu J, Haley KN, Treadway JA, et al. Immunofluorescent labeling of cancer marker Her2 and other cellular targets with semiconductor quantum dots. *Nat Biotechnol* 2003, 21:41–46.

Wu Y, Lopez GP, Sklar LA, Buranda T. Spectroscopic characterization of streptavidin functionalized quantum dots. *Anal Biochem* 2007, 364:193–203.

Wu Y, Campos SK, Lopez GP, Ozbun MA, Sklar LA, et al. The development of quantum dot calibration beads and quantitativemulticolor bioassays in flow cytometry and microscopy. *Anal Biochem* 2007, 364:180–192.

Zhang Y, So MK, Rao J. Protease-modulated cellular uptake of quantum dots. *Nano Lett* 2006, 6:1988–1992.

Permissions

The contributors of this book come from diverse backgrounds, making this book a truly international effort. This book will bring forth new frontiers with its revolutionizing research information and detailed analysis of the nascent developments around the world.

We would like to thank Ingrid Schmid, Mag. pharm., for lending his expertise to make the book truly unique. He has played a crucial role in the development of this book. Without his invaluable contribution this book wouldn't have been possible. He has made vital efforts to compile up to date information on the varied aspects of this subject to make this book a valuable addition to the collection of many professionals and students.

This book was conceptualized with the vision of imparting up-to-date information and advanced data in this field. To ensure the same, a matchless editorial board was set up. Every individual on the board went through rigorous rounds of assessment to prove their worth. After which they invested a large part of their time researching and compiling the most relevant data for our readers. Conferences and sessions were held from time to time between the editorial board and the contributing authors to present the data in the most comprehensible form. The editorial team has worked tirelessly to provide valuable and valid information to help people across the globe.

Every chapter published in this book has been scrutinized by our experts. Their significance has been extensively debated. The topics covered herein carry significant findings which will fuel the growth of the discipline. They may even be implemented as practical applications or may be referred to as a beginning point for another development. Chapters in this book were first published by InTech; hereby published with permission under the Creative Commons Attribution License or equivalent.

The editorial board has been involved in producing this book since its inception. They have spent rigorous hours researching and exploring the diverse topics which have resulted in the successful publishing of this book. They have passed on their knowledge of decades through this book. To expedite this challenging task, the publisher supported the team at every step. A small team of assistant editors was also appointed to further simplify the editing procedure and attain best results for the readers.

Our editorial team has been hand-picked from every corner of the world. Their multi-ethnicity adds dynamic inputs to the discussions which result in innovative outcomes. These outcomes are then further discussed with the researchers and contributors who give their valuable feedback and opinion regarding the same. The feedback is then collaborated with the researches and they are edited in a comprehensive manner to aid the understanding of the subject.

Apart from the editorial board, the designing team has also invested a significant amount of their time in understanding the subject and creating the most relevant covers. They scrutinized every image to scout for the most suitable representation of the subject and create an appropriate cover for the book.

The publishing team has been involved in this book since its early stages. They were actively engaged in every process, be it collecting the data, connecting with the contributors or procuring relevant information. The team has been an ardent support to the editorial, designing and production team. Their endless efforts to recruit the best for this project, has resulted in the accomplishment of this book. They are a veteran in the field of academics and their pool of knowledge is as vast as their experience in printing. Their expertise and guidance has proved useful at every step. Their uncompromising quality standards have made this book an exceptional effort. Their encouragement from time to time has been an inspiration for everyone.

The publisher and the editorial board hope that this book will prove to be a valuable piece of knowledge for researchers, students, practitioners and scholars across the globe.

List of Contributors

Arash Zaminy
Department of Anatomy & Cell Biology, Shahid Beheshti University of Medical Sciences, Tehran, Iran

Ambrus Kaposi, Gergely Toldi and Gergő Mészáros
First Department of Pediatrics, Semmelweis University, Hungary

Balázs Szalay
Department of Laboratory Medicine, Semmelweis University, Hungary

Gábor Veress
Analytix Ltd., Hungary

Barna Vásárhelyi
Department of Laboratory Medicine, Semmelweis University, Hungary
Research Group of Pediatrics and Nephrology, Hungarian Academy of Sciences, Hungary

P. S. Canaday and C. Dorrell
Oregon Health & Science University, USA

Jacques A. Nunès, Guylène Firaguay and Emilie Coppin
Institut National de la Santé et de la Recherche Médicale, Unité 1068, Centre de Recherche en Cancérologie de Marseille, Institut Paoli-Calmettes, Aix-Marseille Univ., Marseille, France

Andrea A. F. S. Moraes, Lucimar P. França, Vanina M. Tucci-Viegas, Fernanda Lasakosvitsch, Silvana Gaiba, Fernanda L. A. Azevedo, Amanda P. Nogueira, Helena R. C. Segreto, Alice T. Ferreira and Jerônimo P. França
Universidade Federal de São Paulo – Unifesp, Brazil
Universidade Nove de Julho – Uninove, Brazil
Universidade Estadual de Santa Cruz – Uesc, Brazil

Annelie Pichert, Denise Schlorke, Josefin Zschaler, Jana Fleddermann, Maria Schönberg, Jörg Flemmig and Jürgen Arnhold
Institute for Medical Physics and Biophysics, University of Leipzig, Leipzig, Germany

Tomás Lombardo, Laura Anaya, Laura Kornblihtt and Guillermo Blanco
Laboratory of Immunotoxicology (LaITo), Hospital de Clínicas San Martín, University of Buenos Aires, Argentina

Sinéad B. Doherty and A. Brodkorb
Teagasc Food Research Centre Moorepark, Fermoy, Co. Cork, Ireland

Irena Koutná, Pavel Šimara, Petra Ondráčková and Lenka Tesařová
Masaryk University/ Centre for Biomedical Image Analysis, FI Veterinary Reasearch Institute/ Department of Immunology, Czech Republic

Paula Laranjeira, Andreia Ribeiro, Sandrine Mendes, Ana Henriques, M. Luísa Pais and Artur Paiva
Histocompatibility Center of Coimbra, Portugal

Chung-Jen Chiang
Department of Medical Laboratory Science and Biotechnology, China Medical University, Taichung, Taiwan

Jai-Sing Yang
Department of Pharmacology, China Medical University, Taichung, Taiwan

Yun-Peng Chao
Department of Chemical Engineering, Feng Chia University, Taichung, Taiwan

Li-Jen Lin
School of Chinese Medicine, China Medical University, Taichung, Taiwan

Wen-Wen Huang, Jing-Gung Chung, Shu-Fen Peng and Shu-Ren Pai
Department of Biological Science and Technology, China Medical University, Taichung, Taiwan

Chi-Cheng Lu and Jo-Hua Chiang
Department of Life Sciences, National Chung Hsing University, Taichung, Taiwan

Minoru Tsuzuki
Nihon Pharmaceutical University, Saitama, Japan
Tsuzuki Institute for Traditional Medicine China Medical University, Taichung, Taiwan

Dimitrios Kirmizis
Medical School, Aristotle University, Thessaloniki, Greece

Fani Chatzopoulou, Eleni Gavriilaki and Dimitrios Chatzidimitriou
Laboratory of Microbiology, Aristotle University, Thessaloniki, Greece